MATHEMATICS AND ITS TEACHING IN THE SOUTHERN AMERICAS

with an introduction by Ubiratan D'Ambrosio

SERIES ON MATHEMATICS EDUCATION

Series Editors: Mogens Niss *(Roskilde University, Denmark)*
Lee Peng Yee *(Nanyang Technological University, Singapore)*
Jeremy Kilpatrick *(University of Georgia, USA)*

Published

Vol. 1   How Chinese Learn Mathematics
Perspectives from Insiders
Edited by: L. Fan, N.-Y. Wong, J. Cai and S. Li

Vol. 2   Mathematics Education
The Singapore Journey
Edited by: K. Y. Wong, P. Y. Lee, B. Kaur, P. Y. Foong and S. F. Ng

Vol. 4   Russain Mathematics Education
History and World Significance
Edited by: A. Karp and B. R. Vogeli

Vol. 5   Russian Mathematics Education
Programs and Practices
Edited by A. Karp and B. R. Vogeli

Vol. 7   Mathematics Education in Korea
Volume 1: Curricular and Teaching and Learning Practices
Edited by Jinho Kim, Inki Han, Joongkwoen Lee and Mangoo Park

Vol. 9   Primary Mathematics Standards for Pre-Service Teachers in Chile
A Resource Book for Teachers and Educators
By P. Felmer, R. Lewin, S. Martínez, C. Reyes, L. Varas, E. Chandía, P. Dartnell, A. López, C. Martínez, A. Mena, A. Ortíz, G. Schwarze and P. Zanocco

Vol. 10   Mathematics and Its Teaching in the Southern Americas
Edited by H. Rosario, P. Scott and B. R. Vogeli

Forthcoming

Vol. 3   Lesson Study
Challenges in Mathematics Education
Edited by M. Inprasitha, M. Isoda, B.-H. Yeap and P. Wang-Iverson

Vol. 6   How Chinese Teach Mathematics
Perspectives from Insiders
Edited by L. Fan, N.-Y. Wong, J. Cai and S. Li

Vol. 8   Mathematical Modelling
From Theory to Practice
Edited by N. H. Lee and D. K. E. Ng

Series on Mathematics Education Vol. 10

MATHEMATICS AND ITS TEACHING IN THE SOUTHERN AMERICAS

with an introduction by Ubiratan D'Ambrosio

Edited by

Héctor Rosario
University of Puerto Rico, Puerto Rico

Patrick Scott
Inter-American Committee on Mathematics Education, USA

Stafford Library
Columbia College
1001 Rogers Street
Columbia, MO 65216

Bruce Vogeli
Columbia University, USA

 World Scientific

NEW JERSEY · LONDON · SINGAPORE · BEIJING · SHANGHAI · HONG KONG · TAIPEI · CHENNAI

Published by

World Scientific Publishing Co. Pte. Ltd.
5 Toh Tuck Link, Singapore 596224
USA office: 27 Warren Street, Suite 401-402, Hackensack, NJ 07601
UK office: 57 Shelton Street, Covent Garden, London WC2H 9HE

Library of Congress Cataloging-in-Publication Data
Mathematics and its teaching in the Southern Americas / edited by Héctor Rosario (University of Puerto Rico, Puerto Rico), Patrick Scott (Inter-American Committee on Mathematics Education, USA), Bruce R. Vogeli (Columbia University, USA).
    pages cm. -- (Series on mathematics education ; volume 10)
  Includes bibliographical references and index.
  ISBN 978-9814590563
  1. Mathematics--Study and teaching--South America. I. Rosario, Héctor, editor. II. Scott, Patrick, 1947-   editor. III. Vogeli, Bruce R. (Bruce Ramon), editor.
  QA14.S63M38 2015
  510.71'08--dc23
                                      2014022288

British Library Cataloguing-in-Publication Data
A catalogue record for this book is available from the British Library.

Front cover image: De Visu/Shutterstock.com

Copyright © 2015 by World Scientific Publishing Co. Pte. Ltd.

All rights reserved. This book, or parts thereof, may not be reproduced in any form or by any means, electronic or mechanical, including photocopying, recording or any information storage and retrieval system now known or to be invented, without written permission from the publisher.

For photocopying of material in this volume, please pay a copying fee through the Copyright Clearance Center, Inc., 222 Rosewood Drive, Danvers, MA 01923, USA. In this case permission to photocopy is not required from the publisher.

Printed in Singapore by Mainland Press Pte Ltd.

Contents

Preface	*Half a Century Later!*	ix
Introduction	*Traversing the Path of Mathematics Education in the Southern Americas*	xi
	Ubiratan D'Ambrosio	
Chapter 1	**ARGENTINA: A Review of Mathematics Education through Mathematical Problems at the Secondary Level**	1
	Betina Duarte	
Chapter 2	**BOLIVIA: An Approach to Mathematics Education in the Plurinational State**	31
	A. Pari	
Chapter 3	**BRAZIL: History and Trends in Mathematics Education**	57
	Beatriz S. D'Ambrosio, Juliana Martins, and Viviane de Oliveira Santos	
Chapter 4	**CHILE: The Context and Pedagogy of Mathematics Teaching and Learning**	89
	Eliana D. Rojas and Fidel Oteiza	

| Chapter 5 | COLOMBIA: The Role of Mathematics in the Making of a Nation | 115 |

Hernando J. Echeverri and Angela M. Restrepo

| Chapter 6 | COSTA RICA: History and Perspectives on Mathematics and Mathematics Education | 151 |

Ángel Ruiz

| Chapter 7 | CUBA: Mathematics and Its Teaching | 193 |

Otilio B. Mederos Anoceto, Miguel A. Jiménez Pozo, and José M. Sigarreta

| Chapter 8 | GUYANA: The Mathematical Growth of an Emerging Nation | 223 |

Mahendra Singh and Lenox Allicock

| Chapter 9 | HAITI: History of Mathematics Education | 239 |

Jean W. Richard

| Chapter 10 | HONDURAS: Origins, Development, and Challenges in the Teaching of Mathematics | 265 |

Marvin Roberto Mendoza Valencia

| Chapter 11 | MÉXICO: The History and Development of a Nation and Its Influence on the Development of Mathematics and Mathematics Education | 295 |

Eduardo Mancera and Alicia Ávila

| Chapter 12 | PANAMÁ: Towards the First World through Mathematics | 321 |

Euclides Samaniego, Nicolás A. Samaniego, and Benigna Fernández

Chapter 13 PARAGUAY: A Review of the History
of Mathematics and Mathematics Education   347
Gabriela Gómez Pasquali

Chapter 14 PERÚ: A Look at the History of
Mathematics and Mathematics Education   363
*César Carranza Saravia and
Uldarico Malaspina Jurado*

Chapter 15 PUERTO RICO: The Forging of
a National Identity in Mathematics
Education   381
*Héctor Rosario, Daniel McGee,
Jorge M. López, Ana H. Quintero, and
Omar A. Hernández*

Chapter 16 TRINIDAD and TOBAGO:
Mathematics Education in the Twin
Island Republic   405
Shereen Alima Khan and Vimala Judy Kamalodeen

Chapter 17 VENEZUELA: Signs for the Historical
Reconstruction of Its Mathematics
Education   439
*Fredy Enrique González
(Translated by Nathan C. Ryan)*

Epilogue   A Half Century of Progress   461
Patrick Scott

Index   465

Preface

Half a Century Later!

The country-by-country reviews of mathematics education in the Americas included in the 1966 *Report of the Second Inter-American Conference on Mathematical Education* were collected by Professor Howard F. Fehr between the First Inter-American Conference held in Bogotá, Colombia in 1961 and the second held in Lima, Perú in 1966. These brief reviews have remained the only united source of information about the mathematics educational policies and practices across the hemisphere.

The purpose of this anthology of essays by national experts in *Mathematics and Its Teaching in the Southern Americas* is twofold. First, political changes early in the twenty-first century imply educational reforms that should not occur in isolation. Hence, national experts must be well-informed about regional developments in mathematics and its teaching as they plan their own nations' futures. Second, external contributors should be aware of the influence that past interventions have had on local policies and practices in mathematics education.

The editors of this volume have strived to accomplish Dr. Fehr's goal of authenticity: Latin American and Caribbean experts carried out the review and editing process. Translations, where required, were done by native English speakers, some of Latin heritage. In particular, we thank Professor Nathan Ryan[1] for general discussions and help with translations, as well as Professor Jonathan D. Farley,[2] a proud son of the Caribbean, for editing the introduction.

Despite our efforts as editors to assure author independence and national integrity, errors occur that are entirely our responsibility. To the generations of readers who will refer to *Mathematics and Its Teaching in the Southern Americas*, we offer our regrets and apologies. Moreover, we owe

---

[1] Bucknell University.
[2] Morgan State University.

the contributing authors our most sincere appreciation and admiration for their exceptional efforts in describing their nations' activities in mathematics accurately. We badgered them ruthlessly about deadlines, attributions, and details of interpretation. Dr. Fehr would have been impressed with the results of their efforts. We certainly are.

Héctor Rosario   Patrick Scott   Bruce Vogeli

Dedicated to the memory of Dr. Howard Franklin Fehr (1902–1982), Professor of Mathematics at Teachers College, Columbia University from 1948 to 1967, and a founding trustee of the Inter-American Committee on Mathematics Education.

Introduction

Traversing the Path of Mathematics Education in the Southern Americas

Ubiratan D'Ambrosio

Experts from different countries of the region prepared this collection of essays, detailing aspects of the evolution of mathematics education in their respective countries prior to, during, and after the colonial period.

The concept of country is the result of a total reformulation of the territorial demarcation of the lands in the Americas after the Europeans took possession of different regions in the name of their monarchs. The Spanish conquest of the Americas began in 1492, with the arrival of Christopher Columbus (c. 1450–1506) to San Salvador Island (Watling Island) or Samana Cay, both in the Bahamas. From there, it is said that he sailed to Hispaniola, a major Caribbean island shared today by Haiti and the Dominican Republic. A few years after this first arrival, navigators explored and conquered the entire region. It was decisive when Hernán Cortés (1485–1547), in the period 1519–1921, conquered the Aztec empire, followed by the much longer conquest of the Yucatan Peninsula. In 1540, the Viceroyalty of New Spain was established. This was followed, in 1542, by the creation of the Viceroyalty of Perú, after the difficult conquest of the Inca Empire by Francisco Pizarro (c. 1471–1541) in 1529. Some two hundred years later, Spain established the viceroyalties of New Granada (1717) and Río de la Plata (1776), with Bogotá and Buenos Aires as their respective capitals.

Brazil, with an area approximately equal to 47% of the total area of South America, was encountered in 1500 by the Portuguese. Contrary to the situation in Spanish America, with its initial bounty of silver and gold, Brazil was at first a disappointment to its "mother country", which decided to establish the colony mainly to avoid its takeover by France, Spain,

England or Holland. Brazil was a relatively poor colony until 1659, when gold and diamond were discovered.

The colonial administrative organization followed, in some ways, cultural ties of the native population, respecting traditional territorial organization and families, language, and means of production. During the colonial period, groups of economic, military and political interest began to organize movements, which eventually led to independence. The origin and common interests of these groups gave rise to countries, designing a political map, which is, basically, what we have today.

This resulted in different developments of mathematics education. In this introduction I will not discuss the political currents and the motivations that shaped the creation of countries. Neither will I comment on individual chapters dealing with specific countries. Instead, I suggest reflections about some common ground and will try to give a general historical view of the formation of the Americas.

Early accounts, among them those by the Dominican friar Bartolomé de Las Casas(1474–1566) and the German mercenary Hans Staden (c. 1525–1579), give accounts of the behavior of the natives. Las Casas wrote *General History of the Indies* from 1527 through 1561. The main body of this book is a visual account of the encounter of Europeans with the natives, generally describing atrocities attributed to the natives. In 1557, Staden published the *True Story and Description of a Country of Wild, Naked, Grim, Man-eating People in the New World*, reporting on the period he was a captive of the Tupinambás, in the Northern coastal area of Brazil, now Maranhão. He described the natives' life and customs, including cannibalism, and the wars among tribes. The book was published in a richly illustrated edition, which became a best seller.

A later chapter of the *General History of the Indies*, the *Apologetic Summary History of the People of These Indies,* became a book on its own. In this contrasting work, Las Casas wrote an early ethnographic account of the Taíno and other native cultures of the islands in the Caribbean. Most of these cultures became extinct, although the Taíno Genome Project shows evidence of its undeniable presence in the genetic pool of the Caribbean peoples (Via, et al., 2011). In the book, Las Casas discusses the cultural level of the indigenous peoples of the Americas, claiming they were as civilized as, in some cases even more civilized than, the Western civilizations of antiquity. In some ways, he praised their political and social life. In this work, the author combined his own ethnographic observations with those of other writers, and compared customs and cultures between different peoples. The mere titles of two chapters of the *Apologetic* synthesize much of the rationale for conversion of the natives to Christianity: "Indians possessed more enlightenment and natural knowledge of God than the Greeks and Romans" and "The Indians are as capable as any other nations to receive the Gospel". The education of the natives hinged on these remarks.

This is the broad scenario for the studies in this collection of essays, which cover elementary, secondary, and higher education from the period preceding the independence of Spanish, Portuguese, French, and British colonies in the Southern Americas to the current status, offering a vision for the future.

Mathematics, Education, and Mathematics Education

Mathematics education is the conjugation of education and mathematics, which are two categories of doing and knowing, of behavior and knowledge. To discuss mathematics education in the Americas and its specificities, we must understand the evolution of categories of modes of thought. In other words, we must discuss the origin of the disciplines as we know them today. This is an exciting research area.

Education is present in every human group and has two purposes. First, it is a strategy to prepare new generations to integrate into daily life, acquiring norms of behavior, basic values, and knowledge about the most common facts, phenomena, and myths, which are the essence of the culture of that particular social group. Second, but not less importantly, its purpose is to identify and stimulate innate and nurtured characteristics of each individual. Synthesizing these ideas — throughout history and in every civilization — we recognize the educational system of a social group as a collection of strategies, focusing on two goals: (1) to prepare for citizenship, and (2) to prepare for the future, stimulating the development of creative potential.

These are conflicting purposes in the sense that the first aims at submission to accepted norms — as it is expected for citizenship — while the second depends precisely on not conforming to norms, which is intrinsic to the freedom to create. This is my concept of education, as discussed in many of my works.

We cannot avoid the possible contradiction of these two purposes, but a good education cannot favor one to the detriment of the other. Indeed, research points to the ability to manage this conflict as an important factor of progress. However, it is easily recognized that favoring one purpose, disadvantaging the other — usually favoring the first goal while restricting the second — may be a major cause of school evasion and revolt. Thus, a major challenge of education is to find a balance between these two goals, in all levels of education and in every cultural environment. All the papers in this collection illustrate, in the diverse situations studied, that this did occur in some cases during the colonial period and in the consolidation of independence of the countries of the New World.

It is agreed by historians that mathematics is a typical mode of thought of the Europeans, and comprises intellectual instruments for every societal behavior needed in the emerging post-feudal European capitalism. Europeans tried to impose their own mathematical practices, focusing on the intellectual instruments of every societal behavior in the conquered lands. These included strategies for counting time (calendars), locating places (distances), delimiting space (territorial property), attributing pecuniary values of objects (commerce), and for work and labor (salaries). Of course, these strategies were absolutely strange to the local peoples.

Evidence of the importance of commerce appears in early publications. It is important to note that the first non-religious book published in the Americas, very useful for the colonizers, was a book on arithmetic related to mining and the commerce of silver and gold: the *Sumario compendioso de las quentas de plata y oro que en los reinos del Pirú son necessarias a los mercaderes y todo genero de tratantes. Con algunas reglas tocantes al arithmética* by Juan Diez Freyle, printed in México, New Spain, in 1556 (Smith, 1921).

This book and others that followed were written using the codes of European mathematics, particularly arithmetic and algebra. Curiously, this practical book contains an appendix with discussions on how to solve quadratic equations, in the style of al-Khwarizmi's *Algebra*. Other than possible practical applications to commerce, I interpret this as the author's attempt to raise the intellectual level of merchants.

Other books from this period deal with geometry and trigonometry, which reveals the need for mathematical instrumentation to cope with production and daily life in the conquered lands. This hints to the development of a special kind of applied mathematics, stimulated by the complexity of problems related to hydraulics and mining. These two constitute the most important problems in the technological development of México. "Subterraneous geometry" became a major theme in Mexican science. Also very important were the efforts for urbanization that took place in all the colonies

Introduction

Figure 1. Cover of *Sumario compendioso*.

(Catalá, 1994). The book *Comentarios a las Ordenanzas de Minas*[1] (1761) by Francisco Xavier de Gamboa, although published much later, is most representative of these developments.

We may consider geometry and trigonometry, along with arithmetic and algebra, as the intellectual instruments to support the socio-economic-political behavior typical of the emerging European capitalism. It is interesting to note that earlier visitors to the New World from the Old World, as for example the Norse navigators to Greenland and Canada, or the Chinese and Japanese to the Eastern Pacific Coast, did not bring this kind of European mathematics. The simple reason for this is that they did not know European mathematics, as I conceive it. They had their own specific intellectual instruments to support their socio-economic-political behavior and their daily life needs.

The conquest, colonization, and early independent times of the Americas had an enormous influence in the development of world civilization. The cultural encounter of the old and the new affected both worlds. The chroniclers of the conquest tell of different ways of explaining the cosmos and

---

[1] A study of mining law in New Spain.

the creation, and of dealing with the surrounding environment. Religious systems, political structures, architecture and urban arrangements, sciences and values were, in a few decades, suppressed and replaced by those of the conqueror. Yet, much of the remaining original behavior and knowledge of the conquered cultures were and still are outlawed or treated as folklore. However, they are surely integrated into the cultural memory of the descendants of the conquered natives. Much of the extant behavior and knowledge of the peoples of the Americas are easily recognized in everyday life. As I have discussed earlier, a new approach to historiography is fundamental for the world history of science and technology, recognizing the contributions of non-Western cultures.[2]

Much of the ideas I present in this introduction have been published elsewhere.[3] It is difficult, indeed impossible, not to repeat myself. I could not avoid using much of the material recently written for my chapter in the *Handbook on the History of Mathematics Education*, edited by Alexander Karp and Gert Schubring, to be published by Springer in 2015.

The Conquest and Early Colonial Times

Spanish and Portuguese colonizers were soon followed by the French, the English, and the Dutch, who also established colonies in the region. The extractive and agricultural economies implanted in the colonies required a human workforce. The resource to slavery was natural. Yet, it is obviously difficult to keep the local population, which is familiar with the environment, under slavery. Much easier was the resource of slaves from other regions. However, with these groups, the colonizers faced the same problems they had trying to engage the natives in their new styles of religious, socio-political, and economic organization. Forced immigrants to the new lands minimize these forms of resistance to the conqueror. Soon, the colonies received immigrants from different parts of the world: many from Africa, in the condition of slaves, some Asiatic, and many Europeans. The immigrants brought new forms of coping with the environment, of dealing with daily life, and new ways of reasoning and learning. The result was the emergence of a synthesis of different forms of knowing and doing, and of explaining and coping with facts and phenomena of the new lands. These new forms

---

[2]See a discussion on these issues in D'Ambrosio (2000).
[3]D'Ambrosio (1977, 1995, 1996, 2000, 2006).

were generated in different communities and were available to workers and to the people. All the new forms of knowledge and behavior were absolutely interrelated, synthesizing cultural forms brought by outsiders and those retained by the natives. Particularly, language became an obstacle to education, and continues to be so.

In the Americas, the variety and peculiarity of the expositions of cultures and the specificity of the migrations reveal an effort of the colonizer to transfer, with minor adaptations, the forms of social, economic, and political organization and administration prevailing in the metropolises, including schooling and scholarship (academies, universities, monasteries). The new institutions in the Americas were based on those styles, mostly under the influence, and even control, of religious orders.

All this, which took place during most of the sixteenth, seventeenth, and eighteenth centuries, occurred while new philosophical ideas, new sciences, new ways of production, and new political arrangements were flourishing in Europe. The cultural facts produced in Europe were assimilated in the colonies under specific, mostly precarious, conditions. Indeed the colonies were the consumers of some of these new cultural facts. There is a clear coexistence of cultural goods, particularly knowledge, produced locally and those produced abroad. This is clearly shown in the various essays about specific countries.

Although this transmission is a question affecting the relations between academia and society in general, hence between the ruling elites and the general population, the dynamics of the relation between social strata is particularly important for understanding the role of intellectuals in the colonial era. Thus ethnomathematics comes as a fundamental instrument for historical analysis.[4]

The information on the pre-Columbian civilizations has relied mostly on the chroniclers of the conquest, who described many facets of the conquered peoples, such as the Maya stelae and writing, the Peruvian *quipus*, the Aztec daily life, and, in general, their vision of the universe. These chronicles were interpreted in order to justify the "civilizatory" mission during the establishment and consolidation of the colonies. These interpretations were biased and, as it is easy to understand, failed to identify and recognize any form of mathematical knowledge in these cultures. The emergence of new scholarship allows for a new reading of the chroniclers and a new understanding of the complexity of the knowledge systems and societies

---

[4]See D'Ambrosio (1995, 1996).

of pre-Columbian civilizations. Scholarship on the history of mathematics of the traditional cultures of the New World is intensifying.

Although in all the countries of the region current mathematics is fully integrated with world mathematics, there are demands, of a social and psychological nature, that justify much attention to the traditional roots of native mathematics. For the pre-Columbian period, sources are available mainly for the Aztec, Maya, and Inca civilizations. To look into other civilizations, mainly in the northern and southern prairies, and of the tropical cultures in the Amazon basin, an enlarged concept of sources is needed, mainly drawn from anthropology and ethnomathematics.

Much finer divisions of the civilizations, taking into account both political and cultural specificities, and new methodologies, are needed for a special study of pre-Columbian mathematics, similar to what occurs when we study traditional African cultures.

Education in the pre-Columbian cultures was different in the various regions of the continent. The imposition of the priorities of the conquerors for production and services profoundly affected the existing models of education. The heterogeneity of cultures makes it impossible to refer to a pre-Columbian educational model. Basically, in the lowlands we had cultures of hunters-gatherers, where initiation rites were central in the preparation of new generations, and more complex cultures in the highlands. Education focused on clearly demarcating the roles of women, in caring for children and crops, and of men, concerned with hunting and war. There were strategies to identify those that would be the ruling class of chiefs and shamans. In the highland cultures, such as the Aztecs and the Incas, education was more elaborated.

There was a clear effort made by the colonial regimes to ignore or obliterate any sense of the history and the achievement of the native cultures. Today, we face the difficult task of reconstructing the histories of these cultures, both looking into the chronology of the events and understanding the important migratory currents that shaped their developments. However, our knowledge of the pre-Columbian period is still very incomplete. Of course, this leads us into the conceptual framework of what might be considered, in today's categories, as science, technology, and mathematics in pre-Columbian times.[5]

The Incas, whose capital was Cuzco, now a city of modern Perú, manipulated a complex texture of knotted strings, called *quipus*, either to record

---

[5] For a brief introduction to this theme, see D'Ambrosio (1977).

Figure 2. Accountant/treasurer in charge of the *quipus*, and a *yupana* in the left corner (from the book of Poma de Ayala, Murra, *et al.* 1987).

narrative or as an accounting device. Their real use in the Inca civilization is not yet clarified, but the *quipus* clearly contained qualitative and quantitative information. For calculation, the Incas also used a device that was a combination of an arithmetical table and abacus, called the *yupana*. Both are clearly identified in the well-known picture of Poma de Ayala (Murra, *et al.*, 1987), a chronicler of the sixteenth century.

A detailed analysis and classification of Peruvian *quipus* is due to the mathematician Marcia Ascher and the anthropologist Robert Ascher, with the 1981 publication *Code of the Quipu*, which became a classic in the field (Ascher & Ascher, 1997). An important source for understanding pre-Columbian cultures is the current practices of extant native communities. Of course, these practices are the result of the cultural dynamics of the encounters, but reveal much of the knowledge and behavior of their ancestors.[6]

---

[6] An important survey of pre-Columbian mathematics of extant cultures is the book by Michael P. Closs (1986), *Native American Mathematics*.

In the colonial times that ensued, we see the exploitation of the lands and resources, and of the peoples of the conquered regions. The colonizers changed the means of production with their traditional European agricultural and mining techniques with which they exploited the native production, mainly in metallurgy. The native religions were simply destroyed. Food habits were also considerably modified. Wheat, rice, coffee, citrus trees, sugar cane, and bananas were all grown for export.

In the sciences in general, Latin America was the recipient of scientific advances. In the process of cultural dynamics, this knowledge was modified and adapted to a new reality. This kind of peripheral science was maintained throughout independence. Indeed, the colonial style and submission of the native population, with land distribution determined by the conquerors and the jurisdiction of the colonizers, was kept after independence. Independence was a movement impregnated with republican ideas led by creoles — not necessarily an aspiration of the native populations — throughout the Americas. Even after independence, education followed the model of the former imperial system. Colonial science, considered as contributing to the mainstream of scientific development, was at best very modest.

Spanish America and Portuguese America followed parallel but distinct historical paths. While Portugal never allowed the establishment of universities in its colony — so that the prevailing form of education consisted of *colégios*, secondary schools organized by the Jesuits — there were numerous universities founded in the Spanish colonies. Gregorio Weinberg explains the reason for this difference in educational structures as the dissimilar ethos of the colonial enterprise (Weinberg, n.d.). While the Portuguese had exclusively economic interests, the Spanish aimed, from the early periods of conquest, to establish complex societies in the New World, with stable social and Christian cultural values as their foundation. Some historians refer to this as the *Reconquista* of the New World for Christianity, echoing the period (711–1492) that sought to end the Islamic rule in Spain on behalf of Christianity.

From Past to Present

Mathematics education as an academic discipline is relatively new. We may trace its origin to the transition of the nineteenth to the twentieth century, mainly as a consequence of the rivalry between European empires in the emergence of new scientific and technological advances and an intense industrialization. Mathematics was an essential component of education for these advances. Also, this period brought new perceptions of the mind and the teaching/learning paradigm. The recognition of mathematics education as a discipline was established in 1908 at the International Congress of Mathematicians held in Rome, when CIEM/IMUK/ICMI[7] was created. South America was present with a participant from Argentina. In 1928, in Bologna, when CIEM/IMUK/ICMI was reformulated, Argentina was again present. However, I have not found any relevant influence of the CIEM/IMUK/ICMI in Latin America.

At the 1950 International Congress of Mathematicians (ICM), held in Cambridge, USA, there were many participants from Latin America, but no relevant presence of mathematics educators. In the 1954 General Assembly of the International Mathematical Union (IMU), it was formally decided to revive the International Commission on Mathematical Instruction (ICMI). An executive committee was appointed, with Henri Fehr as honorary president, Albert Châtelet as president, and Heinrich Behnke, who became president in 1955, as secretary. It is reported that Alberto Sagastume and José Babini were delegates from Argentina. It is important to observe that no Latin American country participated in the Comparative Study of Mathematics Education, commissioned to ICMI by IMU in 1958.

Since their early days, international mathematical organizations were concerned exclusively with academic mathematics and mathematics education as a mission to convey academic mathematics to the new generations as part of a conservative school tradition.

The idea of considering cultural context in mathematics education was not contemplated by CIEM/IMUK/ICMI in 1908. I believe the first reference to cultural issues as a factor to be considered in mathematics education is due to the distinguished Japanese algebraist Yasuo Akizuki in 1960. He wrote:

> Oriental philosophies and religions are of a very different kind from those of the West. I can therefore imagine that there might also exist different modes of thinking even in mathematics. Thus I think we should not limit ourselves to applying directly the methods which are currently considered in Europe and America to be the best, but should study mathematical instruction in Asia properly.

---

[7]Comission Internationale de L'Enseignement Mathématique/Internationale Mathematische Unterrichtskommission.

Such a study might prove to be of interest and value for the West as well as for the East.

I see this as an important moment of the overture of ICMI to ideas coming from other cultural contexts. Indeed, I recognize this as the first opening of a space for ethnomathematics in academic mathematical reflections. Regrettably, the remark of Akizuki did not draw attention from his contemporaries. It took about 20 years for the importance of such remarks to enter the academic mathematics scenario.

In 1959, Marshall H. Stone became the president of ICMI. He announced a meeting, in Bogotá, Colombia in 1961, under the aegis of ICMI and with the financial support of UNESCO, OAS (Organization of American States), and other organizations, to discuss questions related to mathematics education that were common to all countries in the region. The organizing committee of the meeting was formed by Marcelo Alonso (Cuba), José Babini (Argentina), Howard Fehr (United States), and Leopoldo Nachbin (Brazil). In this meeting, which became known as the First Inter-American Conference on Mathematics Education, the Inter-American Committee on Mathematics Education (IACME) was created.[8]

This was the decisive moment of the insertion of Latin America in the international movement of mathematics education. Since then, twelve Inter-American Conferences on Mathematics Education have been held.

When Hans Freudenthal was president of ICMI (1967–1970), he championed the realization of an International Congress on Mathematics Education (ICME), with the support of the French government and UNESCO, in Lyon, France 1969. This was done entirely without any connection with the IMU (Lehto, 1998).

The tone was the New Math, and the influence of the Bourbaki group was felt in Latin America through Jean Dieudonné. Mathematical content, as opposed to pedagogical considerations, was the dominating issue. This characteristic would prevail in the next three conferences.

The program of the 5th IACME, held in Campinas, Brazil, was organized around only three plenary talks and four plenary panels. Invitees then reflected upon what were regarded as important research directions in Latin America. The majority of speakers were mathematics educators from Latin American countries. There was some criticism about the way the reform of mathematics teaching was carried out in the 1960s, and the

---

[8]http://centroedumatematica.com/aruiz/libros/Historia_CIAEM/iacme.html

panelists expressed ideas of a methodological nature. They also stressed the importance of research in the teaching (didactics) of mathematics.

I consider the Campinas meeting as a turning point in mathematics education in Latin America, not only by giving a predominant role to educators from Latin America, but by proposing innovation, and discussing and questioning the modern trends in education which began to be considered as central to mathematics education.

Subsequent IACMEs have been important accounts of innovative actions in mathematics education in Latin America and the establishment of new areas of interest in newly-created research centers. It is difficult to give an overall picture of the state of research in mathematics education in all of Latin America. This is superbly done, for each country, in the essays of this volume.

The ties to Iberian countries are worth noting, which resulted in the creation of a series of conferences (CIBEM, *Conferencias Iberoamericanas de Educación Matemática*) and the establishment of FISEM (*Federación Iberoamericana de Sociedades de Educación Matemática*).[9] Regrettably, internal conflicts still prevail that have consequences for rivalries in the regional organizations.

Mathematics educators from Latin America have been very active in the International Study Group on the Relations between the History and Pedagogy of Mathematics (HPM) and the International Study Group on the Psychology of Mathematics Education (PME) since their inception in 1976. Also mathematics educators from Latin America have a significant presence at the International Organization of Women in Mathematics Education (IOWME), the World Federation of National Mathematics Competitions (WFNMC), and the International Study Group for Mathematical Modeling and Applications (ICTMA).

The Future

In considering the current status, I cannot avoid mentioning the disastrous effects of standardized tests. They are discussed in most of the essays in this collection. I believe these tests are a gross mistake, responsible for lowering standards of education and social justice in all countries. I have commented on this in several of my writings and will not discuss this more in this essay.

---

[9] http://www.fisem.org

Instead, I will address the path I think we are taking as an international community.

My main concerns are of a political nature. We must remember the events of 1960s. That decade was a landmark in the history of modern civilization. No doubt, it was the most active in terms of reflections, theories and practices about mathematics education, as seen in the essays of this volume. However, there are practically no references — except for Puerto Rico and México — to the political, scientific and technological scenarios of the period at the regional or international levels.

In the 1960s, the scientific bases for bio-, nano- and cyber technologies, as well as for artificial intelligence and the conquest of space, were developed. World politics was in a period of turmoil. In synthesis, this was the decade when the Vietnam War escalated and spread to practically all of Southeast Asia. It was the decade of the construction of the Berlin Wall by the German Democratic Republic (GDR, East Germany) in 1961, of the Cuban Missile Crisis in 1962, of the killing of John F. Kennedy in 1962, and of Martin Luther King Jr. in 1968. Never was the Cold War so close to becoming the most destructive war ever. In Latin America, military dictatorships flourished in many countries and guerrilla movements burgeoned. A similar situation took place in Africa and other regions of the world. In June 1967, the third Arab-Israeli War erupted. This involved Israel, Egypt (then the United Arab Republic), Jordan, and Syria. Although short in duration, the consequences were enormous. In May 1968, civil unrest in France, followed by strikes and the occupation of factories and schools, carried the message of disapproval of the capitalist system and the perception that new generations were being prepared to serve in the armed forces in the name of national interests and international rivalries. The younger generations seemed on their way to the same disgrace, suffering, and destruction of the generation of their parents; they were very sensitive to these political movements. The unrest and protests, sometimes violent, spread through the entire world and repression ensued, with different mechanisms and strategies. The consequences of 1968 were particularly felt in the various levels of schooling. Regrettably, the intense movements of change in mathematics education, particularly the New Math proposals, ignored these events.

A new moment of intense political turmoil started in the final decade of the twentieth century. The big change resulting from the fall of the Berlin Wall in 1990 marked the beginning of a new geography of Europe, with geopolitical, cultural and socio-economic characteristics. We entered the twenty-first century under the lasting impact of the events of September 11, 2001, in New York City. Many other events affected the entire world. Among them, we highlight the invasions and civil wars in Afghanistan, Iraq, Eastern Europe, North Africa and elsewhere in the world. All of these events have had, and continue to have, great repercussions across the globe, including in Latin America.

How many of the consequences of this world scenario are being considered by mathematics educators? It is naive to say that this does not affect mathematics education. As an example, school attendance is disrupted in many cities under conflict. Millions of children and teenagers who study mathematics are affected, including children and teenagers outside the conflict zones, mainly because of security measures and control mechanisms. Students are subjected to permanent impact of television and internet news. There is a generalized fear as well as mistrust. All this affects the emotional and the imaginary, hence the cognitive capabilities of students of all ages, and particularly of children and teenagers. They all study mathematics.

We recognize the enormous dynamics of mathematics education as a discipline during the 1960s and 1970s. Although there are only minor, if any, hints of the influence of the student unrest in some aspects of the efforts of mathematics educators, the future of mathematics education must consider all these issues. We must ask how this affects mathematics education — the profound changes in the daily life and perturbing threats to the survival of civilization. This asks for a new look into education, particularly in mathematics education. I quote the 1998 interview of Mikhail Gromov in the *Notices of the AMS*:

> We must do a better job of educating and communicating ideas. The volume, depth, and structural complexity of the present body of mathematics make it imperative to find new approaches for communicating mathematical discoveries from one domain to another and drastically improving the accessibility of mathematical ideas to non-mathematicians.

I paraphrase these remarks and say that we must drastically change and open our communication about the major issues affecting our world today, hence our school practices. I do not know how this affects mathematics education, but I feel it will have an impact. The history of mathematics shows how intimate the evolution of mathematics and the world scenario has been. Clearly, there is an implication of the evolution of mathematics for mathematics education. All the technological advances must be taken into account. It is very important for mathematics educators to be concerned with these implications. A good opportunity is to bring the developments of the project "Mathematics of Planet Earth 2013" (MPE2013), which will now continue as "Mathematics of Planet Earth" (MPE), to the attention of mathematics educators. This may imply some different ways of considering formalism and rigor in mathematics education. The future of mathematics education, particularly in Latin America, must face these issues.

There is still much to be done. I see the publication of this collection of essays as a major effort in a positive direction. I am sure it will have a great impact in the development of mathematics education in Latin America. It will bring together information and ideas from all the countries of the region and I also hope this will be an opportunity to eliminate regional rivalries and foster camaraderie for inter-American and international cooperation.

Bibliography

Akizuki, Y. (1960). Proposal to ICMI, *L'Enseignement mathématique*, t.V, fasc.4, 1960; pp. 288–289.

Ascher, M., & Ascher, R. (1997). *Mathematics of the Incas: Code of the Quipu*. New York: Dover Publications.

Baumont, M. (1949). L'Essor industriel et l'impérialisme colonial. Paris: PUF (col. Peuples et civilisations, v. 18).

Catalá, J. S. (1994). Ciencia y Técnica en la Metropolización de América, Theatrum Machinae, Madrid.

Closs, M. (Ed.) (1986). *Native American Mathematics*, Austin: University of Texas Press.

D'Ambrosio, U. (1977). Science and Technology in Latin America During the Discovery. *Impact of Science on Society*, *27*(3), 267–274.

D'Ambrosio, U. (1995). Ethnomathematics, history of mathematics and the basin metaphor, histoire et epistemologie dans l'education mathématique/history and espistemology in F. Lalande, F. Jaboeuf, & Y. Nouaze (Eds.), *Mathematics Education* (Actes de la Première Université d'Eté Europeenne, Montpellier, 19–23 juillet 1993). IREM, pp. 571–580.

D'Ambrosio, U. (1996). Ethnomathematics: an explanation, vita mathematica, in R. Calinger (Ed.), *Historical Research and Integration with Teaching*, pp. 245–250. Washington DC: The Mathematical Association of America.

D'Ambrosio, U. (2000). A historiographical proposal for non-western mathematics, in H. Selin (Ed.), *Mathematics Across Cultures. The History of Non-Western Mathematics*, pp. 79–92. Dordrecht: Kluwer Academic Publishers.

D'Ambrosio, U. (2006). A concise view of the history of mathematics in Latin America, *Ganita Bharati* (Bulletin of the Indian Society for the History of Mathematics), *28*(1–2, June December), 111–128.

D'Ambrosio, U. (2008). ICMI and its influence in Latin America, in M. Menghini, F. Furinghetti, L. Giacardi, & F. Arzarello (Eds.), *The First Century of the International Commission on Mathematical Instruction (1808–2009). Reflecting and Shaping the World of Mathematics Education*, pp. 230–236. Collana Scienze e Filosofia, Istituto della Enciclopedia Italiana, Roma, 2008.

Dobb, M. (1954). *Studies in the Development of Capitalism*. London: Routledge & Kegan Paul.

Gromov, M. (1998). *Notices of the AMS*, *15*(7), 486–487.

Lehto, O. (1998). *Mathematics Without Borders. A History of the International Mathematical Union*. Berlin: Springer-Verlag.

Murra, J. V., et al. (1987). *El primer nueva corónica y bein gobierno*, Felipe Guamán Poma de Ayala, John V. Murra, Rolena Adorno, Jorge L. Urioste, México: Siglo Veintiuno (Ed.), 1987.

Poma de Ayala, & Guaman, F. (1987). *Nueva crónica y buen gobierno* (orig.1615), Edición, Introducción y Notas de John V. Murra, Rplena Adorno, Jorge L. Urioste, Siglo XXI, México.

Smith, D. E. (1921). The first work on mathematics printed in the new world. *The American Mathematical Monthly*, *28*(1), 10–15.

Via, M., Gignoux, C. R., Roth, L. A., Fejerman, L., Galanter, J., et al. (2011). History shaped the geographic distribution of genomic admixture on the island of Puerto Rico. *PLoS ONE* *6*(1): e16513. doi:10.1371/journal.pone.0016513

Weinberg, G. (n.d.). *Modelos Educativos en la Historia de América Latina*, UNESCO/CEPAL/PNUD, Buenos Aires: A-Z Editora.

About Ubiratan D'Ambrosio

Ubiratan D'Ambrosio earned his doctorate from the University of São Paulo, Brazil in 1963. Since 1972 he has been on the faculty of the State University of Campinas (UNICAMP), São Paulo, where he became an Emeritus Professor of Mathematics. He retired in 1994 and currently he is Professor of the Graduate Program in Mathematics Education at Universidade Anhanguera de São Paulo (UNIAN). D'Ambrosio has been a Fellow of the American Association for the Advancement of Science (AAAS) since 1983, and is also a Member of the *Académie Internationale d'Histoire des Sciences*. He received the Kenneth O. May Medal of History of Mathematics, granted by the International Commission of History of Mathematics (ICHM/IUHPS/IMU), in 2001, and the Felix Klein Medal of Mathematics Education, granted by the International Commission of Mathematics Instruction (ICMI/IMU), in 2005.

Chapter 1

ARGENTINA: A Review of Mathematics Education through Mathematical Problems at the Secondary Level

Betina Duarte

Abstract: We propose a review of the teaching of mathematics in Argentina based on the presence and evolution of problem-solving in textbooks for the secondary school. We also take into account the changes introduced by educational policies in curricular documents. The period selected (1950–2000) has significant influence on the current situation of teaching.

Keywords: History of mathematics education; mathematics education in Argentina; mathematics education; math problem solving; secondary school; math textbooks; curricular documents.

Historical Overview

While we have chosen a timeframe to discuss mathematics education in Argentina, we believe it is necessary and appropriate to provide a brief historical background to contextualize our proposed analysis.

We can surmise that the colonial education promoted and sustained largely by religious orders arriving to Río de la Plata during the midsixteenth century, established the first ideas about the teaching of mathematics.[1] Among the various groups who came to this region were the

---

[1] Known as the *Ratio Studiorum*, the Jesuit educational model was developed as a curriculum to be implemented in all schools of the Society. The education system comprised several levels of learning. In the first, the teaching of basic mathematics, first letters, and Christian doctrine were introduced. The objective of forming an erudite man of faith shaped the pedagogical approach.

Jesuits who had greater presence through the establishment of institutions dedicated to education: education or nursing homes, colleges and universities. One of those, created in 1613 under the name Colegio Máximo, with the aim of granting bachelor's and doctoral degrees, and train teachers, was next transformed into the Universidad de Córdoba del Tucumán in 1621. Its teaching structure emulated the University of Salamanca. It was administered by Jesuits until they were expelled in 1767. This was the first university in Argentina, now known as National University of Córdoba.

Buenaventura Suárez, a priest, astronomer, geographer, and mathematician (1679–1750), established an astronomical observatory. His calculations were used to select the exact locations of the thirty Jesuit missions in Paraguay (Stacco, 2011).

The same order, settled in what is now the center of our country, founded in 1662 the San Ignacio School in Buenos Aires, a great intellectual and cultural center. Similarly in 1685 they created the Monserrat School, an institution that prepared students for college. For many years it trained the future leaders of the region. Both schools — today depending on their respective national universities — contributed to the formation of those who gestated the emancipatory project of our country.

In 1773, the Viceroyalty propelled the creation of a school of applied mathematics by French engineer Joseph Sourryères of Souillac. The shortage of students forced the establishment to reorient their activities. Shortly after, around 1799, and thanks to the initiative of Manuel Belgrano, it transformed into the Nautical Academy, which included the teaching of arithmetic, geometry, spherical and rectilinear trigonometry, cosmography and differential and integral calculus.

His first 26 students used the texts of Etienne Bezout, *Cours de mathématiques à l'usage du Corps Royal de l'artillery* (Paris 1770–1772) and Benito Bails, *Principles of Mathematics* (Madrid, 1789–1798).

Manuel Belgrano was a central figure to ensure the presence of mathematics since the late viceroyalty to the early period of emancipation in the region. In this second stage, the teaching of mathematics was thought necessary for the instruction of the army because it was "the more useful science for a military man" and the more efficient means to form "smart military in the art of defense" (Arata and Mariño, 2013). This included the teaching of arithmetic, plane geometry, rectilinear trigonometry and practical geometry. Instruction was carried out in the School of Mathematics established in September 1810 (Gaceta de Buenos Aires, 1910).

The University of Buenos Aires, founded in 1821, had an Exact Sciences Department since its inception. Under the initiative of Felipe Senillosa, of Spanish origin and trained at the Academy of Engineering of Alcalá de Henares, Lagrange's trigonometry texts, Lacroix's *Arithmetic*, De Monge's *Descriptive Geometry* and Poisson's *Principles of Mechanics* were taught. This evidences the early influence that the French school of mathematics had at the beginning of the University of Buenos Aires, now the largest public university in Argentina. In 1872, driven by members of the Faculty of Exact Sciences, the Argentina Scientific Society was created to "promote, in particular the study of the mathematical, physical and natural sciences" (Stacco, 2011).

Before the end of the nineteenth century, three new universities were created: the Tucumán University (1875), La Plata University (1882), and Santa Fe University (1889). Discussion on a possible educational model, the role of the State, teaching contents, the conditions to be met by teachers, teaching methods and funding sources, were developed largely in the Pedagogical Congress of 1882 and crystallized by the 1420 Law, also known as Láinez Law, in 1884.

So far we have tried to present some historical milestones of education in Argentina, particularly in mathematics education, to give a context to our proposal. These facts do not deny the earlier presence of *pueblos originarios* (native peoples) in our region who, in one way or another, constructed mathematical notions. There was a mathematical knowledge developed by those who inhabited these regions that were neglected or destroyed by the effect of the arrival of the Spaniards. In order to make clear some kind of testimony to these cultures, it is appropriate to say that, for example, the Mapuche people, from the Patagonia region, "developed an oral decimal number system" to meet their need to count and, therefore, limited to numbers less than 10,000 (Belloli, 2009). Also, the north territories populated by Kolla and Aymara communities, significantly influenced by the Inca culture, used *yupanas* to count and *quipu* to record information. The numbering system of these communities was also decimal. The geometry was present in their clothing designs and the construction of their homes.

Foundational Ideas

The teaching of problem-solving in mathematics has attracted the attention of various educational communities in Argentina: teachers, educators, researchers, school heads, government officials and their respective technical teams, and textbook publishers. Some of them are also interested in problem-solving as a way to teach mathematics. These are two different matters — different approaches to the same topic, namely, how math problems are conceived in the classroom. The two views indicate different conceptions about the teaching of mathematics.

Some of the questions regarding the first topic we have indicated center around the problems themselves. How is the solving of problems taught? What is the goal? Based on what mathematical topics? Does mathematics begin and end with the solving of problems?

The questions that arise when considering the possibility of teaching mathematics on the basis of problem-solving brings to light the need to think about classroom management and the link between the problems and the mathematical knowledge one intends to teach. How do math problems exist within a secondary school classroom? Who presents them? Who solves them? How is a class organized around the solving of a problem? What part of the student's problem-solving task should be left to the individual, and what part should be dealt with in the collective space of the classroom? How can interaction among students be generated while problems are being solved?

To answer whether problem-solving can be a way of teaching mathematics, we must accept that problem solving has certainly been and continues to be a powerful driver of mathematical output. One has only to look at the colossal production that resulted from Fermat's marginal note on his copy of Diophantus' *Arithmetica*. We can also point to the 23 problems posed by David Hilbert at the International Congress of Mathematicians in Paris at the beginning of the twentieth century (Stacco, 2011).

Nevertheless, turning away from the production of mathematics to the teaching of mathematics, when research in math education establishes the existence of a strong link between the mathematics teaching and learning process and the mathematical production of students, an impulse is given to the idea of considering problem-solving as a process capable of being interwoven in the learning process. Research on this topic (G. Brousseau, M. Artigue, R. Charnay, J. Kilpatrick and A. Schoenfeld, among others) has had repercussions in Argentina.

The contributions of these researchers confirm the need for problems to be considered in interaction with the subject that resolves them. In addition, they promote the idea that problem-solving gives meaning to the mathematical notions in play. We will take these approaches as a starting-point for the study of the presence of problems in the textbooks we have selected.

Our proposal in this chapter consists of reviewing some of the history of the presence of problem-solving in the teaching of mathematics in Argentina. To do so, we use two points of reference: (1) textbooks and teaching manuals — books that have defined secondary school teaching in the country for many generations of Argentines, and (2) curricular educational policy documents. We have considered it advisable to restrict the period to be considered. Hence, we propose reviewing two eras: the period between the 1950s and 1970s inclusive, and the period from the return of democracy until the end of the 1990s. We highlight certain historical landmarks that represent significant changes in the way mathematics has been taught, as well as changes in the customs and traditions of the publishers of such texts.

An Anachronistic Example

We have chosen to include an example from outside the timeframe we have adopted, as well as from outside the secondary school system that concerns us, because of its originality. In 1901, a book written by a "normal professor"[2] named Alfredo Grosso under the title of *Arithmetic Exercises and Problems for Use by Standard Schools*, dedicated a chapter to methodological considerations regarding "studying together". He points out to students the advantages of sharing the task of problem-solving with peers, based on the description of the studying process of two hypothetical children (from the fourth grade of primary school)[3]:

> When it came to problem-solving, they acted as follows: Each one, separately, solved the problems in his own manner, and once solved, wrote the result on the blackboard; if the answers were the same, they checked their calculations to verify whether their solutions, despite arriving at the same result, were the same or not. If the

---

[2] At the time, primary school teachers graduated from Normal Schools, some kind of special high schools, hence the denomination.

[3] The web page of the National Teachers' Library has a section with scanned images of 19th and 20th century mathematics textbooks. http://www.bnm.me.gov.ar/e-recursos/medar/exposiciones/matematicas/pdf_matematicas/011759.pdf.

results differed, each explained his answer, discussing it until arriving at a solution they considered to be exact. If there was no agreement, they submitted the problem with their own solution (Grosso, 1907).

At first sight we find the validity and currency of such words stressing the importance of *interaction among peers* and the *clarifying of strategies by the students*. Images of classrooms in the early twentieth century show students organized in a manner (the classic wooden bench pattern) that is not conducive to group study.[4] Such a structure would only begin to be seen at the beginning of the 1970s. At the start of the twentieth century, classrooms were designed for individual work by students at all levels of learning.

A second reading of the paragraph in question allows us to infer that the problems to which the author refers have a "single" correct, numerical answer, which serves as a starting point for consideration of procedures or "solutions" in the form of calculations based on what is indicated next, reinforcing the idea that mathematics is about performing calculations. It therefore states, albeit implicitly, that (a) mathematical problems have "a result or solution", (b) such a result derives from numerical calculations, and (c) desirable solutions in mathematics are precise, exact, and unique.

These inferences seem to describe the concept of mathematics that has been transmitted through the teaching of mathematics, at least since the start of the twentieth century. In this chapter we document the scope of this approach.

First Stage: Textbooks from the 1950s to the 1980s

Publisher Kapelusz brought out the first edition of a series of books for all secondary school years in their different categories (regular high school, commercial and technical) in 1940, all of them written by Celina Repetto,[5] Marcela Linskens[6] and Hilda Fesquet.[7] A dozen mathematics textbooks

---

[4]Grosso's notes went on to indicate that it was advisable not to form study groups of more than three students.
[5]Professor of Secondary School Mathematics and Physics Education and Doctor in Physics and Mathematical Sciences at the University of Buenos Aires.
[6]Professor of Secondary School Mathematics Education.
[7]Teacher of Education and Professor of Secondary School Mathematics Education.

were written and re-written. In the early 1950s these textbooks were widely used in the country's various types of secondary schools. Some ran into over 20 editions through the end of the 1970s.[8] In the mid-1960s, they incorporated the proposals for change formulated for Latin America[9] with the introduction of the teaching of the new math.

The tables of contents of these books all have something in common. At the end of each chapter came an identical heading in bold print: **Applied Exercises and Problems**.

The term "exercises" is applied principally to the performance of mathematical operations that can result in *simplification* (which involves simplification in mathematical terms but also factorization, operating and re-writing for reduction) of expressions or the solving of equations, whereas the term "problems" is reserved for premises expressed by means of verbal constructions "in words" which always include "the data" and come with a request to find the answer to some question.

The term "applied" for both exercises and problems indicates the intention: the student will be required to bring into play the theoretical notions developed in the chapter to be able to solve them. This in turn reflects one conception of the teaching of mathematics: theory precedes practice. Teaching is responsible for communicating the notions that mathematics has developed and coined, and the student is responsible for putting them into practice.

The texts have a generic target that can be either the student or the teacher. Indeed, in the period to which we refer many teachers used these textbooks to deliver their lectures (outlining the theory that the student could read in the textbook), as well as to indicate the exercises that the student needed to solve. Students used the textbooks to do the exercises proposed in class for "practice" and also to go over concepts that had not been properly understood.

The use of textbooks in secondary schools during that period was similar in the case of many disciplines. Teachers used them as their classroom text or as support for the development of their classes, and students used them as study material. The books themselves did not have space for students to write in them, and the sale of used textbooks was common practice.

---

[8] They can still be found in second-hand bookstores today.

[9] One of the organizations promoting changes in the teaching of mathematics for Latin America was the Inter-American Committee for the Teaching of Mathematics (CIAEM) since its first conference in 1961.

In the case of the math texts we have just described, they did not contain any "teacher guide" or a prologue directed to either the teacher or the student. There was no indication of the teaching intention in any of their chapters. Seldom was a footnote used, usually addressed to the teacher, to reveal some very specific intention of the authors.

This is the case of a footnote in the text of *Aritmética 2*, in which the authors alert at the start of the **Applied Exercises and Problems** section that:

> Exercises such as these, lengthy and painstaking to solve, are intended solely to develop in the student the habit of observation and meticulous and well-organized work. Teachers who do not consider them to be necessary, or who can replace them with others for the same purpose, may ignore them.[10]

There follow 40 exercises to practice operations with whole and rational numbers, increasing in order of quantity and complexity of the operations proposed. We show here one of the problems found a little over halfway into the proposed exercises to elucidate both the explanation of the authors and the way this teaching intention was conceived.[11]

$$\sqrt{\frac{\sqrt[3]{(-2)^5}\left(\frac{1}{2}\right)^{-1} + \frac{\left(2-\frac{1}{2}\right)^{-2}}{\left(\frac{3}{4}\right)^{-1}} + \sqrt{\frac{\frac{1}{2}+\left(-\frac{1}{2}\right)^{-2}\left(\frac{1}{10}\right)^{-1}}{2}}}{\frac{\left(-1+\frac{2}{5}\right)^{-2} \div \frac{1}{3}}{1-\frac{1}{4}}\left(-\frac{1}{36}\right)\left(-\frac{6}{5}\right)^{-1}}}$$

To solve this problem, students must have the capacity to resolve, simplify and operate a significant number of expressions and similar equations[12] in which they can put into practice their calculating skills, time and time again.

---

[10] This footnote is contained in *Aritmética 2*, in the chapter headed **Review of operations with whole and rational numbers**, page 46.

[11] This exercise has been chosen as a representative example of the full collection of proposed exercises.

[12] Another topic is that of **repetition** as a teaching strategy. This matter exceeds the scope of our coverage in this chapter, but we should say that the repetition of operations, actions, and tasks was part of a scenario for which many professions and jobs were noted at the time.

Operations in the field of whole numbers

In general terms, the chapters focusing on the study of numbers (natural, whole, rational, irrational, complex) and the operations that secondary schools were responsible for presenting to the respective groups of numbers present a practice orientated to calculate, operate, and obtain results. There is a significant absence of *intra-mathematical* problems that could problematize such notions. At successive stages of the practice, students are required to practice operations such as[13] (a) removing parentheses and distinguishing terms of a numerical expression to be able to calculate it, (b) dividing and multiplying by a unit followed by multiple zeros, and (c) representing fractions as decimals and/or decimals as fractions.

The first chapter of *Arithmetic and Algebra for the Fourth Year of Secondary School and Teacher Education*[14] is dedicated to radicals.[15] Beginning with the definition of powers, the text goes on to provide definitions (for example, of nth roots, nth powers) and theorems (e.g. "The nth root of a product is equal to the product of the nth roots of each of the factors, as long as the operations are possible.") in which the hypothesis and thesis of each theorem are detailed and their demonstration is explained. The corresponding inverse theorems are also presented.

In over 68 pages, *Teoremas y Ejemplos* (Theorems and Examples) deals with matters such as the uniqueness of the root when the index of the root is odd (under the heading of "Rule of Signs"), simplification of radicals, reduction of radicals to a minimum common index, extraction of factors from radicals, introduction of factors within radicals, similar radicals, sum of similar radicals, multiplication and division of radicals and rationalization of denominators (a procedure that requires three cases to be considered).

The exercises to be applied for this chapter, a total of 547 (contained in the aforementioned 68 pages), require the application of this set of rules explained in the form of the presentation of theorems and their respective

---

[13] To enumerate the tasks and activities that we will analyze we take the same chapter on "Review of operations with whole and rational numbers" as this same topic will be used in the analysis of the texts from the second stage.

[14] Third printing of the 13th edition, January 1964.

[15] The following chapters deal with: Powers of fractional exponents, Logarithms, Arithmetical progressions, Geometric progressions, Complex numbers, Second degree equations with an unknown, and conic sections.

demonstrations. As from there, students are expected to exercise these properties of calculations with radicals applying the distributive properties when required, so as to be able to simplify lengthy expressions (mostly numerical expressions, in a few cases with the appearance of letters). Each statement specifies the property to be applied for the set of exercises. For example, the indication could be: "Resolve the following exercises applying the reciprocal property of the distributive with respect to multiplication and division" proposing to work with expressions of the type:

$$\sqrt[3]{-\frac{3}{5}m^{-2}} \cdot \sqrt[3]{\frac{1}{15}m^{-5}n^2} \cdot \sqrt[3]{\frac{1}{5}m^{-2}n^{-2}}.$$

As can be imagined, the mechanization of operations is one of the goals of the work. Evidently there is a cultural value in the algorithms that mathematics has achieved as an efficient way of proceeding in the face of the need to operate numerically. What is not shown is where the need for such numerical operations comes from. Beyond numerical operations, the chapter offers no connection to any other mathematical or non-mathematical area. We could conclude on the basis of the text that knowing mathematics is the same as knowing how to deal with numbers. The display format of this knowledge in the "hypothesis-thesis-proof" style keeps one intention: to account for a finished formulation of knowledge, show the structure of mathematical knowledge, and inform the reader of the need to understand mathematical formulation rules and justification of knowledge.

We are aware that we are dealing with a period that predates the presence of calculators in the classroom, which will later lead to a questioning of such calculations and an imbalance between the mathematical practices that the various levels of education decide to accept. Until the end of the 1960s, secondary students were taught manual processes for the calculation of square roots that were abandoned in the 1970s, in the same way as logarithm tables fell out of use in the face of the presence of calculators in the early 1980s.

As our interest focuses on the presence of mathematical problems in the teaching project, in the rest of this section we will present two examples in which mathematical problems are visible in these textbooks. These examples will be looked at again later in the next stage of the analysis.

A well-known problem concerning geometric progressions

The chapter dealing with geometric progressions begins with a definition: "Given three or more numbers in a given order, they are said to form a geometric progression if each one is obtained multiplying the previous number by a fixed number, which is known as the common ratio," and it is followed by a couple of examples of geometric progressions. The text details a series of formulas enabling determination of the initial value, the nth term and the common ratio. It also explains how to interpolate the terms included within two elements of the progression and how to find the product and the sum of the n terms of a finite progression. It also states as a theorem that "the product of each pair of terms equidistant from the extremes is equal to the product of the extremes of the progression".[16] All this theoretical development is accompanied by examples of progressions in which these properties are identified.

As the chapter progresses, it is observed that both the sum and the product of the terms of a progression grow strikingly large and the classic problem of the chessboard is introduced. The text reads:

> Sissa, who invented the game of chess, so pleased his emperor with his invention that the ruler offered him whatever reward he desired. Sissa requested that he should receive the grains of wheat resulting from placing one grain on the first square of the chessboard, two on the second, four on the third, eight on the fourth and so on successively until all 64 squares had been used. The emperor considered this to be a derisory request, and was greatly surprised when he became aware of the result (Repetto, Linskens and Fesquet, 1962).

The text then goes on to "explain" the problem as follows:

> In effect: The number of grains of wheat asked for by Sissa is the sum of the terms of a geometric progression in which:
>
> $$\begin{cases} a = 1 \\ q = 2 \\ n = 64 \end{cases}$$

---

[16] *Arithmetic and Algebra for Secondary School and Teacher Education*. 4th Year. Chapter 5: Geometric Progressions.

Then:
$$S = \frac{a(q^n - 1)}{q - 1} = \frac{1(2^{64} - 1)}{2 - 1} = 2^{64} - 1.$$

And performing the corresponding operations, one arrives at:
$$S = 18446744073709551616 - 1.$$

We therefore see that the solution to the problem, as outlined in the text, consists of:

(a) Recognizing that the request detailed in the problem is the sum of all the terms of the progression.
(b) Recognizing the elements in the progression, that is to say: the first term, the number of terms, and the ratio.
(c) Applying the formula for the sum of a progression for $n = 64$ to calculate the number of grains that the emperor should pay.

A problem outlined in such a way undoubtedly has a purpose — application of a theory. The student's task consists of identifying the mathematical objects disclosed in the text and the properties that have been demonstrated to be linked to them, so as to place them in action in the case of the specific problem. Here, knowing the need to find a geometric progression, it is evident that the squares of the chessboard are responsible for each of the terms and that the number of grains of wheat needed will be obtained from the elements of the progression.

Equations of the second degree

Chapter 7 of the same textbook is titled "Equations of the second degree with an unknown" and starts out by explaining that "for an equation with an unknown to be of the second degree, it is necessary for that unknown to include a squared term as one of its components and that there should be no other square greater than that unknown" (*ibid.*, p. 248). Several examples of complete and incomplete quadratic equations are presented below. Development continues with the following proposal:

(a) Search for roots in the various cases in the absence of one of its terms.
(b) Search for roots of the "fully reduced" equation (expressed as $x^2 + px + q = 0$).
(c) Search for roots of the "fully general" equation (expressed as $ax^2 + bx + c = 0$).

There follow various "application examples" in which different types of equations are solved (including fractional equations), for which at some stage the solving of the exercise requires the search for roots of a quadratic equation.

Concluding the chapter's explanations of the theory, some of the properties of the roots are established in terms of the sum and product of the roots and their relationship with the coefficients of the full general equation.

The application problems presented later in the textbook are basically various types of equations to be solved. Out of a total of 201 exercises, 41 are designed to resolve the "fully reduced" equation $x^2+px+q = 0$, while 40 exercises deal with finding the solution to the "fully general" equation $ax^2 + bx + c = 0$. The rest are reserved for other equations that eventually lead to the resolution of some type of quadratic equation or the use of the indicated properties.

Following this exposition of multiple quadratic equation problem-solving strategies, this part of the chapter presents several problems under the heading of "problems resolved by means of second degree equations with an unknown" declaring at this point the resource — or linking theory — that will be used to produce a solution.

This section is then developed on the basis of four cases.

(1) "Given the sum and the product of two numbers, calculate those numbers." These problems are therefore presented as an application of the properties of the roots that were proposed previously.
(2) "Calculate numbers that meet certain conditions that differ from those set in the previous case." An example is provided for this case in which the product of two consecutive natural numbers is 68 when it is "diminished by 42", generating the equation $x(x + 1) - 42 = 68$.
(3) "Applications to geometry." Three problems are presented in this section: (i) calculation of lengths in successive segments that arise when dividing a segment into an extreme and mean ratio, which requires the use of proportions; (ii) calculation of the base and height of a rectangle when the area and perimeter are known; (iii) calculation of the sides of a right-angled triangle when it is known that its dimensions are consecutive numbers. The latter requires the use of the Pythagorean Theorem.
(4) "Applications to physics." Here, a problem is presented involving notions of uniformly variable motion, initial velocity and acceleration.

These variables are presented to the student with the following statement:

Remember that the formula for the distance traveled by a vehicle in a uniformly accelerated motion is:

$$e = v_0 t + \frac{1}{2} a t^2.$$

In this problem we propose to calculate t when e, v_0 and a are known (*ibid.*, p. 297).

Based on this statement, the equation is reorganized so that the quadratic equation is visible depending on t as a fully general equation, and numerical data are provided to find the time at which a vehicle has traveled a given distance.

Once again, the theory is something external that is presented to the student initially, and the student's activity consists of discovering the application of the theory to the problems presented — in a more explicit manner in this example.

This enumeration of cases of application problems is a sign of a resource often mentioned in math teaching: the idea of "types of problems", as mentioned by Charnay, which are explained or shown. Thereafter, students are expected to know how to solve them by means of minor changes in their data or presentation.

We should make it clear that we are not claiming that textbooks have unilaterally given rise to a theory concerning how mathematics should be taught; it is just that in the case of our analysis they provide evidence of how math teaching was viewed in that period.

In these textbooks, *exercises and application problems* are presented in the context of the treatment of a topic with specific indications, and as we have shown, with details as to what part of the theory needs to be applied, or what type of problem-solving strategy is to be considered. There is no point at which students must decide for themselves what part of the notions they have already seen they must apply to solve the problems they have been presented with (mathematical problems do not precede theory in these texts). No exercises or problems can be found for which it would have been necessary to use notions dealt with in different chapters of the book, that is to say, where the topics taught have been grouped. The compartmentalization of knowledge proposed by educational organization (then, but also today) is therefore evident.

We have chosen these Kapelusz textbooks because of their widespread use and huge print runs. We could also have used the well-known texts by Tapia as a source, and we would also mention the slightly less representative texts from Cabrera and Médici. The observations based on these texts or other similar ones would be basically the same, however.

In the mid-1970s, the seizing of power by the last de facto government that our country endured, among many very painful consequences, resulted in the silencing of ideas. School textbooks were not exempt from this reality, and therefore during this period the status quo was preserved as far as the type of textbooks being produced, unless they were removed from the public eye altogether.[17]

Second Stage: The Return of Democracy and the 1990s

Until 1995 the education system was governed by Law 1420 (in force since 1884) that established a mandatory seven years of basic instruction from entry to primary school at age 6 to leaving primary school at age 12.

The restoration of democracy in 1983 provided an impulse for the renewal of the regulatory framework of education at the national level, and a National Pedagogical Congress was convened in the second half of the 1980s (Tedesco and Tenti, 2001). Federal Education Law 24,195, passed in 1993, increased the length of obligatory schooling through the inclusion of an initial level (five years) and the extension of primary education (as well as its restructuring) for a further two years in a new cycle known as EGB (Rivas, 2010). These changes were modified by further changes in 2006 following the sanction of the National Education Act. This legislation extended mandatory education to the end of secondary school, a total of 12 years of mandatory primary and secondary education (without taking into account initial education).

During this period publishing flourished, not only because of the presence of new publishing companies specializing in textbooks, but also because of the frequency with which texts were changed. Their authors continued to come from the field of mathematics education, and in some cases interdisciplinary teams were formed with the participation of teachers. Publishers imposed standards for the creation of material that exceeded the

---

[17]The banning of books affected certain school texts, although none of those we have mentioned here.

scope of publishing itself, so it became necessary to "negotiate" the hard core of the texts produced. This had several implications, and is one possible explanation for the movement of authors between publishers and the latter's entry and exit from the school text publishing business. There were also significant changes in the authors specializing in this task.

We will take a few texts representative of a large portion of publishing output as examples to show material typical of the changes that took place.

Whole numbers

First and second year textbooks deal with whole number operations, in sequence with primary education. In many of the texts analyzed, problems with an extra-mathematical context help grant significance to whole numbers. For example, students are asked to record daily changes in cities where temperature records are negative. An apartment building elevator provides a reason to register changes among its users (floors traveled up or down).

To develop problems in a mathematical context, mental arithmetic and the use of estimates is encouraged to put in play the characteristics of multiplication and addition operations in the field of whole numbers, while at the same time stimulating the student's argumentation strategies. It is about operating, but also about comparing, estimating and validating. We speak about "estimating" because a sound estimate makes it possible to argue the distance or closeness of numbers without the need to provide the exact value of the result.

Exploration and the formulation of conjectures is also encouraged (explain the largest set of whole numbers m, n and p, for which the expressions (a) $12 \cdot m + 75$, (b) $12 \cdot n^2 + 75$, and (c) $12 \cdot p^3 + 75$ are negative in each case).

This small sample of activities for students is evidence of *extra* and *intra-mathematical* problems in the study of whole numbers and their operations. Texts reflect a change, where problems "problematize" calculations.

Geometric progressions

All the chapters of the 4th and 5th year textbooks by Barallobres and Sassano propose an approach to each of the topics they deal with that is based on the analysis of an initial problem. In Chapter 1 of the 4th year textbook, in which real numbers and successions are studied, it is proposed under the heading of "Sequences" that consideration be given to a situation in which a ball is dropped:

> A ball is allowed to fall from an initial height of 15 meters onto a concrete floor. Every time it bounces it reaches 2/3rds of the previous height (Barallobres Gustavo, 1997).

The text continues by explaining that this is a mathematical model, and like all models, it is an ideal model. A Cartesian graph (which contains a drawing of the ball as its bounces) is provided so that the student can see on the x axis the number of bounces, and on the y axis the height of the ball.

The problem proposes that the student calculate the heights of the first bounces, and assign them a mathematical name: "Let us call this height $f(1) = 10$." Calculation of the height reached by the ball through to the tenth bounce support the production of a formula, and students are asked whether they can anticipate the height of the tenth bounce even before these calculations are shown. Similarly, they are asked whether it is possible that at some moment the ball will stop bouncing, and why that would happen. As a result, the mathematical model that is proposed enters into dialogue with the situation offered to present it. We transcribe the following paragraph of the text:

> You will also have noticed that $f(n) = \frac{2}{3} \cdot f(n-1)$ and that $f(n-1) = \frac{2}{3} \cdot f(n-2)$ and so on. You will have arrived at the general formula:
> $$f(n) = \left(\frac{2}{3}\right)^n \cdot 15.$$
> This formula is also sometimes written as:
> $$a_n = \left(\frac{2}{3}\right)^n \cdot 15.$$

This last equation gives rise to the explanation in more formal terms of what is understood by a sequence, what the limit to a sequence means, and the notation that mathematicians use for it.

Although we have provided a very small extract from the book, we think it is sufficient to give an idea of the new shape of mathematical problems in this school textbook: the problem is intended to contextualize the mathematical objects that are presented (Duarte, 2011). The problem becomes a way to attach the mathematical object to some aspect of the students' reality.

It is also true that this manner of introducing students to the topic delays the presentation of the entire set of mathematical properties that can be studied in relation to the object. We also note that these proposals take up school time that is always limited as a result of the developments proposed by the design of the curriculum. We understand that this could be considered to be a weakness of these didactic proposals, and although we are aware that the "orthodox" form for axiomatic presentation of a set of mathematical notions could be viewed as a more economic use of time, it does at the same time give the impression of the transmission of a set of notions that the students will then soon forget.

Quadratic functions

The textbook "*Modelos Matemáticos para interpretar la realidad*" (Mathematical Models for Interpreting Reality) develops subjects for the 4th and 5th years of secondary school. Chapter 3 of this publication by Editorial Estrada presents a series of problem situations that provide a context for treatment of the quadratic function. Although the problem is headed as "Carpet Sale", the description of the situation does not involve data on the sale of goods, but rather on the measurements of the sides of some square carpets and their corresponding areas. "In a furnishing store, carpet coverage areas are calculated according to the measurement of one side, for a series of square carpets with sides measuring between one and five meters."

Tables are provided with some measurements in whole numbers, and data are represented on a Cartesian graph. This leads to presentation of the quadratic function form: $f(x) = a \cdot x^2$.

The next situation-problem deals with a monkey hanging from a tree branch 7.2 meters above the ground that lets itself fall to fetch some fruit. Reference is made to the formula for free-falling objects, explaining that the initial velocity is null because the monkey simply lets go, and goes on to mention the use of gravitational acceleration. This generates a function of the form $f(x) = a \cdot x^2 + c$. Several questions go on guiding the interpretation of the time and height in the formula. Finally, a graphic representation of the data that have arisen outlines the design of the quadratic function of the situation.[18]

As a result, five new situations-problems give rise to different concrete cases of the quadratic function so as to present the complete formula, as well

---

[18] We use the expression "situation" in a broad sense. We do not refer to the definition of situation in terms of the Theory of Situations.

as the canonical equation of the parabola. Problems are used throughout the whole chapter to provide a reference framework for the characteristics of the quadratic function being studied.

Undoubtedly a significant change has taken place in these mathematics textbooks. From their early pages, students are invited to solve problems related to the ideas that are being explained.

We have noted certain new traits in these three groups of examples that we have discussed based on school texts from this second period. Texts now have the student as their reader. Some of these texts are accompanied by a separate text for the teacher (we will not be analyzing this aspect here). These texts for teachers explain the purpose of the activities developed, and in some cases there is even a description to assist teachers in the design of activities for their classrooms. A minority include references to curricular documents.

Problems contain questions taken from real life, although students are advised that the model is "an ideal". These questions illuminate certain mathematical issues (such as the heights eventually reached by that ball, or the way in which carpet areas increase, helping to demonstrate the novel aspect of the quadratic model compared to the linear model).

We have found proposals for problem-solving without explicit details of the area of knowledge of references, and there are also problems that are proposed as an overall conclusion to one or more chapters. In these cases, students are challenged to discover the resources to resolve them as they cover various topics.

However, it is also true that textbooks cannot deal with classroom management, or student output and procedures. There is a group of classroom matters that remain without treatment, which the teacher needs to find a way to resolve. This involves decisions that are necessary even when considering the way in which the text itself can be used.

The freedom that publishers have in the design of their textbooks has left its mark. During this period of analysis, publishers changed their textbooks every year, and as can be imagined, most of the time the only change that could be detected was one of style rather than content.

Problem-solving activity in curricular documents

The first curricular guidelines only began to be provided as from the Federal Education Law. That is why we have no curricular guidelines or designs for the first stage we have chosen. In the period from the 1960s to the 1980s the Ministry of Education used to issue the programs for the subjects to be taught in secondary schools. As a result, the "Program of fourth year subjects for the secondary cycle of the baccalaureate[19] of the 1956 plan, with new math topics in effect for 1978", contains the same thematic units we have found in the index of contents of the textbooks we mentioned in the section on the period from 1950 to 1980, namely, real numbers, complex numbers, elementary functions, sequences, notions of plane analytical geometry, combinatorics, notions of statistics and probability, and a summary of solid geometry.

Luis Santaló,[20] an Argentine researcher and mathematician with a particular interest in the teaching of mathematics, confirms that the rise of the new math during the 1960s resulted in certain "exaggerations and deviations. The main defect was that often form was preferred over substance, and mathematics was transformed into a science of definitions" (Santaló, 1992).

On the matter of the formalizing of mathematical notions in secondary school, Santaló recalled that "rigor is a function of age". Just as Euclid when writing his "Elements" relied on mathematics developed by Thales, Pythagoras and Menaechmus, so did Bourbaki build on the constructions of Cantor, Borel and Lebesgue. On this basis, Santaló claimed that "the foundations of mathematics always come well after mathematics itself, and the investment that this law has sought to make in teaching is somewhat inexplicable" (*ibid.*, p. 15).

In the absence of documentation from public policy spheres, we take into account the contributions of those who, like Santaló, have been active participants in the production of proposals for the teaching of mathematics in secondary schools in this first period, from 1950 to 1980.

Referring to problem-solving, Santaló has indicated his firm conviction of its pertinence, echoing the words of Polya[21] in 1968, "Problems can even be considered as the most essential part of a book, and problem-solving by students is the most essential element of their mathematical education" (*ibid.*, p. 22).

Santaló differentiates these problems from routine exercises that apply rules, and stresses the advantages of proposing that students solve problems

---

[19]This is the old denomination for the fourth year of high school.

[20]Santaló represented South America on the Inter-American Committee for the Teaching of Mathematics (CIAEM) active as from 1967.

[21]Conference delivered at St. Augustine, Trinidad in 1968, as per Santaló's citation.

in their math class as long as they "awaken in students an interest in learning about the mathematical tools needed to solve them, and to generalize problems for situations that are more complex or more real".

As we have indicated, the changes introduced by the Federal Education Law in 1993, and later by the National Education Law in 2006, were numerous (and on some matters almost contradictory), and one of them was the incorporation of a planned curriculum and recommendations on teaching for all disciplines, including mathematics.

The text serving as a source of new proposals for secondary school teaching (under the name of CBC: Common Basic Contents, dated 1997) includes a few comments on the teaching of mathematics in the previous period:

> Until the 1960s the teaching of mathematics had as its basic aim the transmission of contents considered to be essential by means of procedures that emphasized the mechanical learning of algorithms for the solving of problems.
>
> Subsequently so-called newmath was introduced, based on the set theory, which did not achieve the expected educational success. We believe that this was due on the one hand to an excessive formalism that was introduced simultaneously with the new concepts that blocked the constructive and intuitive spirit that we consider should predominate in all the learning processes of this science, and on the other, to an attempt to justify the incorporation of these concepts by linking them in an unnatural manner to the notions traditionally being taught to students at their different levels. ("Sources for Curricular Transformation", p. 16).

This document mentions the recommendation by the Federal Education Council in relation to math problems that proposes "acquisition by the student of specific mathematical thought processes designed to solve problems linked directly to real-life situations".[22]

This same text makes special mention of the matter of problem-solving[23] developing the notion of a problem and the uses of the problems developed by R. Charnay[24] among other matters.

---

[22] Attachment I to Recommendation 26/92 issued by the Federal Council for Culture and Education.
[23] Developed by mathematics education researcher Irma Saiz.
[24] Problems for the construction of new knowledge or problem situations for the use of knowledge already studied, or reinvestment problems for extension of the

In this area, the document for Mathematics Curriculum Updating for the first cycle of the EGB issued by the City of Buenos Aires Curricula Department in 1996 offers a concept of problem that is in harmony with the definition we considered at the beginning of this chapter:

What is a problem?
A problem is understood to be any situation that leads students to bring into play the knowledge they possess while at the same time including some degree of difficulty, making such knowledge insufficient and forcing the search for solutions in which new knowledge is generated by modifying (enriching or rejecting) earlier knowledge. We refer to problems that assist learning. This differs from the idea that problems are just an opportunity to apply things that have already been learned.

Problem-solving plays a key role in learning. Problems encourage the construction of new learning and provide the opportunity to make use of knowledge acquired previously.

This last sentence shows us that there is an intention to differentiate the knowledge that students already possess from the reworking that would take place as a consequence of confrontation with the new problem.

The text goes on to state, "Nevertheless, problem-solving is only part of the teaching project. It will be essential to reflect on other aspects and consider the role of the teacher, an absolutely core factor if the educational objective is to be achieved."

This leads to the document dealing with the matter of the organization of the class that is required so that such productive student activity can take place.

We consider that these ideas make it necessary to distinguish between (a) the problem statement and the problem itself, and (b) the solving of a problem and the management of a class where the problem is developed, as we indicated at the start of this chapter.

Although the changes introduced by the two laws were not always in the same direction, there is consensus about the importance of the presence

---

field of use of a notion already studied, or transference problems for joint use of various categories of knowledge, or problems for integration or synthesis, to determine the state of knowledge, or problems for evaluation, and lastly, problems to place the student in a situation requiring investigation, or open problems.

of mathematical problem-solving at secondary school level. The curriculum plan for the province of Buenos Aires for the first year of secondary school[25] states that:

> Math is basically problem-solving, whether problems come from within or outside mathematics, and it therefore plays a pivotal role in teaching.
>
> It should be pointed out that solving problems on its own is insufficient: for the construction of knowledge that can be transferred to new situations there is a need to reflect on what has been done, with teacher intervention to establish links between what has been constructed and scientific knowledge.
>
> In the ESB[26] the situations outlined should go beyond the application of concepts. Analysis should be made of the use of knowledge as a tool to resolve problems from the point of view that helps to recognize the need for generalizations and enables consideration of the notions constructed as mathematical objects.

We note the indication of the need to consider a work stage with the students that extends beyond the solving of the problems that are presented. At the same time, the problems can come from mathematics itself (intra-mathematical) or from the outside world (extra-mathematical). Lastly, mention is made of the need to conceive problems beyond the "theory-problems-application" format.

The text also states that mathematical problems are a way of producing knowledge: "Teachers will be responsible for organizing learning situations that represent challenges that students recognize they are capable of accepting, generating interest in the problem solving, which will enable them to build new knowledge."

All sources quoted subscribe to the idea that there is a need to ensure mathematical problems enable students to deploy a genuine work of mathematics. At the same time, it is obvious that the possession of an arsenal of powerful problems statements is no more than an optical illusion: a problem statement does not encompass an activity that can be deployed in class. The class, the students, their strategies, knowledge, and the object of the lesson

---

[25] Directorate General of Culture and Education, "Curricular Design for Secondary Education Grade 1 ESB", Government of the Province of Buenos Aires, 2006.
[26] Basic Secondary Education.

are just a small sample of the many decisions that a teacher must make when teaching mathematics based on problem-solving. It is the teachers themselves who need to be able to solve genuine mathematical-didactic problems when embarking on their teaching project using this method. Problems in which there are scenarios of uncertainty, unforeseen and unpredictable (in relation to the output of their students) with a consequent need to take decisions on the move.

Furthermore, group work appears as a fundamental instance in the exploration, review and solving of problems. Working in small groups, interaction among students, and making explicit the thought processes that this involves encourages analysis of errors, reviews of reasoning and concepts used in solving a problem, the preparation and contrasting of hypotheses and conjectures, and the understanding of the nature of mathematical knowledge (Sadovsky, 2005). The aim is to encourage reflection on the study process itself as well as to establish a record of the work done, relating it to past work.

Another aspect derived from the proposal to base work on problems, and which implies a significant change regarding the traditional manner of teaching mathematics has to do with the use of time (Felix, 2013, p. 16). The curricular proposal for mathematics assigns time for exploration as part of the learning process, time for discussion, and time for review and reworking, which means extending the time that is traditionally dedicated to the teaching of a subject. The space assigned to error as a progress driver, also implies enabling space and time for successive tests and reformulations. This not only comes into conflict with teacher practices and ingrained work habits, but also clashes with institutional schedules.

Conclusions

Analysis of secondary school mathematics textbook content in Argentina identifies two distinct periods in relation to aspects such as the concept of mathematics teaching, the concept of mathematics, the roles of teachers and students, and last, the role of math problems themselves in the teaching and learning of mathematics.

In the first stage we have indicated that mathematics teaching is presented as the act of communicating results: definitions, properties, theorems, demonstrations. Teaching is responsible for communicating the theory, and the student for using it to solve problems. Mathematics problems thus become *an opportunity to apply* notions from the fields of the discipline.

We have also found in these texts evidence of a concept of knowledge: mathematics is a discipline that produces ideal objects with rules and properties for formulation and treatment. These rules can be communicated in a systematic, orderly manner on the basis of definitions, properties, propositions and theorems. Knowledge is communicated starting from the simple in order to then approach the complex.

In the second period, the textbooks we have analyzed show a change of course, according to which problems (occasionally extra-mathematical but also intra-mathematical) are presented as a resource to begin discussion on a given field of mathematics (we chose for this chapter the topics of geometric progression and quadratic function). These books reflect a substantial change in which students are invited to give meaning to mathematical notions in play because of their capacity to produce knowledge in relation to a problem.

Nevertheless, this educational purpose is not always achieved with the same effectiveness. The great disparity in the output of school texts in this second stage has made it possible for there to be some very different proposals. We have chosen textbooks written by math teachers dedicated to mathematics teaching research. These books are representative of significant changes (both in the number of texts that reflect them and the way they are used), although they coexist with other proposals.

As from the 1990s, Argentina began to experience the introduction of state policies to achieve changes in education, by issuing documents on curricular design. The effectiveness of curricular policy as a tool for change has been attenuated because it tends to be limited to discussion and legislative proposals that are not put into practice. Many textbooks reflect the changes proposed, but this does not mean that such changes are being seen in Argentine classrooms. Sadovsky mentions that "the practice of teaching obeys reasons that cannot be changed at will" (Sadovsky, 2005).

The changes that we have verified in the design of teaching material in textbooks have taken place in the context of publishing policies that have been "liberated" to the effects of market forces, leading to proposals with few real differences between them. Such differences are hard for teachers to identify without a detailed analysis of the material that is generated every year. These changes destabilize teaching work, and consequently lead teachers to become indifferent to such materials.

We believe we are going through a period of high instability and great mobility as regards to the concept of math teaching carried out on the basis of problem-solving. We consider that the current time of instability is necessary for a real change in the face of a model that has remained in force unaltered for an extended period.

Bibliography

Arata, N., & Mariño M. (2013). *La educación en Argentina: una historia en 12 lecciones*. Buenos Aires: Ediciones Novedades Educativas.

Babini, J. (1949). *Historia de la ciencia argentina*. Buenos Aires: Ed. Fondo de Cultura Económica de México.

Belloli, L. (2009). Algunos aportes al conocimiento de la numeración mapuche. *Revista Electrónica de Investigación y Educación en Ciencias*, *4*(2), 1–6. Retrieved from http://reiec.sites.exa.unicen.edu.ar/ano4-nro-2

Brousseau, G. (2007). *Iniciación al estudio de la teoría de las situaciones didácticas*. Buenos Aires: Libros del Zorzal.

Charnay, R. (1994). Aprender (por medio de) la resolución de problemas. In Saíz, I., & Parra, C. (Eds.), *Didáctica de Matemáticas. Aportes y Reflexiones*. Buenos Aires: Editorial Paidós Educador.

Chemello, G. (1997). El cálculo en la escuela: ¿las cuentas son un problema?. In Gotbeter, G. (Coord.), *Los CBC y la enseñanza de la matemática*. Buenos Aires: Editorial A-Z.

Dassen, C. (1924). *Evolución de las ciencias en la República Argentina 1872–1922*. Tomo IV. Matemáticas. Buenos Aires: Sociedad Científica Argentina.

Duarte, B. (2011). *Cuestiones didácticas a propósito de la enseñanza de la fundamentación en matemática. La función exponencial, el razonamiento matemático y la intervención docente en la escuela media*. Doctoral Thesis. Max Von Buch Library. Universidad de San Andrés. Buenos Aires.

Felix, B. (2013). *Investigación colaborativa. Aportes para la enseñanza de la matemática*. Thesis for Degree in Education Sciences. Max Von Buch Library. Universidad de San Andrés. Buenos Aires.

García Venturini, A. (2004). *Los matemáticos que hicieron la historia*. Buenos Aires: Ediciones Cooperativas.

Grosso, A. (1907). *Ejercicios y problemas de aritmética para uso de las escuelas comunes*. Librería Económica. Buenos Aires.

Hanflin, M. (1996). Problemas para enseñar, problemas para aprender. In Hanflin, M., Savón, S., Sessa, C., Camuyrano, M., Crippa, A., & Guzner, G., *Matemática: temas de su didáctica*. Buenos Aires: Programa Prociencia. Conicet.

Junta de Historia y Numismática Americana (1910). *Gaceta de Buenos Aires: 1810–1821*. Reimpresión Facsimilar. Buenos Aires: Compañía Sudamericana de Billetes de Banco.

Kilpatrick, J. (1987). Problem formulating: Where do good problems come from? In Schoenfeld, A. (Ed.), *Cognitive Science and Mathematics Education*. New York: Routledge.

Polya, G. (1945). *How to solve it*. New Jersey: Princeton University Press.

Polya, G. (1957). *Mathematics and plausible reasoning*. New Jersey: Princeton University Press.

Rivas, A. (2010). *Radiografía de la Educación Argentina*. Buenos Aires: Fundación CIPPEC. ISBN 978-987-1479-21-4. Retrieved from http://www.cippec.org/files/documents/Libros/Radiografia-edu.pdf

Sadovsky, P. (2005). *Enseñar matemática hoy: miradas, sentidos y desafíos*. Buenos Aires: Libros del Zorzal.

Santaló, L. (1992). *La enseñanza de la matemática en la escuela media*. Proyecto Cinae. Centro de Investigación y acción educativa. Buenos Aires. Second edition. First edition by Editorial Docencia (1986).

Stacco, E. (2011). *200 años de la Matemática en la Argentina*. Retrieved July 15, 2013 from http://inmabb-conicet.gob.ar/publicaciones/iti/iti99.pdf

Schoenfeld, A. (1994). *Mathematical thinking and problem solving*. New York: Routledge.

Tedesco, J. C., & Tenti, E. (2001). La reforma educativa en la Argentina. Semejanzas y particularidades. Technical report for *Alcance y resultados de las reformas educativas en Argentina, Chile y Uruguay*. Ministries of Education of Argentina, Chile and Uruguay. Advisory group for University of Stanford/BID. IIPE-UNESCO. Buenos Aires.

Vilanova, S., et al. (2001). La Educación Matemática. El papel de la resolución de problemas en el aprendizaje. *OEI-Revista Iberoamericana de Educación*. Espacio de los lectores. From http://www.rieoei.org/deloslectores/203Vilanova.PDF and from http://www.rieoei.org/deloslectores_Didactica_de_las_Ciencias_y_la_Matematica.htm

Curricular Documents Consulted

Cuenya, H., Fava, N., Gysin, L., & Saiz, I. (1996). *Fuentes para la transformación curricular. Matemática*. Ministerio de Cultura y Educación de la Nación. República Argentina.

Dirección General de Cultura y Educación (2006). *Diseño curricular para la Educación Secundaria 1°ESB*. Gobierno de la Provincia de Buenos Aires.

Dirección de Currícula de la Ciudad de Buenos Aires (1996). Documento de Actualización Curricular. Matemática. Primer Ciclo. EGB.

Ministerio de Cultura y Educación de la Nación. Consejo Federal de Cultura y Educación. (1997). *Contenidos Básicos Comunes para la educación Polimodal.* Retrieved July 23, 2013 from http://www.me.gov.ar/consejo/documentos/cbc/polimodal/present.pdf

Textbooks Consulted

Aragón, M., Camuyrano, B., & Net G. (2000). *Matemática I. Modelos matemáticos para interpretar la realidad.* Buenos Aires: Editorial Estrada.

Barallobres, G., & Sassano, M. (1997). *Matemática 4.* Buenos Aires: Aique Grupo Editor.

Barallobres, G., & Sassano, M. (1997). *Matemática 5.* Buenos Aires: Aique Grupo Editor.

Cicala, R., Díaz, B., Franco, E., Kaczor, P., & Schaposchnik, R. (1999). *Matemática I. Polimodal.* Buenos Aires: Editorial Santillana.

Cicala, R., Díaz, B., Franco, E., Kaczor, P., & Schaposchnik, R. (1999). *Matemática II. Polimodal.* Buenos Aires: Editorial Santillana.

Repetto, C., Linskens, M., & Fesquet, H. (1952). *Aritmética para primer año de las escuelas de comercio.* Buenos Aires: Editorial Kapelusz.

Repetto, C., Linskens, M., & Fesquet, H. (1957). *Aritmética para segundo año de las escuelas de comercio.* Buenos Aires: Editorial Kapelusz.

Repetto, C., Linskens, M., & Fesquet H. (1960). *Aritmética para segundo año de las escuelas normales y nacionales.* Buenos Aires: Editorial Kapelusz.

Repetto, C., Linskens, M., & Fesquet, H. (1960). *Aritmética para segundo año del ciclo básico.* Buenos Aires: Editorial Kapelusz.

Repetto, C., Linskens, M., & Fesquet, H. (1962). *Aritmética y Algebra 4.* Buenos Aires: Editorial Kapelusz. Thirteenth edition.

Repetto, C., Linskens, M., & Fesquet, H. (1964). *Geometría del espacio.* Buenos Aires: Editorial Kapelusz.

Repetto, C., Linskens, M., & Fesquet, H. (1968). *Aritmética y Algebra. Tercer año del ciclo básico.* Buenos Aires: Editorial Kapelusz.

Repetto, C., Linskens, M., & Fesquet, H. (1968). *Geometría 3.* Buenos Aires: Editorial Kapelusz.

Santaló, L. (1993). *Matemática 1.* Serie Horizonte. Buenos Aires: Editorial Kapelusz.

Santaló, L. (1993). *Matemática 2.* Serie Horizonte. Buenos Aires: Editorial Kapelusz.

Santaló, L. (1993). *Matemática 3.* Serie Horizonte. Buenos Aires: Editorial Kapelusz.

About the Author

Betina Duarte is a professor at the Mathematics Department in the Universidad de San Andrés. She also teaches in a graduate program for teachers at the Universidad Pedagógica (UNIPE), where she chairs the Mathematics and Experimental Sciences Department. Her doctoral thesis addressed the problem of the teaching of mathematics at the secondary level using the foundation argument as a theoretical framework. Professor Duarte is currently focused on research about the use of technology, mostly as related to the problem of teachers' orchestration and the students' instrumental genesis.

Chapter 2

BOLIVIA: An Approach to Mathematics Education in the Plurinational State

A. Pari

Abstract: This article explores the history of mathematics education in different socio-political periods and the current status of research in this field. The chapter is divided in three periods: Pre-Hispaninc, Colonial, and Republican. Although research is scarce, there have been signs of interest in mathematics education in the Plurinational State of Bolivia.

Keywords: History of Mathematics Education; Plurinational State of Bolivia.

Life is good for two things, learning mathematics and teaching mathematics.

Siméon Denis Poisson

Mathematics Education in Bolivia, as a field of research, can be described as incipient. However, it is possible to find sporadic samples of interest in the teaching of mathematics during the last century. Even though the Bolivian contribution to international scientific production can be described as scarce and marginal, Bolivian universities have been able to increase the personnel and attention dedicated to research. Still, the context is not favorable as the State does not have a clear policy with respect to higher education and does not have funds to dedicate to research and innovation (Rodríguez & Weise, 2006).

Bolivia is a land of contrasts, not only for the nature of its geography, its soil and its climate, but also for its history, its politics and the condition in which its population lives. Its riches are considerable, but the majority of its population is poor. It has soil and a variety of climate sufficient to assure that everyone is properly fed, yet many are undernourished. Today, at the dawn of the twenty-first century, Bolivia continues to be a State of contrasts and paradoxes. It is a developing State with marvelous possibilities that struggles to look to the future while its feet are chained to the past and its heart is divided.

Within this framework, Bolivian mathematics education has been and continues to be determined by institutional factors, politics, and popular social movements. One of the difficulties faced in presenting this document is the lack of information from mathematics education as a field of research.

The History of Mathematics Education in Bolivia

Some important aspects of the political history and history of education in Bolivia are presented below in order to situate Mathematics Education in the country. Three historical periods will be considered: Pre-Hispanic, Colonial (sixteenth and seventeenth centuries) and the Republic (nineteenth and twentieth centuries).

The Pre-Hispanic Period

The Pre-Hispanic era covers a period of time in the history of the Americas that ended with the arrival of the Spanish on the American continent. The conquistadors and colonizers from the Old World found peoples in South America with various levels of cultural development. Specifically, on the Andean Plateau various cultures were developed: Viscachani, Pucara, Wankarani, Chiripa, Mollo, Kolla Tiwanaku and Inca. The most advanced at the time of the Spanish arrival was the Inca. A study of these cultures is beyond the scope of this chapter. Nevertheless, here is a brief mention of Tiwanakota culture "whose remains can still be found as megalithic monuments engraved with inscriptions that no one has been able to translate (Elbarquero, 2006; Kolata, 1993).

The Tiwanakota spoke Aymara, a language that is still spoken by many Bolivians. Not much is known about the origins and duration of the Tiwanakota culture (Diez de Medina, 1986).

The earliest available written records (Kolata, 1993) are those left by the Spanish conquistador and chronicler Pedro Cieza de León (1549) who wrote "Chronicles of Perú and a New Spain" after his visit to

Tiwanaku. It contains interesting historical and geographical data. Later in the nineteenth century information on the arts and literature was added.

In the twentieth century, various foreign archeologists worked in Bolivia. In 1934 the North American Andean archeologist Wendell Benner discovered the monolith that is now named after him. The Austrian Arthur Posmansky (1873–1945) published a classic four-volume work: *Tihuanacu: The Cradle of American Man*.

Among Bolivian archeologists, it was Carlos Ponce Sanginés (1925–2005) who became the necessary reference for Pre-Hispanic studies (Ponce, 2000). He initiated Bolivian archeology in the mid-1950s. He was the founder and first director of the National Institute of Archeology (1971–1982), as well as of the Center for Archeological Research on Tiwanaku (1957–1975).

The archeological evidence from Tiwanaku indicates important advances in astronomy, architecture, technology and mathematics. The Bolivian mathematics teacher Jaime Alfonso Escalante Gutiérrez used to say to his students in Garfield High School in Los Angeles that "The Aymaras knew math before the Greeks and Egyptians" (Pari, 2011; Schraff, 2009). Tiwanaku is a rich archeological site that still hides secrets within a defined area of time and space (Diez de Medina, 2005). It also contains a vast artistic depository of textiles and ceramics.

In 2000 UNESCO declared Tiwanaku to be a Cultural Patrimony of Humanity. Currently there is interest in recovering the cultural practices of different groups in order to be able to understand them. In that context various works should be mentioned: *The Tiwanaku: Portrait of an Andean Civilization* by Alan Lewis Kolata (1993); *Mathematics Was Born in Tiwanaku, Bolivia: The reason for the natural numbers* by Fidel Rodríguez Choque (2008); *Tiwanaku Capital of Mystery: Five Meditations and Two Tales* by Fernando Diez de Medina (2005/1986), and *The Mathematics of the Puerta del Sol of Tiwanaku* by Jorge Emilio Molina Rivero (2000).

The other Pre-Hispanic culture that developed in the Andean region was the Inca. Education in the Incan Empire went through a period of formation in which the teaching of the children of the nobles was carried out by wise men called "Amautas". It was an oral, practical and experimental education.

There were specialists who were like statisticians or accountants who were called "Quipucamayos". They preserved data in knots of colored thread called "quipus". The houses where the Quipucamayos met were called "Yachaywasis" and commoners were not allowed; thus, education was very class-oriented. The first school for the children of the nobles was created by the leader called "Inca Roque" in Cuzco.

The basic Incan educational principles that still exist in some indigenous villages in the Bolivian highlands and valleys were based on three fundamental themes:

- *ama sua* (Don't be a thief).
- *ama llulla* (Don't be a liar).
- *ama kjella* (Don't be lazy).

These norms were indisputable values for all the population and were followed as models for behavior (OEI, 1988, p. 3).

In 1532 the Incan Empire was invaded and conquered by the Spanish under the leadership of Francisco Pizarro.

Period of Conquest and Colonization (Sixteenth and Seventeenth Centuries)

The first Spaniard to arrive in what today is Bolivia was Diego de Almagro. In 1535 he arrived at Lake Titicaca and the Valley of La Paz. When Almagro died in 1538, Francisco Pizarro sent his brother Gonzalo to conquer the provinces of Charcas and Collao, which had not been colonized by the Almagro expedition.

Pedro Anzúrez founded Chuquisaca in 1538. Potosí emerged in 1546, La Paz in 1548, and Cochabamba in 1574. Meanwhile, colonization from the River Plate region to what is today Bolivia led to the founding of Santa Cruz in 1561. However, it was the discovery of silver in Potosí in 1545 that led that city to become a key point for Spanish exploitation in the Americas. During the colonial period the mines of Potosí were the principal producers of silver in all the Americas.

The immense deposits of silver led the colonizers to focus on mineral exploitation. However, the riches of the mines, ethnic inequalities, and rivalries among the conquistadors caused a turbulent history in High Perú (now Bolivia) during the sixteenth, seventeenth and eighteenth centuries.

Among the most important institutions in the Spanish governing of the Americas were the Royal Courts ("Reales Audiencias"). The highest legal authority in High Perú was the Royal Court of Charcas in the department of Chuquisaca. Better known as the Court of Charcas, it was created by decree of King Felipe II on September 18, 1559:

> The Court of Charcas is created in 1559, although one of the first documents indicating a need for said Court is the "Agreement of the Council of the Indies on the advisability of locating the Court in the Villa de la Plata", on April 20, 1551 (A. G. I., Estado 140, Caja 7, Leg. 31): The principal goal of Your Majesty is to govern those new territories in the Indies, providing them with an abundance of justice, because that is how the Christian religion is founded and our faith kindled and the natives are well-treated and receive adequate instruction in them, and that is what experience has taught us ..." (González & Bravo, 2008, p. 1042).

According to the historian, journalist and ex-President of Bolivia Carlos D. Mesa Gisbert (May 3, 2013), "Charcas had a greater significance in the seventeenth and eighteenth centuries than the Republic of Bolivia at the beginning of its independence." The conquest had material as well as spiritual motives, even though one of the main aims of the Spanish monarchs was the spread of the Catholic religion. In fact, a church was constructed in every city or town that was founded.

Colonial education was for the children of creoles, mestizos, rich businessmen, and land holders who received lessons in reading and writing in their homes from teachers called home tutors ("leccionisantes"). There was no free obligatory elementary education during the colonial period nor was there tutoring for the majority of the indigenous population. There was no teaching program and the Catholic Church had a monopoly on instruction in the hands of priests from various orders. Parochial classes were implemented. Thus, the only educational action was catechistic, which negated any possibility of valuing local cultures.

The first schools

The first educational institution in Bolivia was founded by Father Alfonso Bárgano in 1571. In 1599, Bishop Alfonso Ramírez founded the Saint Christopher Seminary School in Chuquisaca, which was also known as the Red School because its emblem was a red medallion. Another school was founded on February 22, 1621 by the order of Francisco Borja, Viceroy Prince of Esquilache. It was called the Santiago School. On April 10 of the same year its name was changed to Saint John the Baptist or the Blue School because of its emblem (OEI, 1988).

Education continued to be classist and only children of the gentry could attend. That is, women and indigenous children were excluded. In 1792, Brother José San Alberto, Archbishop of La Plata created the School for Poor Children in Chuquisaca, also known as Saint Albert's School (OEI, 1988).

The emergence of the Bolivian university

In 1623 the first Bolivian university was created in the department of Chuquisaca. At the time, Chuquisaca was the center of economic, social and intellectual power. It was also the home of the Saint John the Baptist School (the Blue School) (Serrudo, 2006, p. 56):

> In 1623, enjoying the privileges and prerogatives and immunities of royal schools, the "Blue School" was allowed to offer bachelor's, master's and doctoral degrees in the arts, theology, canons and laws, to be accepted by any university (OEI, 1988, p. 4).

On March 27, 1624 the University of St. Francis Xavier of Chuquisaca was created by Father Juan Frías de Herrán, provincial head of Jesuits in Perú. It was given that name to honor Father San Francisco Xavier, the new Apostle of the Indies, so his name could protect the achievement of the students (Serrudo, 2006, p. 56).

In line with the paradigms of the times, the teaching objectives of the University of St. Francis Xavier were those of the Jesuits and the Catholic Church in general: spreading the gospel to "save souls and eradicate idolatries" (Serrudo, 2006, p. 56). Initially the university prepared lawyers based on three scholastic disciplines. It functioned regularly until 1767, when the Jesuits were expelled from the university. After the expulsion of the Jesuits, the university entered a second stage of development and the Carolina Academy was created in 1778 to provide legal practicums for the graduates (OEI, 1988, p. 4). The Academy gave new life to the university and by 1780 it had reached the same level as the University of Salamanca. Nevertheless, it was not until 1798 that the Spanish government recognized it as a Royal and Pontifical official institution with all the privileges of the University of Salamanca (OEI, 1988, p. 4).

The university education was criticized for its failure to apply science to the development of the region. However, pure and exact sciences were not necessary for the means of production at the time. Moreover, there was no evidence of the utility of the incipient science (Serrudo, 2006, p. 56).

The Period of the Republic (Nineteenth and Twentieth Centuries)

The nineteenth century represents the final stage of colonial period and the beginning of the Republic of Bolivia. The Court of Charcas, called High Perú, achieved independence from Spain and its autonomy from the viceroyalties of Perú and River Plate. Political instability and economic collapse ensued (Mesa, 2012). Thus, republican life began for Bolivia on August 6, 1825, with its capital in Sucre. With the exit of Field Marshall Antonio José de Sucre, there was a political crisis until Andrés de Santa Cruz assumed the presidency (1829–1839). According to Mesa (2012), it was during that period that Bolivia had the most solid state, possibly the only one in its history. This period is bound to the personality of Andrés de Santa Cruz.

However, by the end of the nineteenth and beginning of the twentieth century, Bolivia was living with grave post-revolutionary consequences and a crisis of identity and economic devastation (Lozada, 2004). It had a population of about two million inhabitants, but only about 7% had an elementary education. Bolivia approached the twentieth century bloodied by wars and international treaties.

On the other hand, the decline in the mining of silver had to do with changes in monetary policies and increased industrialization, which made tin more attractive in the powerful countries. This generated not only changes in production, but in power as well, and led to a civil war in Bolivia — called the Federal War — from 1898 to 1899. This civil war pitted La Paz against Sucre in trying to become the capital to begin the new century. It was an encounter between the conservative forces with an oligarchic ideology representing the silver mining interests in the south (Sucre and Potosí) against the liberals with a federalist orientation in the north (La Paz and Oruro) who defended the interests of tin miners. In this context, Bolivian history should be viewed from three dimensions: economic, political and social (Mesa, 2012). Beginning at that time, the center of development and political decisions was in the west with La Paz at the center. This persisted until the consolidation of production in the east in the 1990s (El Diario, 2011).

The emergent ideology attacked the old regime and broke the power of the Catholic Church declaring that education would be non-denominational and that the indigenous population would no longer have to provide domestic service to the parish churches. As Lozada (2004) stated, "In the rural areas of the country the liberals promoted the founding of educational entities dedicated to indigenous education and in the cities they assumed the creation of a national identity as the principal task" of the State.

The National Revolution had begun by the middle of the twentieth century, particularly in 1952, led by Víctor Paz Estenssoro. Political opposition and social movements continued into the twenty-first century. The most notorious confrontations were the Water War in 2000 and the Gas War in 2003, concluding with the election of Evo Morales as President (Neso, 2013). These are the most important events or phenomena of republican life in Bolivia. Education in general — and mathematics education in particular — has been determined and even limited by the complex circumstances with which the country has lived.

Elementary and secondary education

In the early years of the republic, elementary education as such did not exist and there was not even a single printing press in the country (Aramayo, n.d.). Nevertheless, Simón Bolívar was very interested in public education and assigned his own teacher, Simón Rodríguez, as Director General of Public Education (OEI, 1988). Bolívar and Rodríguez initiated school legislation in December of 1825 with Decree 11:

> ... education is established as the first duty of the government and it should be Uniform and General ... an elementary school should be established in every department's state capital ... to receive children of both genders ... (OEI, 1988).

By the Law of January 9, 1827 elementary, secondary and central schools were established:

Reading, writing, religion, moral doctrine, and agriculture should be taught in elementary schools by the method of mutual teaching. In addition to elementary schools, every provincial capital should establish secondary schools to improve reading and writing, religion, moral doctrine, Spanish grammar, **the four rules of arithmetic**, agriculture, industry and veterinary medicine [emphasis added]. In addition to the aforementioned Department school, there should also be central schools to thoroughly teach arithmetic, grammar, drawing and design. These schools should admit only those students judged by their teachers to have demonstrated aptitude.

On the other hand, in the cities of La Paz and Potosí — where there were mines — schools of mineralogy were created to teach ***geometry*** and ***subterranean architecture***, elements of chemistry and mineralogy, and the art of extracting and melting all kinds of metals, for which they constructed small laboratories.

In the capital of the Republic, in addition to the disciplines taught in the departmental schools, history of literature, ***complete mathematics***, chemistry, botany, painting, sculpture, engraving and music should be taught. They also considered establishing a literary center called the National Institute.

This plan was at the time without a doubt a great step for education. Some of the provisions were implemented and others remained on paper, but it did give guidelines to which Bolivian education could aspire. Nevertheless, the teaching methods were teacher-centered, with little or no student participation. In other words, it was based on rote learning.

Educational reforms in Bolivia

Throughout the twentieth century, Bolivian education went through various transformations or reforms. The first reform that should be mentioned was an unsuccessful attempt in 1874 at decentralization that was called the municipalization of education. At the beginning of the twentieth century, liberal governments proposed the modernizing of education and were supported by foreign experts such as Georges Rouma.[1] In the 1930s, a

---

[1] Rouma was a Belgian pedagogue born in Brussels in 1881, who died there in 1976. He graduated as an elementary teacher in 1900. Nine years later he received the title of Doctor in Social Sciences and his dissertation was entitled *The Graphic Language of the Child*. Beginning in 1902 he collaborated with Ovidio Decroly, a medical doctor, in the study of language difficulties among children. When he was contracted by the Bolivian government, Rouma was a young researcher just

model of indigenous education was developed in Warisata. In the 1950s not only were Bolivian natural resources nationalized, but education received its first Bolivian Education Code in 1955. The code is a document that unifies the whole system while at the same time divides it into urban education on the one hand and rural education on the other. In 1994, a Law of Education Reform, referred to as Law 1565, was approved. Currently the Avelino Siñani–Elizardo Pérez Law of 2010 is in the process of being implemented.

The Bolivian education code (1955)

It was during the presidency of Victor Paz Estenssoro that the first Education Code was approved. The code was supported by Article 157 of the 1947 Constitution that indicated that education was the highest function of the State:

> Education is the highest function of the State. Public teaching will be organized according to the system of the unified school. The obligation of the school system in general is from seven to fourteen years. Elementary and secondary instruction is free (Art. 157).

Until the National Revolution, education in Bolivia was a monopoly of a minority in service of foreign interests that exploited the country's riches. Large sectors of the population remained uneducated and unable to contribute significantly to the development of the nation.

In the 1960s and 1970s there was a significant reform in the teaching of mathematics at the international level that affected almost all countries. The contents of that reform are well known: introduction of set theory, modern symbolism, the elimination of Euclidean geometry, introduction to algebraic structures and axiomatic systems, the algebraization of trigonometry (Barrantes & Ruiz, 1998).

George Papy is associated with the New Math in Latin America (Pérez, 1980, p. 1). However, New Mathematics had preceded its introduction into elementary and secondary classrooms. The Bourbaki group as early as 1950

---

25 years old with pedagogical training in the best ways to understand modern education, therefore his pedagogical proposals were for an education that took into account psychological, physical and social elements of the socio-cultural environment in which children lived (Iño, 2010, p. 26).

had realized a synthesis of new tendencies in mathematical language (Papy, in Pérez, 1980, p. 1).

The New Math was implemented almost everywhere. It arrived in Bolivia with the support of the Simón Patiño Pedagogical and Cultural Center that assumed the challenge of promoting it throughout the country.

A team composed of Belgian educators and teachers from various districts in the city of Cochabamba was organized to implement it. This team took charge of preparing teachers to join in a multiplier effect in the nine departments. This movement was not very successful and ended in 1985 without achieving the goals it had proposed (Grigoriu, 2005, p. 86).

One of the major difficulties was that not all teachers were well-prepared, particularly in rural schools. Later there were various attempts at reforms and counter-reforms. Behaviorism was the prevailing pedagogical model.

Educational reform of 1994

Bolivian education was reformed in 1994, almost 40 years after the National Revolution had implemented the Bolivian Education Code. Among the contributions of the Education Reform Law of 1994 were including communities in the educational process, and recognizing intercultural and bilingual education. The Bolivian educational system was structured into four levels: initial, elementary, secondary and tertiary. The initial level has two cycles, elementary has three, and secondary has two (Ley de Reforma Educativa, 1994). The elementary level became an open institution at least in aspirations if not always in reality (Grigoriu, 2005).

This educational reform emphasized elementary education and teacher preparation:

> The curricular focus at the elementary level gives attention to diversity, satisfying basic learning needs of the population, and gives attention to society's emerging problems. Based on these principles, there is a focus on an orientation toward the development of competencies, preparation for work, bilingualism, and the integration of students with special needs (Grigoriu, 2005, p. 58).

Within this framework, the mathematics contents of elementary and secondary are organized as follows: number and operations, data processing, spatial sense and geometry, and measurement (Grigoriu, 2005). Previously mathematics content had been organized as arithmetic, algebra, geometry and trigonometry.

The government has published and distributed textbooks in indigenous languages and Spanish for the elementary level. At the secondary level, there has not been much support in terms of textbooks. At the end of the twentieth century the textbooks of the Cuban mathematician Aurelio Angel Baldor were still being used throughout Bolivia, particularly in rural areas.

Information on the preparation of mathematics teachers in normal schools and universities will be presented below.

The Bolivian university

As in the majority of South American countries, the history of the Bolivian university begins in the colonial era with Spanish influence on academic aspects and with a profound Catholic vocation (Serrudo, 2006). The Bolivian educational system is made up of public, private and special institutions (Ley de Educación, 2010: capítulo 1, artículo 1).

Public universities receive their funding from the State General Treasury. Their academic and administrative actions are based on their autonomy, as well as on a co-government by faculty and students, but the latter has at various times been abolished to be re-established in the interest of democracy (Serrudo, 2006). Although private universities receive no financing from the government, their regulations, programs and plans of study must be approved by the government. They can grant academic diplomas, but Titles by National Decree are granted by the State (Constitución Política del Estado Boliviano, Art. 188°). Special are not considered private and are instead actually part of the National System of Bolivian Universities and are recognized by the Executive Committee of the Bolivian University. On the other hand, private universities belong to the National Association of Private Universities (Márquez, 2004).

The universities created in the nineteenth century were the Higher University of San Andrés (La Paz), the Higher University of San Simón (Cochabamba), the University of Gabriel René Moreno (Santa Cruz), the Autonomous University of Tomás Frías (Potosí), and the Technical University of Oruro (Oruro) (CEUB, 2011, p. 10).

The first university created during the colonial period was the Higher, Royal and Pontifical University of San Francisco Xavier in Chuquisaca in

1624, with Jesuit influence. It was two centuries before the second Bolivian university was established, following the Supreme Decree of November 30, 1830, signed by President Andrés de Santa Cruz. The first university founded under the Supreme Decree was called the Junior University of La Paz. In 1831, the Constituent Assembly declared it to be the Higher University of San Andrés of La Paz. In Cochabamba, during the same government in 1832, the Higher University of San Simón was created. It had its origins in the Academy for Legal Practicums.

On December 15, 1879, during the government of Hilarión Daza, the fourth university district was created. On January 11, 1880, the University of Santa Cruz was established and in 1911 it became the Gabriel René Moreno University.

In 1892, under the presidency of Mariano Baptista and via a Law Decree, university districts were created for Oruro and Potosí, and subsequent regulations created the Autonomous University of Tomás Frías and the Technical University of Oruro.

Following the creation of those universities in the nineteenth century, several more public universities were created in the twentieth and twenty-first centuries: University of Juan Misael Saracho in Tarija in 1946, Mariscal José Ballivian Technical University of Beni in 1967, XX Century National University in 1993 with a branch in the Mining District of Llallagua in the department of Potosí, and, finally, the Public University of El Alto in 2000 in the city of El Alto, La Paz.

There are several universities that are classified as "special". The Catholic Bolivian University in La Paz was created in 1966. The "Mariscal José Ballivian" Military Engineering School, also in La Paz, was founded in 1950. In 1985 the Universidad Andina Simón Bolivar with branches in Sucre and La Paz was incorporated to provide graduate studies. In 2009 the Police University Mariscal José Antonio de Sucre was incorporated.

Beginning in the second half of the twentieth century, private universities began to emerge. However, it was in the 1990s that there was a virtual explosion of such institutions and their contribution cannot be denied.

Of all the public universities, only three offer degree programs in mathematics: the Higher University of San Andrés in La Paz, the Higher University of San Simón in Cochabamba, and the Autonomous University of Tomás Frías in Potosí. Of the more than 50 private universities, only the University of Simón I. Patiño in Cochabamba offers a degree in mathematics.

The founding of the Normal School for Teachers and Private Tutors

During the government of Ismael Montes (1904–1909) the Minister of Instruction, Daniel Sánchez Bustamante, recognized the need for institutions to prepare future elementary and secondary teachers. Therefore, in 1909 he founded the Normal School for Teachers and Private Tutors for the Republic in Sucre. The Belgium mission led by Georges Rouma was responsible for the new normal school. According to Iño (2010), the professors of mathematics and physics were Constan Lurquin, professor of mathematics; José Maria de Araujo, professor of physics; and Mariano Oropeza, professor of arithmetic, geometry and algebra.

This normal school was a boarding school for youths who wanted to be the future teachers of Bolivia. They were selected by the government and had all expenses paid (room, board, clothing, textbooks, health costs, and personal hygiene articles) during the four years of study in which they attended 35 hours of classes per week (Lozada, 2004).

The admission requirements were:

- Be between 16 and 20 years old.
- Pass an examination on **elementary notions of arithmetic**, Spanish, ethics, history, geography, and handwriting.
- Evidence good behavior before an examining board through a school certificate or other means deemed acceptable by the board.
- Pass an oral examination before an examining board composed of the President of the university in which the candidate resided, the Principal of the Primary School for Boys, and the National Inspector of Instruction.

Initially, it prepared elementary teachers and later secondary teachers as well. Beginning in 1937 it had two sections: elementary and secondary. When it began to prepare secondary teachers its Secondary Department had four sections: Philosophy and Letters, Mathematics and Physics, History and Geography, and Biological Sciences.

In 1917 the Normal School of La Paz was also created under the leadership of Georges Rouma and the Belgium mission. According to Lozada (2004), Rouma created the Normal School of Sucre while the government for the first time promoted rural indigenous education with the creation in 1931 of the Normal School of Warisata.

The Normal School of Warisata

The Normal School of Warisata was founded in 1931 during the government of Daniel Salamanca. One of its protagonists was Elizardo Pérez Gutiérrez who had graduated from the Normal School of Sucre in 1914 (Iño, 2010). There were many details in the vicissitudes that needed to be overcome in creating this school. Its creation marked a new era and a new path for the education of the masses in Bolivia. Warisata was a unique experience, not only for Bolivia, but for the entire American continent (Lozada, 2004). It was the major revolutionary expression in the field of education and it was the exploit that liberated the indigenous population. It helped to bring about the revolution of April 1952, the Agrarian Reform, and liberation from the federal yoke. These events would not have been possible without this decisive transformation of the indigenous mind.

The background for Warisata can be found in the founding of clandestine schools by Avelino Siñani, himself indigenous, at a time when formal education for the indigenous population was prohibited. The intellectual Elizardo Pérez joined with Siñani and together they established indigenous schools and began to plan a normal school for indigenous teachers. In 1917, Pérez was an educational inspector in La Paz. In that capacity, he visited the humble school in the district of Warisata. His visit would have had no transcendence had he not happened upon a private school run by Siñani, clear evidence of the importance of private schools in Bolivia. Pérez and Siñani found there mutual support for their ideas.

The Education Code of 1955 assured the existence of both urban and rural normal schools. In the Education Reform of 1994, the normal schools were transformed to Higher Normal Institutes. Finally, with the Avelino Siñani–Elizardo Pérez Law of 2010, the name became Higher School for Teacher Preparation. Also created was the Complementary Teacher Preparation Program for in-service teachers to meet the need to transform the education system with support from teachers in the framework of the Socio-Communal Productive Educational Model that was designed to contribute to the creation of a multinational state. The degree offered by the Complementary Teacher Preparation Program is a bachelor's degree equivalent to that offered by the Higher School for Teacher Preparation (Ministerio de Educación, 2013).

The idea of complementary teacher preparation programs had arisen in the 1990s in some universities to allow access to a bachelor's degree.

However, the courses stressed general theoretical matters with little emphasis on pedagogical material. They had even less focus on teaching methods for specific scientific disciplines.

Universities offer bachelor's degrees in pedagogy and a few offer bachelor's degrees in mathematics though usually oriented to pure and applied sciences. Mathematics teachers are prepared only in the normal schools, which are now called Higher Schools for Teacher Preparation. This exclusiveness was and continues to be a factor that limits the participation in teacher preparation of the few mathematicians that do work in Bolivian universities. Also, the professional preparation of mathematics teachers is very heterogeneous and there definitely is no degree in mathematics education.

Mathematics in Bolivian universities

The degree program in mathematics arrived in Bolivia much later than those in law and medicine. Additionally, only four Bolivian universities offer mathematics degrees: three public and one private.

The Higher University of San Andrés created the degree program in mathematics in 1967 in the Faculty of Pure and Natural Sciences. The Autonomous University of Tomás Frías began to offer a similar degree in the Faculty of Pure Sciences in 1972. In the Higher University of San Simón, it was created in the Faculty of Science and Technology in 1985. At the beginning of the century, the private University of Simón I. Patiño established a degree in mathematics. It also hosted the XXVI National Congress of the Bolivian Society of Mathematics in 2013.

The professional profile of the mathematician in Bolivia has been described as follows: "The mathematician is a person who dedicates himself to mathematics."

- Creates mathematics (Research)
- Disseminates mathematics (mathematics teaching)
- Applies mathematics to other areas of knowledge (Expert advice to other disciplines)

The Higher University of San Simón has bachelor's degrees in mathematics and engineering mathematics. The first prepares professionals in pure mathematics and the second has an emphasis in applied mathematics. The other universities just have a bachelor's degree in mathematics. The program lasts ten semesters divided into three cycles: basic, intermediate and orientation (Reunión Sectorial de Carreras de Matemáticas, 2008). To facilitate learning, mathematics has been structured into the following areas: algebra, analysis, geometry and topology, and applied mathematics.

Rimer Zurita Orellana (2003) has done a comparison study of university systems in Bolivia and Switzerland. He based his comparison on the academic programs and stated:

> Although it appears that the list of courses is more extensive in Bolivia than in Switzerland, often more than one course in Bolivia is needed to cover just one course in Switzerland. Many of the courses in Bolivia, particularly those in the first year, are courses in the last year of high school in Switzerland. Another difference is that often in Bolivia not all the material in a course is actually covered, while in Switzerland it is all covered. Finally, in Switzerland, in the first three years they cover all the topics — even additional one — that are in the five year program in Bolivia.

Similar informal observations can be made about university programs in Spain. Although no rigorous study has been carried out, the observation is valid for reflection. Similar observations can be made about high school education in the two countries.

According to data from the universities that have degree programs in mathematics and from the Bolivian Society of Mathematics, there are no more than about 50 professional mathematicians with degrees ranging from bachelor's to doctorate in Bolivia (Portal UMSS). As Efraín Cruz has said:

> There are very few of us in the department with doctorates, but a good group is now engaged in doctoral study ... the research that has been done is mostly reviews of the literature, but we hope that soon we will receive support from the international community.

> This number, only taking teaching into consideration, is insignificant in relation to the actual needs of the higher education system. The pressing needs for professional mathematicians in specific fields where mathematics is applied, and even more for the transference of universal mathematics knowledge to meet the needs of the countries, are not being adequately met.

> The gradual placement of primary and secondary teacher education programs from higher normal schools to universities will require the recruitment of professional mathematicians.

> Society has yet to understand the need for professional mathematicians in various multidisciplinary projects. However, the

situation is improving, which means that in the near future more mathematicians will be needed.

The permanent expansion of advances in science and technology demands that developing countries such as Bolivia need professionals who adequately transfer and assimilate such knowledge. The bachelor's degree in mathematics prepares students to study at the master's and doctoral levels as well as to work in specific areas of applied mathematics (Cruz, personal contact).

It is very important that some mathematicians have an interest in mathematics education, especially those who have made teaching mathematics their major activity. Also, all Bolivian mathematicians should work for the consolidation of the Bolivian mathematics community. That surely would benefit the country.

The mathematics program at the Higher University of San Andrés publishes *The Bolivian Journal of Mathematics* (*Revista Boliviana de Matemáticas*) and *Fascicles* (*Fascículos*). Porfirio Suñagua, head of the mathematics program in the Faculty of Pure and Natural Sciences at the Higher University of San Andrés from 2001 to 2004, in the prologue to the *Bolivian Journal of Mathematics* #3 indicated that the first *Scientific Journal* of the program was edited in 2000 and the first *Fascicle* in 2002.

> ... in 2000 the program edited its first scientific journal as a result of the research projects presented at the World Mathematical Year. The next year, volume 2 was published. In 2002, the first *Fascicle* published all the research papers from the week-long 35th anniversary of the mathematics program (Suñagua, 2004).

The Bolivian Society of Mathematics

Since its founding in 1991, the Bolivian Society of Mathematics has organized congresses, seminars and colloquia at the regional and national levels. In July of 2013, the 16th Kurt Gödel Bolivian Congress of Mathematics was held (Andia, Los tiempos, July 25, 2013).

Even though the Bolivian mathematics community is not particularly strong, the Society has been able to attract national and international speakers to its congresses (from Germany, Argentina, Brazil, Chile, Costa Rica, United States, Spain, Switzerland, México, Perú, etc.). Many academic contacts have been generated that have permitted many students and some professors to be able to pursue graduate studies. The Bolivian Society of Mathematics has also programmed parallel sessions on mathematics education directed at teachers of mathematics at different levels.

The Bolivian Society of Mathematics Education was created in 1995 to promote professional advancement and improvement of mathematics teaching (Grigoriu, 2005). As of 2005 it had 200 official members and each year sponsors a National Congress for an average of 600 Bolivian mathematics teachers. In 1997 it hosted the 4th Southern Cone Meeting on the Teaching of Mathematics and the 4th Iberoamerican Congress on Mathematics Education (IV CIBEM) in 2001 (Grigoriu, 2005).

Distinguished teachers of mathematics

Detailed descriptions of Bolivian teachers are scarce. The various movements and reforms were full of prescriptions, but there are no profound descriptions of teaching. There is no doubt that there are many Bolivian mathematics teachers whose teaching practices should be documented. Among them are Santiago Conde, Begoña Grigoriu, and Humberto Giacoman. However, a history of Bolivian education must mention the famous teacher of mathematics Jaime Alfonso Escalante Gutiérrez (1930–2010), who came to prominence for his work at Garfield High School in Los Angeles, California. His story was immortalized by Hollywood in *Stand and Deliver* (1988), starring Edward James Olmos.

Jaime Escalante was born in La Paz on December 31, 1930, to a family of educators. His parents were teachers in a school in Achacahi, so Jaime grew up among Aymara children. His mother, who had high aspirations for her children, took them to the city on the first opportunity she had (Pari, 2011; Scraff, 2009; Mathews, 1988).

Upon finishing high school at the Colegio San Calixto, Escalante enrolled in the normal school that had been created by Georges Rouma in La Paz, but he did not agree with the teaching methods that were being used there (Pari, 2011). Before finishing his pedagogical studies, he was invited by one of his teachers, Humberto Bilbao, to study mathematics and physics. Later, several institutions competed for his services (Pari, 2011; Mathews, 1988). He went to Puerto Rico for graduate studies in sciences and mathematics. He returned to Bolivia, but because of the political and economic instability in Bolivia, he decided to emigrate to the United States.

Upon arriving in the United States, with his limited English and without a teaching credential recognized by the State of California, he had to start over. For admission to Pasadena City College, he finished the two-hour placement test in mathematics in 20 minutes with a perfect score (Mathews, 1988). He studied English, electronics, and mathematics at night while working during the day. He first worked in a restaurant, and then at Burroughs Corporation. Upon earning his teaching credential, he took a job at Garfield High School, even though it meant earning much less than at Burroughs. This was a decision that would change his life forever — and the lives of those he touched. His success at this new position earned him international recognition. He was awarded the Presidential Medal for Excellence in Education by President Ronald Reagan in 1988, the Andrés Bello Inter-American Prize for Education by the Organization of American States (OAS) in 1992, and 10 honorary doctorates. More information about his success at Garfield — and on the prizes and recognitions he received — can be found in Mathews (1988), Schraff (2009), and Pari (2011).

Escalante returned to Bolivia in 2000 intending to share his experiences with his Bolivian colleagues. Nevertheless, authorities at the Ministry of Education and elsewhere in the government gave him no support or recognition. He is an example of the phrase from Jesus "No prophet is accepted in their own land". He did manage to present various seminars and workshops that were very well attended by teachers from different levels. For Escalante, mathematics was defined in four words: concept, language, process and application (Pari, 2011). Teachers need to understand the discipline, know how to motivate, manage personal relationships, but motivating is the most important (Pari, 2011). Students need to learn responsibility, honesty and a positive attitude; the most important is a positive attitude to mathematics.

The Avelino Siñani–Elizardo Pérez Law

At the end of the twentieth and beginning of the twenty-first centuries, there were political changes in Bolivia. The country with its multiethnic composition became known as the Plurinational State of Bolivia (Estado Plurinacional de Bolivia) (NCPEB, 2008).

On December 20, 2010 the Avelino Siñani–Elizardo Pérez Educational Law was passed. This legal regulation initiated a new process of educational transformation in Bolivia, enshrined in the new constitution that insists that

> Every person has a right to receive an education at all levels that is universal, productive, free, integral and intercultural, without discrimination (Ley Avelino Siñani–Elizardo Pérez, Art. 1, Part 1).

According to Mario Yapu[2] (in Mayorga, 2012, p. 53), the law begins with an ideological reflection, a vision of society, and not with a serious analysis of the technical terms of what the Educational Reform of 1994 was. The foundations of education appear in a list of adjectives: decolonizing, liberating, community, democratic, universal, unique, diverse, plural, integrating, inclusive, intercultural, productive, scientific, etc.

This torrent of terms has generated certain doubts and questions such as this one from Carlos D. Mesa[3] (2012): "New education or new adjectives?" And this is the question from many Bolivians, because the law has not specified the technical aspects and the contents.

The teachers ask: What are the new contents that we are going to teach in this new paradigm, this new educational model? (Yapu, in Moyano, 2012, p. 54).

Perspectives on Mathematics Education

After the creation of the Bolivian Society of Mathematics in 1991, teachers of mathematics saw the need to create the Bolivian Society of Mathematics Education in 1995.

The Bolivian Society of Mathematics, through its national congresses, has allowed mathematicians and mathematics teachers to learn about opportunities for graduate study in mathematics and mathematics education. There has also been the emergence of several fields of research: control theory, dynamic systems, matrix theory, mathematics education, etc.

---

[2]Yapu is a sociologist and anthropologist who received a doctorate in sociology from the Catholic University of Louvain in Belgium. He is a professor and researcher on research methodology and topics in education. He is the author of various books and Academic Director at the University for Strategic Research in Bolivia.
[3]Investigative journalist and former President of Bolivia.

Both societies have seen the need to promote the consolidation of the Bolivian mathematics community so that professionals in mathematics and mathematics education can generate a focus characterized by the sociocultural dimension of Bolivian society, as well as initiatives aimed at teacher development and research in ethnomathematics.

It is important to look at the past in order to understand and give a correct interpretation of the present, but we should not pontificate on history nor remain with our feet trapped and our hearts divided by the past. Of course, there are many difficulties to be overcome and challenges to be met. To do so we need the collaboration of all who are involved (teachers, principals, the community, government officials and students) for the sake of achieving a quality educational system for the preparation of mathematics teachers. We need to create graduate courses, particularly in the universities that can rely on the departments of mathematics and help teachers with the interest and desire to achieve disciplinary and instructional qualifications specific to mathematics.

The MEMI program (Improvement of the Teaching of Mathematics and Informatics) began in Cochabamba in 1992 (Grigoriu, 2005). Sponsored by the Faculty of Technology at the Higher University of San Simón, it had support from the Freudenthal Institute of the Utrecht University in The Netherlands. Currently, MEMI has its own installations in the university including classrooms equipped with computers, projectors, etc. It continues to develop activities and supports teachers by organizing courses, workshops and seminars related to the teaching of mathematics (Zegarra, 2013, personal contact).

The initiative of Jaime Escalante upon his return to Bolivia led to many seminars and workshops throughout Bolivia and in other Latin American countries. He awakened the interests of many teachers by showing that it is possible to enjoy the learning of mathematics.

Another proposal to consider is that of Eduardo Valenzuela Siles,[4] who has offered to donate basic software licenses for the Universal Data Manager (ADM).

> Bolivian scientist Eduardo Valenzuela Siles offered advanced mathematical software that facilitates the teaching and learning of teachers and students with complex exercises in physics, calculus, algebra and more to the Ministry of Education.

---

[4]Valenzuela is a Bolivian scientist living in Germany.

There are many more initiatives of teachers who work in Bolivia and abroad, but they are not documented. They simply happened and are often forgotten. This work invites mathematicians, mathematics teachers and researchers to document the mathematics teaching experiences of educators and institutions. We have a commitment to our country and to our profession.

Bolivian society is beginning to understand the importance of the development of mathematics, its teaching and its learning. On the one hand, professionals are earning graduate degrees in mathematics abroad. On the other, political authorities are advancing innovative initiatives from a constructivist and socio-productive community approach, with an emphasis on ethnomathematics.

To some, it would seem as if teachers are bewildered by this new approach. However, once the initial shock is overcome, teachers would be better prepared to develop school mathematics in a way different than the traditional one. This clearly shows that we must work for the innovation of mathematics teaching, and for the improvement of the initial and continued training of teachers — goals of mathematics education.

Conclusions

This chapter consists of two parts. First, we make an approach to the history of mathematics education in Bolivia, framed in the socio-political and educational events in the country. Then, we describe issues related to research in mathematics education and its development. We have divided the history in three periods: Pre-Hispanic, Colonial, and the Republican.

Although it is true that the creation of mathematics degrees in Bolivian universities happened in the second half of the past century, it is also true that they show a sustained quantitative and qualitative growth at the state level. In the 1990s the Bolivian Society of Mathematics and the Bolivian Society of Mathematics Education were created. All of this favors and facilitates the development and consolidation of the Bolivian mathematics community (mathematicians, teachers of mathematics and students).

Furthermore, the innovative initiatives contemplated by the Education Reform of Bolivia need the active participation of the Bolivian mathematics community so that the development of mathematics will correspond to social and local needs, as well as to the needs of the Plurinational State of Bolivia.

We hope that this chapter will not only show the scarce production in mathematics education research in Bolivia, but that it will also help us to reflect upon and discuss creative means by which to attain equity in education. The Plurinational State of Bolivia has the need to realize educational possibilities that have yet to be discovered and probed.

Bibliography

Acebo, R. (2012). *Crisis de la Educación.* Monographic. Retrieved from http://es.scribd.com/doc/148365849/MONOGRAFIA

Aramayo, J. (n.d.). Educación en Bolivia desde el Incario hasta la actualidad. Retrieved September 27, 2013 from http://www.eabolivia.com/bolivia/5822-educacion-en-bolivia-desde-el-incario-hasta-la-actualidad.html

Barrantes, H., & Ruiz, A. (1998). *Historia del Comité Interamericano de Educación Matemática.* Bogotá: Academia de Ciencias Exactas, Físicas y Naturales. Retrieved from http://www.cimm.ucr.ac.cr/ciaem/?q=es/node/37

CEUB (2011). Modelo Académico del Sistema de la Universidad Boliviana. La Paz, Comité Ejecutivo de la Universidad Boliviana. Retrieved September 23, 2013 from http://www.uatf.edu.bo/web_unidades/dsa/descargas/NUEVO_MODELO_ACADEMICO_EN_LA_SUB_2011.pdf

Código de Educación Boliviana (1955). Retrieved October 14, 2013 from http://www.lexivox.org/norms/BO-DL-19550120.xhtml

Conclusiones de la reunión sectorial de Carreras de Matemáticas (2008). Universidad Mayor de San Simón. Cochabamba, 11 al 14 de noviembre.

Constitución Política del Estado (1947). Retrieved October 14, 2013 from http://www.lexivox.org/norms/BO-CPE-19471126.xhtml

Diez de Medina, F. (2005/1986). *Tiwanaku Capital del misterio.* La Paz: Rolando Diez de Medina.

El Diario (2011). Historia de Bolivia. Retrieved October 14, 2013 from http://www.eldiario.net/bolivia/

Elbarquero (2006). Tiwanaku: la capital de una civilización perdida. Retrieved from http://www.viajeros.com/diarios/tiwanaku/tiwanaku-la-capital-de-una-civilizacion-perdida

González, L., & Bravo, C. (2008). Fundación y límites de la Real Audiencia de los Charcas. Retrieved October 10, 2013 from http://www.americanistas.es/biblo/textos/08/08-071.pdf

Grigoriu, B. (2005). Educación Matemática en Bolivia, *Revista Iberoamericana de Educación Matemática, 1,* 55–88. Retrieved September 27, 2013 from http://www.fisem.org/www/union/revistas/2005/1/Union_001_011.pdf

Kolata, A. (1993). *The Tiwanaku. Portrait of an Andean Civilization.* Peoples of América. Retrieved from http://www.lavoisier.fr/livre/notice.asp?ouvrage=1807600

Iño, W. (2010). A más de cien años de la fundación de la "Escuela Normal de Profesores y Preceptores de la República", en Sucre (1909). XXIV

Reunión Anual de Etología–RAE. Retrieved from http://200.87.119.77: 8180/musef/bitstream/123456789/513/1/421-433.pdf
Ley de Reforma Educativa (1994). Retrieved October 20, 2013 from http://www.oei.es/quipu/bolivia/Ley_Reforma_Educativa_1565.pdf
Ley de Educación "Avelino Siñani–Elizardo Pérez" (2010). Retrieved October 10, 2013 from http://www.oei.es/quipu/bolivia/Leydla%20.pdf
Lozada, B. (2004). *Formación Docente en Bolivia*. La Paz: Multimac S.R.L. Retrieved October 10, 2013 from http://www.cienciasyletras.edu.bo/public aciones/Textos/pdf/La%20formacion%20docente%20en%20bolivia.pdf
Márquez, S. (2004). La educación superior universitaria en Bolivia, in *Digital Observatory for Higher Education in Latin America and the Caribbean*. Available at http://www.iesalc.unesco.org.ve/
Martínez, R. (1910). Anuario de Leyes y Resoluciones Supremas de 1909. Compilado por Ricardo Martínez Vargas. Tipografía La Unión: La Paz.
Mathews, J. (1988). *Escalante: The Best Teacher in America*. New York: Henry Hold and Company.
Mayorga, J. A. (2012). Entrevista a Mario Yapu sobre la última Reforma Educativa en Bolivia: los desafíos de la nueva Ley Avelino Siñani-Elizardo Pérez. *Propuesta Educativa*, 38-Año 21-Noviembre, 2012, Vol. 2, pp. 49–58.
Mesa, C., Mesa, J., & Gisbert, T (2012). *Historia de Bolivia*. La Paz: Editorial Gisbert, (8ª Ed).
Ministerio de Educación, Estado Plurinacional de Bolivia (2013). Matemática. La reconstrucción sociocultural de la matemática. Programa de Formación complementaria de Maestros y Maestras. PROFOCOM. http://profocom.minedu.gob.bo/index.php/material/material_subs_participante/1
Molina, J. (1991). *Los enigmas geométricos en Tiwanaku*. La Paz: Hisbol.
Molina, J. (2000). *La matemática calendárica de la Puerta del Sol en Tiwanaku*. La Paz: Hisbol.
Neso, N. (2013). La guerra del agua y la guerra del gas- los movimientos sociales de Bolivia y la llegada de evo Morales. *Revista de ciencias Sociales de la Universidad Iberoamericana*. 15, 207–232. Retrieved from http://www.ibero.mx/iberoforum/15/pdf/ESPANOL/8.%20DOSSIER%20 IBEROFRORUM%20NO15.pdf
Nueva Constitución Política del Estado (2008). Retrieved October 18, 2013 from http://pdba.georgetown.edu/Constitutions/Bolivia/constitucion2009.pdf
Organización de Estados Iberoamericanos OEI (1988). Breve reseña histórica del sistema educativo, en *Sistemas Educativos Nacionales-Bolivia*. Retrieved from http://www.oei.es/quipu/bolivia/boli02.pdf
Organización de Estados Iberoamericanos OEI (1997). *Sistemas Educativos Nacionales–Bolivia*. Retrieved September 27, 2013 from http://www.oei.es/quipu/bolivia/boli02.pdf
Organización de Estados Iberoamericanos (OEI) (2009). 24 *Olimpiada Iberoamericana de Matemáticas*. Retrieved from: http://www.oei.es/oim/reporteOIM09.pdf

Pari, A. (2011). *Historia de Vida y Metodología de la Matemática de Jaime Alfonso Escalante Gutiérrez.* Salamanca: Universidad de Salamanca. Doctoral thesis.

Pari, A. (2012). El boliviano Jaime Alfonso Escalante Gutiérrez, profesor de matemáticas en los Estados Unidos, in *Formación de Élites y Educación Superior en Iberoamérica (ss. XVI–XXI).* X Congreso Iberoamericano de Historia de la Educación Latinoamericana, José Maria Hernandez Diaz (Cord), 229–239. Salamanca: Universidad de Salamanca.

Pérez, A. (1980). Las matemáticas modernas: pedagogía antropología y política. Entrevista a Georges Papy. *Perfiles Educativos. 10,* 41–46.

Ponce, C. (2000). *Tiwanaku y su fascinante desarrollo cultural: ensayo y síntesis arqueológica.* La Paz: Cima.

Rodríguez, F. (2008). *La Matemática nació en Tiwanaku.* La Paz, Publicrafic Digital.

Rodríguez, G., & Weise, C. (2006). *Estudio superior universitario en Bolivia: estudio nacional.* UNESCO. Retrieved from http://unesdoc.unesco.org/images/0014/001489/148999s.pdf

Schraff, A. (2009). *Jaime Escalante: Inspirational Math Teacher.* Berkeley Heights, NJ: Enslow Publishers, Inc.

Suñagua, P. (2004). Prólogo. *Revista Boliviana de Matemáticas #3.* Retrieved October 15, 2013 from http://bibmat.umsa.bo/bibventas/rev3res.pdf

Serrudo, M. (2006). Historia de la Universidad en Bolivia. *Revista de la Educación Latinoamerica, 8,* 49–64. Retrieved September 29, 2013 from http://www.redalyc.org/pdf/869/86900804.pdf

Zurita, R. M. (2003). Proyecto MEMI Universidad Mayor de San Simón-Bolivia. Retrieved from http://www.cesip.org/es/enlaces-bdd/trabajos/bolivia/2003/zurita_memi.pdf

About the Author

Abdón Pari Condori has worked as a mathematics professor in Bolivia and Spain. He earned his undergraduate degree (licenciatura) in mathematics education at the Universidad Peruana Unión, a master's degree in mathematics at the Universidad Católica del Norte in Chile, and a doctorate in mathematics education from the Universidad de Salamanca. He is a member of the Bolivian Society of Mathematics (SoBolMat) and is the representative for Bolivia before the Inter-American Committee of Mathematics Education (CIAEM).

Chapter 3

BRAZIL: History and Trends in Mathematics Education

Beatriz S. D'Ambrosio, Juliana Martins, and Viviane de Oliveira Santos

Abstract: This chapter begins with a brief history of mathematics education in Brazil, contextualized within the socio-political and educational events in the country. Research in ethnomathematics and mathematical modeling provide the foundation for curriculum reform and teacher preparation initiatives that strive to achieve greater social justice and equity for Brazilian children.

Keywords: History of mathematics education; ethnomathematics; mathematical modeling; mathematics education; social justice; equity.

Brazilian mathematics education is a young field of research, establishing itself in the latter half of the twentieth century. In this chapter we describe the history of mathematics education in Brazil, from colonial times to more contemporary developments. During this journey we will highlight a few aspects that characterize Brazilian mathematics education today and finally conclude the chapter with a discussion of current trends, including new directions in teacher education and aspects of curriculum reform.

The History of Brazilian Mathematics Education

In order to situate the history of mathematics education in Brazil, we have included in our discussion some important aspects of the political and educational history of the country. Our discussion is organized around the same periods used by D'Ambrosio (2011) to describe the history of mathematics in Brazil. These are: pre-Colombian/Cabral: from pre-history to the first settlements; the conquest and colony (1500–1822); Empire (1822–1889); First Republic (1889–1916) to the beginning of modernity (1919–1933); Modern Times (1933–1957); and contemporary developments (since 1957) (p. 19).

From pre-history to the first settlements

Like most countries in South America, the history of Brazil's beginning is characterized by colonialism. Prior to the colonial period, native peoples inhabited the lands that would come to be called Brazil. It is beyond the scope of this paper to bring details of the education of the native peoples. Many different groups with distinct educational and cultural practices constituted the population of the lands found by the Portuguese. Today, efforts exist to recover the cultural practices of the different groups, and with oral histories, to try to understand the history of their educational and other cultural practices. Many cultural anthropologists and historians have engaged in the challenge of writing the history of Brazil prior to the colonial period. Of particular interest to this chapter are the studies of various ethnomathematicians, who focus on the examination of the mathematical practices of different tribal groups within Brazil. Their work invariably leads to inferences about the history of Brazil prior to the colonial period. Ethnomathematics will be discussed in more detail later in this chapter.

The conquest and colony (1500–1822)

In 1500 Brazil was established as a Portuguese colony, and similarly to what occurred in other Portuguese colonies throughout the world, there was little educational investment made in the colony. Unlike in the ruling of the Spanish colonies, the Portuguese did not establish educational institutions, libraries, or printing presses in their possessions, and their behavior in Brazil was no exception. However, in 1808 the Portuguese court was forced to flee from Europe to Brazil, as Napoleon's threats of invasion of Portugal grew stronger. With the court came a new era for Brazil as the colony became the new home of the Portuguese nobility. It was this turn of events that established in Brazil the first educational institutions — created to educate the Portuguese children arriving at the colony — the first institutions of higher education and research, the first libraries, a printing press, and many other institutions that would provide the necessary infrastructure for the permanence of the royal family, and the court, in Rio de Janeiro. With this event, Brazil quickly shifted from the status of a colony to being part of the United Kingdom of Portugal; in fact, Rio de Janeiro became the administrative center of the Portuguese empire.

1500–1808: Until the events of 1808, all responsibility for education had been in the hands of the Jesuits and clergy. In the early years of the colonial period, the only institutions created by the colonial powers were strictly those targeted to commerce and trade. The main business was the storage of "pau-brasil", i.e. the wood found on the colony that was to be shipped to Portugal.

In 1532, the first community was founded — Vila de São Vicente — a port city from which shipments of goods would be sent to Portugal. As of 1549, the Portuguese king Dom João III decided to establish organized government for the colony in Bahia, where the exploitation of sugar cane would be organized and administered; yet, there was no evidence of any efforts to organize education or instruction. Tomé de Souza, of military background, was appointed to govern the colony with the mission of organizing life in the town, the production of sugar, the distribution of lands, and the opening of roads. On his move to the colony, six Jesuit priests accompanied the new Governor General Tomé de Souza and his commission. Niskier (1989) refers to these six religious men as the first educators of Brazil who had the intention of instilling in the natives the Christian faith. Their main goal was to teach the natives the Portuguese language, including reading and writing, as part of their efforts toward catechism. According to Miorim (1998), the clergy dominated Brazilian education for nearly 200 years. Silva (2003) reports that

> ... in some elementary schools the four operations were taught, and that in some art courses, topics in Euclidean Geometry appear. In the year 1605, there were classes in Arithmetic at the Colégio Salvador, in Recife (Pernambuco), and in the city of Rio de Janeiro. Among the topics taught are Ratios and Proportions, as well as Euclidean Geometry (p. 15).

Certainly, the developments in the colony suffered the repercussions of the crises on the mainland. During the years between 1580 and 1640, Portugal was dominated by Spain, leading to instability and unrest in the kingdom. This instability seems to have caused some vulnerability in the colonies and as a result the Dutch seized the opportunity to colonize parts of Brazil, invading and occupying different locations in the north of Brazil. The longest occupation occurred in what is today the city of Recife, and lasted from 1604 to 1679. Meanwhile Portugal struggled to regain its independence from Spain and was finally able to do so, with the awareness that it was necessary to reorganize and improve its armed forces in order to better protect its possessions both on the mainland and its colonies. This awareness resulted in the establishment of military training efforts (*Classes in Fortification and Military Architecture*) in Portugal (1647) and, a few years later, in the three Brazilian regions of Rio de Janeiro, Bahia (*School of Practical Artillery and Military Architecture*) and Maranhão (*School of Fortification*) (1699) (Valente, 1997; Niskier, 1989). It is thus that the first institutions of higher education appear in Brazil to prepare stronger armed forces with the ability to plan, build, and strategize for the protection of the colony. Following closely behind the establishment of these educational organizations are the publications of the first mathematical textbooks written in Brazil: the first appeared in 1744 (The Artillery's Exam) and the second in 1748 (The Bombers' Exam) (D'Ambrosio, 2011), both written by José Fernandes Pinto Alpoim (for more information about this author see Valente, 1997).

Simultaneously, the Jesuits were torn between educating the natives, as described above, and meeting the demands of educating the young Portuguese men who found themselves in the colony, but who intended to go to Portugal for a degree in law and theology, or to France for a degree in medicine. Opting to pay greater attention to the young Portuguese elite, the focus of the Jesuit shifted to secondary education (Carvalho & Dassie, 2012).

There were primarily two events that provoked a change in the direction of the colonial educational system. The first was the growth of the military academies as the official educational establishments in the colony, and the second was the 1759 expulsion of all Jesuits from the colonies and from Portugal. In order to provide for the educational needs of the colony, given the elimination of the Jesuit educational structure, the Reform of Pombal of 1772 established the *Aulas Régias*, a series of lectures in various subjects, including Arithmetic, Algebra and Geometry (Miorim, 1998). Given the colonial status of Brazil, the educational needs of the population were considered minimal, and the focus remained on the preparation of the male children of the colonial elite to pursue higher education in Europe.

1808–1822: The arrival of the royal family and the court to Brazil resulted in many changes to the educational structures existing in the colony. In particular, the creation of multiple institutions of higher education occurred quickly, among them were medical schools, and naval and military academies, which soon became engineering schools for both military persons and civilians. To prepare students to enter these institutions, the lecture structure (*Aulas Régias*) remained stable and intact until 1834. To provide the lecture structures, as well as the institutions of higher education with the textbooks needed, a publishing company was established. All of this development occurred within the first year of the royal family's presence in Brazil. While there was much activity in secondary education, up to this point there was no organized structure for elementary education. According to Carvalho and Dassie (2012) the elite generally found tutors for the early education of their young boys in order to prepare them for participating in the lecture structure.

In 1821, due to political unrest in Portugal and a threat to the throne, King João VI returned to Lisbon and left his son to rule in the colony as Regent. This seemed unacceptable to the population and soon afterwards, in 1822, feeling pressured by the population, the Regent declared independence for Brazil and became the Emperor Pedro I.

The Empire (1822–1889)

Having acquired its independence, the new country of Brazil sought to establish its economic and social autonomy. This historical period was marked with the need for construction of factories, ports, roads, urban development of its cities, and so many other needs that required specialized knowledge. At the time, the only institution capable of preparing individuals to undertake and supervise these initiatives was the military academy, then called the Imperial Military Academy. The development of civil engineering courses required the restructuring and enhancement of the mathematics courses. It was clear too that as an independent country, it no longer seemed viable that the preparation of lawyers, theologians, and other professionals for meeting the demands of the new country should occur in Portugal. Hence the need emerged for creating additional institutions of higher education. Examples of the historical movement were the two law schools created in the cities of São Paulo and Olinda in 1827. Interestingly, a factor that pushed mathematics teaching demanding more structure and organization was the entrance examination to the law programs, which required geometry (Valente, 1997). It was still the case that the preparation for students to attend the military academies and other institutions of higher education was structured as independent lectures. These independent lectures offered to students throughout the provinces would soon be reorganized as an integrated unit in institutions of secondary education, the first of which were inspired on the French Lycees. Coupling the demands for preparing students for higher education and the new laws established during the Empire, institutions of elementary and secondary education were formed, and free public education began to take shape. In 1834 the first *liceus* emerged throughout the province and in 1837, in Rio de Janeiro, the *Colégio Imperial Dom Pedro II* (Imperial School Dom Pedro II) was founded. We will refer to the school as *Colégio Pedro II* throughout the rest of this document. For the first time the teaching of mathematics was organized according to a plan and sequence. Following the French tradition, the sequence of studies was established as Arithmetic, Geometry, and Algebra and developed over an eight-year educational experience.

During this period the imperial government incentivized higher education throughout the country and financed the *Colégio Pedro II* in Rio de Janeiro. All other institutions of primary and secondary education throughout the provinces had to be supported by the provincial governments themselves. Insights into the mathematics taught at the time can be gained from the textbooks written by Brazilian authors. We refer the reader to Valente (1997) for a detailed study of those textbooks.

At the same time that the schools were established and began to provide certain stability in the education of the wealthiest young men in the Empire, the socio-political landscape of the country was one of instability. The clamoring for change and reform reflected the people's dissatisfaction with slavery, illiteracy, access to the educational system, immigration, marriage laws, and the separation of the State and the Catholic Church (Silva, 2003). The socio-political unrest resulted in two dramatic changes in the socio-political life of the country: the abolition of slavery (1888) and the drastic governmental shift from Empire to the First Republic (1889).

For nearly 70 years, during the Empire, the teaching of mathematics from secondary school to higher education was primarily related to the military schools, and the volume of mathematics produced during that time was scholarship that occurred in the *Military School*. Small signs of change are observed toward the end of the nineteenth century, primarily through the emergence of mathematics textbooks written by Brazilian authors. The sequence and breadth of content in the areas of Algebra, Geometry and Trigonometry are what appear to vary from one textbook to another.

First Republic (1889–1916) to the beginning of modernity (1916–1933)

With the proclamation of the Republic, in 1889, the government was reorganized and ministries were established. In particular, we find the Ministry of Instruction, Mail and Telegraphs. The first minister to be named was Benjamin Constant Botelho de Magalhães and his first act was to promote reform of the educational system based on positivist doctrines. The Reform of Benjamin Constant (1890) moved the secondary system away from the classic-humanist focus that characterized it at the time, toward a more scientific focus. In spite of numerous attempts at educational reform beginning with the constitution of 1891, followed by the Reform of Rivadária Correia (1911), the Reform of Carlos Maximiliano (1915) and the Reform of Rocha Vaz (1925), the original Reform of Benjamin Constant was the one to have the most impact on Brazilian education during the Republic. Even as this initiative established programs of study for both the primary schools and the secondary schools, the educational system still had the single goal of preparing students for the professional schools — law, medicine and engineering (Miorim, 1998).

During the first two decades of the twentieth century — referred to by D'Ambrosio (2011) as the beginning of modernity — the work of John Dewey gained momentum, particularly in the United States. His writings reflected new developments of psychology and their implications for rethinking schools and education. In contrast to the existing traditional views of learning at the time, Dewey proposed that the child be the center of the educational process, shifting focus away from the actions of the teacher and emphasizing the importance of an active student in order to promote learning. This characterized the movement of Progressive Education and was referred to in Brazil as the New School Movement. With the end of the First World War the great English investors in the Brazilian economy lost their grasp of the financial and mercantile monopoly they held over Brazil. This resulted in the greater presence in and influence of the United States on Brazilian economy, permeating the educational system as well.

The founding of the Brazilian Association for Education, in 1924, is an important marker of the New School Movement in Brazil. Members of the association were intellectuals who were becoming more and more responsible for instruction in the schools. Adhering to the Progressive Education movement occurring in the US, the association proposed several reform initiatives for the Brazilian educational system. In spite of this movement, mathematics education in Brazil stayed faithful to the French Catholic school models using collections of textbooks that guided instruction. According to Valente (1997), "Slowly the *lessons* begin to create space for *exercises* in the mathematics textbooks" (p. 162, our translation), in an effort to assure enhanced learning opportunities for students.

The French influence reached the *Colégio Pedro II*, in the last decade of the XIX century, through the adoption of the collection *Frère de l'Instruction Chrétienne* (Brothers of Christian Instruction) — FIC for mathematics instruction. These textbooks remained influential in Brazil, with new editions published, until the middle of the 1950s. Through these texts (translated to Portuguese by Raja Gabaglia) the program for secondary mathematics education came to include Arithmetic, Algebra, Geometry, Descriptive Geometry, Trigonometry, Cosmography, Mechanics, Measurement (Valente, 1997). Competition for FIC emerged with the *Frère Téophane Durand* (FTD), a congregation of Marist Brothers founded in 1817 in France, and established in 1897 in Brazil. The FTD publication introduced an innovation in their collection — the teacher's guide. These two collections, FIC and FTD, dominated the teaching of mathematics in Brazil for quite some time.

With the advent of the industrial revolution, European powers turned their attention to the need for reform of their educational systems. In an effort to amplify the discussion of mathematics instruction internationally, the International Commission on Mathematical Instruction (ICMI) was founded in 1908, during the IV International Congress of Mathematicians (ICM). According to this group, the reform of mathematics instruction was urgently needed (Schubring, 2004). The first president of the ICMI was Félix Klein. With his persuasive leadership style, he put forth many ideas about the reform needed in the teaching of mathematics. The analysis in detail of the proposals for reform of mathematics teaching put forth by Klein is beyond the scope of this chapter. Instead, we focus the discussion here on the opportunities that emerged for Brazilians to become aware of and participate in the movement.

Arthur Thiré — a French mathematician invited during the Empire to teach in one of the institutions of higher education of Brazil — moved to Rio de Janeiro and joined the faculty of *Colégio Pedro II*. He is considered the first mathematician to fight for reform in mathematics teaching in

Brazil (Valente, 2004). Under his influence, the faculty of the *Colégio Pedro II* adhered to the international movement in 1912, the year of the V ICM, selecting a delegate from Brazil to the commission. The Brazilian delegate, Raja Gabaglia, attended the 1912 meeting of the commission, during which all participants agreed to prepare a document describing in detail the mathematics education of their countries, to be presented at the 1916 meeting. Due to the First World War, the next meeting of the commission did not occur until 1920.

Gabaglia returned to Brazil after the 1912 meeting and assumed the position of director of the *Colégio Pedro II* until 1914. We will not try to hypothesize in order to explain the reasons why Gabaglia did not disseminate the reform proposed by the commission and all that he learned during the IV ICM. Instead he chose to stay focused on the FIC textbooks with teacher guides from FTD used at the school. The existing evidence shows that no changes occurred in the teaching of mathematics at the school during his leadership, in spite of the activities of Arthur Thiré, pushing for change. The faculty was divided and it was only after Gabaglia's death in 1919, that a few members of the faculty began to work towards reform of mathematics teaching at the *Colégio Pedro II* in an effort to modernize mathematics instruction and align it to the New School Movement and the reform efforts in France. In 1925 Euclides Roxo was appointed director of the school and the reform efforts gained momentum. Still, due to the ongoing resistance and reluctance among the faculty coupled with the legacy of his predecessor, Roxo's efforts were not successful until the later years of the 1920s. According to Dassie (2008), between 1915 and 1922 there is no evidence of any change to the school's mathematics program that might have been initiated by Roxo (p. 41). The voices clamoring for improvement in the teaching of mathematics were only acknowledged during the latter half of the 1920s.

It is within the scenario of the *Colégio Pedro II* that Roxo wrote his book *Lições de Aritmética* (Lessons in Arithmetic). Although not a direct translation, Roxo's book is based on the work of the French mathematician Jules Tannery, *Leçons d'Arithmétique*. The faculty's discussions around this book, beginning in 1923, became the springboard for the changes that occurred in the mathematics programs at *Colégio Pedro II*. In fact, the school's mathematics program was radically modified in 1927 (Valente, 2004). Euclides Roxo's influence on mathematics teaching in Brazil went beyond the *Colégio Pedro II*, due to his book's national success. In 1928 the Brazilian Education Association and the National Department of Education approved his proposal for the reform of mathematics teaching at *Colégio Pedro II* to be the national curriculum. Basically, what was a fragmented curriculum composed of the study of Arithmetic, Algebra, Geometry, and Trigonometry became an integrated curriculum that was to be called simply the study of Mathematics (Valente, 2004, p. 73). For Valente (2005) this is the moment in which we can claim that the discipline of mathematics was created in Brazil. In 1929, affirming the existence of mathematics as a discipline in Brazil, the first Brazilian journal that was fully dedicated to mathematics was founded in the State of Bahia, the *Revista Brasileira de Mathematica Elementar* (Brazilian Journal of Elementary Mathematics) (D'Ambrosio, 2011, p. 67).

This time period closes with the publication of several textbooks that guide the reform efforts throughout the country. According to Dassie (2008), the following texts appear: *Como se aprende mathematica* (How mathematics is learned) by Savério Cristófaro, in 1929; three volumes of *Curso de Mathematica Elementar* (Course in Elementary Mathematics) by Euclides Roxo, also in 1929; and three volumes of *Mathematica* (Mathematics) by Cecil Thiré and Mello e Souza, in 1930 (p. 154). Later Euclides Roxo joined Thiré and Mello e Souza as third author of volumes four and five of their collection, *Mathematica*.

The last few years of this period were of great turmoil in the economic and political scenario of the country. The economy depended on its export of coffee primarily to the United States, but with the Great Depression of 1929 the exports were hugely affected, leading the economy into a state of much instability. The public dissatisfaction with the economic crisis resulted in the Revolution of 1930, when Getúlio Vargas, with the help of the armed forces, deposed the elected president and declared himself president of Brazil. The Minister of Education of Vargas' government, Francisco Campos, was responsible for the educational reform of 1931, *Reforma Campos*. According to Miorim (1998), Campos' Reform was to modernize the content and teaching methods of secondary education. These efforts were highly compatible with the changes in mathematics teaching proposed by Euclides Roxo, and not surprisingly, his proposal was fully adopted by the national reform (p. 94).

Modern Times (1933–1957)

The 24 years encompassed by the Modern Times were marked by many fewer changes in mathematics education than the previous period.

Nevertheless, politically, there were many changes. Vargas' government, known as the Vargas Era, was a dictatorship that lasted until 1945, when Vargas was deposed and exiled until 1951. He then returned to power by way of popular elections. For the next four years, he governed democratically, but not being able to withstand the political pressures, he committed suicide in 1954.

A few reforms and educational laws were passed during this period, continuing to mold the Brazilian educational system and to shape the mathematics education of children and young adults of the time. In terms of mathematics curriculum in the schools, these reforms had little impact. It was mostly laws guaranteeing access to schools, as well as the duration of schooling, that were modified during this period. With the *Reforma Capanema*, schooling became more elitist and bifurcated. Schooling, with the intent of accessing higher education, was an option for a few, while the rest of the population was to be channeled into a professional education system that would prepare the young people for access into the job market.

Probably, the most important development for mathematics education during the modern period was the founding of the University of São Paulo (USP), which unified the many different institutions of higher education into one institution. At this university a new dimension of higher education took shape, as professionals were prepared to teach and to do basic scientific research, such as mathematics. To meet this demand, several researchers from abroad were contracted to teach at USP. As a result, the students in São Paulo became aware of the major new developments and trends in the mathematics of the times (Silva, 2003, p. 137); in particular, the seeds of modern mathematics, which was appearing in Europe, found their way to Brazil. This is how in the decade of the 1950s, a new period for mathematics education began; it was marked by the attempt of several scholars to introduce aspects of the "modern" mathematics into the pre-college levels of education (D'Ambrosio, 1991).

The university community and recent graduates felt compelled to organize themselves for national professional conversations, hence the creation in 1945 of the Mathematical Society of São Paulo. Several years later (1969), during one of the meetings of the professional organization, the Brazilian Mathematical Society was founded, with the goal of stimulating the development of research and teaching in the country. It was primarily during the meetings of these societies that members discussed the insertion of modern aspects of mathematics in secondary schools. In fact, during the First National Congress of Mathematics Teaching, held at the University of

Bahia, members recommended the insertion of modern mathematics topics in the curriculum in light of the experiences that were being developed at the secondary schools tied to the universities — the laboratory schools.

Of course, any effort to determine periods in history is a difficult feat, since periods inevitably overlap. Much activity in mathematics education occurs in the transition from the period, which D'Ambrosio (2011) calls Modern Times, through the new period, which he refers to as Contemporary Times. While these periods may be more clearly separated in the history of mathematics in Brazil, they tend to blend into each other when we try to describe the history of mathematics education.

Contemporary Times (1957–today)

The late 1950s in Brazil were politically troubled times. The country was about to undergo a military coup which would overthrow the existing government and for two decades (1964–1985) consecutively sit five Army Generals as the Presidents of the Republic. Just prior to this military coup (in early 1964), the Ministry of Education and Culture had instituted the National Literacy Program, based on Paulo Freire's pedagogy for adult literacy education. With the coup, all initiatives from the previous government were discontinued, Paulo Freire was arrested and exiled from Brazil, and was allowed to return only in 1980. As exemplified by this act of the military dictatorship, education can be characterized in that era as undergoing a period of great repression, with the privatization of education, exclusion of large portions of the population from quality elementary education, institutionalization of professional education, pedagogy of training, and a general disorientation of the ministry of education due to excessive and confused educational legislation (Ghiraldelli Jr., 1991, p. 163).

The imposition on schools of the pedagogy of training (a North American trend at the time) had the goal of the production of a qualified and competent work force, trained by the secondary school system. This training view of teaching and learning permeated the educational system until the end of the military dictatorship in 1985.

An important aspect of the history of mathematics education in Brazil was the emergence and dissemination of the modern mathematics reform movement mentioned above. As we mentioned earlier, late in the 1950s in the university setting, interest in modern mathematics had emerged as mathematicians explored new areas of research. In 1957 a second National Congress of Mathematics Teaching was held in Porto Alegre. At that

Congress, with a large number of secondary teachers in attendance, there was much discussion of the adequacy of mathematics teaching given the new advances in science and psychology. Proposals for working with topics from modern mathematics in the pre-college curriculum were still modest. At several meetings of mathematicians the topic of introducing modern mathematics into the curriculum were raised. However, it was only in the 1960s that the Modern Mathematics Movement (MMM) gained momentum. Scholars (Miorim, 1998; Fischer, *et al.*, 2007) refer to the founding of the study group *Grupo de Estudos do Ensino da Matemática* — GEEM (Study Group for the Teaching of Mathematics) in São Paulo in 1961, as the historical moment that officially established the MMM in Brazil.

Other study groups emerged throughout the country with similar goals as GEEM; those were to offer courses for mathematics teachers that would prepare them to implement the curriculum supporting the MMM. These courses resembled the Summer Institutes offered in the United States during the same period, and counted on the participation of visitors from abroad such as Marshall Stone (USA), George Springer (USA), George Papy (Belgium), and many others. Since its founding, GEEM has played an important role in the organization of the National Congresses of Mathematics Teaching and, as a result, the focus of the fourth and fifth congresses was related to themes that would support the implementation of modern mathematics in secondary schools. The emphasis was the modern mathematics content of Set Theory, Logic, Abstract Algebra, Vector Spaces and Geometry. It was at this same time that the textbooks of the School Mathematics Study Group (SMSG) were translated to Portuguese and became the only textbook to support the implementation of modern mathematics in secondary schools.

Like many countries throughout the world, Brazil followed a similar journey through modern mathematics. However, during the 1970s, when the rest of the world, in reaction to the difficulties of the MMM moved back towards the teaching of the basics, instead Brazil took a turn towards a socio-cultural perspective on the teaching and learning of mathematics under the influence of the Brazilian mathematics educator Ubiratan D'Ambrosio. According to D'Ambrosio (1986), it was at the Fourth Inter-American Conference of Mathematics Education that he first proposed that the mathematics curriculum consider the socio-cultural context of the learners. This focus on pedagogical action tied to socio-cultural reality was highly controversial, but gained momentum internationally as it shaped into D'Ambrosio's Ethnomathematics Research

Program. Throughout the 1970s and 1980s Freire and D'Ambrosio, two internationally renowned Brazilian scholars, emphasized the improvement of learning by considering and valuing the culture of the learner. Freire and D'Ambrosio — working independently and in different fields — proposed culturally relevant literacy and mathematics education respectively, as critical components of a democratic and equitable education for all Brazilians. Under their leadership and guidance, Brazilian scholars and policy makers developed bold and unique creative visions of teaching literacy and mathematics.

We see 1975 as a critical point for mathematics education in Brazil, as it established itself as an interdisciplinary knowledge domain transcending the content area of mathematics, and integrating within it sociology and politics, psychology and cognitive science, anthropology and history, arts and communication, to mention just a few of the areas included in this new knowledge domain. It was this new and vibrant view of mathematics teaching and learning that could begin to create an educational system that was democratic, with equal opportunities for all the children in the country.

The late 1970s into and throughout the 1980s was a time of great development in mathematics education in Brazil. The major mathematics education journals were founded, mathematics education research centers were created, and graduate programs were established to educate a cadre of leaders to conduct research in mathematics education. Of major impact was the First National Meeting of Mathematics Education held in 1987, in which the mathematics educators of the country were compelled to organize themselves and create the Brazilian Society of Mathematics Education, founded the following year at the Second National Meeting of Mathematics Education. As a result of these initiatives, Brazilian mathematics education became firmly established as a field of study and grew a community responsible for the directions to be taken in the teaching and learning of mathematics.

Brazilian Mathematics Education Today

Brazilian education today is characterized by many of the same maladies of educational systems worldwide. These include an overemphasis on examinations undermining formative assessment of children's mathematical talents, poorly funded schools and school systems, underpaid and devalued teachers and school administrators, unmotivated and apathetic students who are underperforming, and inequities in availability and quality of schools leading to unequal educational opportunities for Brazilian children. Acknowledging that all of these issues are important in characterizing the educational system, and all have great impact on the teaching of mathematics, we have chosen to focus on two serious challenges, issues of equity and of teacher preparation.

Before delving into these discussions we must explain the structure of the educational system in order to guide the reader as to some of the terminology used throughout our discussion. The current system, referred to as *Ensino Básico* [Basic Education], is a sequence of five years of *Ensino Fundamental I* [Foundation I Education] (6 to 10 year olds) early elementary grades; four years of *Ensino Fundamental II* [Foundation II Education] (ages 11 to 14) upper elementary grades; and three years of *Ensino Médio* [Middle Education] (ages 15 to 17) which we will refer to as Secondary Education, the older terminology used to refer to this age group since it is a more usual term used by English language readers. Today Foundation I and II Education levels (9 years of schooling) are compulsory. In 2016 all twelve years of Basic Education will be compulsory. Prior to Basic Education children can attend non-compulsory preschool education. In some states, new legislation has been proposed to require the attendance of children ages 4 and 5 in preschool settings.

Equity

Based on the 2012 UNICEF report titled *Global Initiative for Out of School Children*, several important facts are raised that characterize issues related to equity in the Brazilian educational system.

The 1988 constitution established quality education as a right of all citizens. Since then, the federal government, through the Ministry of Education, has made extensive efforts to achieve the goal of providing quality education for all. While the averages on indicators of success are high, these averages tend to mask the regional inequities still existing in the system. In fact, in spite of the advances in which a large number of people have had a significant improvement in their economic situation, still 20.2% of Brazilians suffer some form of serious deprivation in education. According to the United Nations Development Program educational deprivation is considered serious when no one in a family completes five years of schooling and/or at least one school age child is not attending school (UNICEF, 2012, p. 21).

While Brazil as a whole is close to achieving universal education, the differences between the poor and the wealthy are startling. On average the rich will study roughly 10.7 years, while the poor will average 5.5 years in school (UNICEF, 2012, p. 23). The out-of-school children in Brazil tend to be under 6 years of age and between 15 and 17 years of age. In other words, there is a problem accessing school when very young and a problem staying in school past 14 years of age. Here again numbers are misleading. Reports indicate that 85.2% of adolescents between the ages of 15 and 17 are enrolled in school. Yet, only 50.9% of this age population find themselves in the correct grade, meaning that these children have failed to be promoted to the Secondary Education level. Even more telling is the fact that in the south-east of the country 60.5% are in the correct grades, but in the north only 39.1% of the same age range are in correct grades. Looking at the same age group from yet another angle, we find that only 31.3% of adolescents from the poorer strata of society are in the correct grades in comparison to 72.5% of their colleagues who are wealthier Brazilians. This indicates that access to secondary education of the wealthier Brazilians is more than twice the access of those from less advantaged homes (UNICEF, 2012, p. 24). Furthermore, if we take a look at the data relative to race, we find that 70% of white children finish the first eight years of schooling while only 30% of black children will finish; only 4.3% of the black population 30 years and older have a university degree compared to 18% of their white peers; and a black adolescent between the ages of 12 and 17 is 42% more likely to be out of school than his white peers (UNICEF, p. 47).

It is important to mention that in Brazil there are 56.2 million Brazilians, age 18 and over, who are not in school and have not completed the first nine years of schooling (Foundation I and II). In contrast, almost 4 million Brazilians, who missed the opportunity for schooling at the appropriate age, are enrolled in what is called *Educação de Jovens e Adultos* (EJA) [Education for Youths and Adults], an educational program available nationwide that seeks to educate the population over 18 years of age that never completed a Basic Education. Interestingly, 65.6% of youths and adults enrolled in EJA are trying to complete their first 9 years of schooling, while the others are in the secondary programs (INEP, 2013). Brazil has also seen an increase in the number of students seeking a secondary education program that is professional in nature. Since 2007 there has been an increase of 54% of enrollment in professional programs (INEP, 2013).

Children with disabilities are also highly disadvantaged in the educational system. According to the School Census 47% of children of school age with disabilities are not attending schools (UNICEF, p. 58).

Recognizing that children with disabilities have, for the most part, had little access to quality education, national initiatives are seeking to increase the enrollment of these children in mainstream schools. As a result of these initiatives the number of children enrolled in special schools for children with disabilities has steadily decreased while the number of children with disabilities included in the regular school system has steadily increased (INEP, 2013) as public schools become better equipped to meet the needs of these students. Recent research activities have focused on understanding and supporting the needs of these children and their teachers who struggle to provide a better experience for their newly-included students, particularly in their mathematics instruction. In order for the educational experiences of included children to be successful, it will be important to explore the nuances of teaching mathematics in ways that account for and minimize the impact of their difficulties.

It is beyond the scope of this chapter to analyze the complex societal reasons for such inequities in access to and permanence in the Brazilian educational system. However, it is safe to say that one of the major obstacles to a child's completion of basic education is his or her lack of success in the system. The very high rates of retention in grades due to poor achievement, coupled with the need to work in order to help sustain the family, are two of the major struggles of the children and reluctance of poor families to maintain their children in schools. Of course, mathematical achievement is one of the primary reasons that children fail and consequently abandon school.

Teacher preparation

According to the data presented in the School Census of 2009 only 68.4% of school teachers hold a degree from an institution of higher education, and of these 10% have not been licensed to teach. In fact, it is mostly in the first nine years of schooling that students might be taught by a teacher who does not meet the minimum legal requirements for teaching. The situation is more serious if we look specifically at the teachers working with the indigenous population. A mere 21.2% of teachers in the upper elementary grades and only 51% of teachers of the secondary schools meet the minimum licensure requirements.

Beginning in 2016 Brazilian law will require that secondary education be universal. With this expectation the demand for teachers for secondary school will grow tremendously, creating a new challenge for institutions preparing teachers for the secondary level.

The growing demands for quality teachers, the difficult access to higher education in remote areas, the lack of school professionals with licensure, the low income of teachers throughout the country, are all challenges that require alternative strategies in order to be overcome. Some of the national initiatives for improving the preparation of teachers to meet the needs of the country will be discussed below.

Initiatives and Innovations

Amidst these conditions educators strive to provide the best possible education for the young Brazilian population. In the next few paragraphs we describe some of the innovative initiatives of contemporary mathematics educators in Brazil. It is impossible to report on all of the efforts and innovative activities in schools. We will dwell on a few that seem to have the potential to result in large-scale innovations. In particular, we describe three innovations that influence the movement in preparing teachers of high quality and strive to provide an adequate number of teachers for the growing demands of the Brazilian educational system. We then describe specific work towards innovation in mathematics teaching that characterize much of the mathematics education research in Brazil. In particular, we highlight two specific innovations: ethnomathematics and mathematical modeling. The goal of both of these innovations is to increase student motivation and interest in mathematics by developing mathematics that is related to students' socio-cultural realities.

Mestrado Profissional em Matemática em Rede Nacional (ProfMat)

The goal of the *Mestrado Profissional em Matemática em Rede Nacional* (ProfMat) [Professional Master's Program in Mathematics in a National Network] is to enhance the mathematical content knowledge of practicing teachers. It is coordinated by the Brazilian Mathematical Society and implemented through a national network of institutions of higher education. In ProfMat (n.d.), we find that this federally funded program attends to one of the long term goals of the *Plano Nacional de Educação* (National Education Plan, 2010) for the decade 2011–2020, which reads: *"Goal 16: Graduate fifty percent of the practicing teachers with a master's degree and provide all with a guarantee of continuing education in their content area."* (p. 18, authors' translation).

The ProfMat graduate program is offered nationally and requires participants to attend class meetings both online and face-to-to face on weekends at one of the associated Support Centers. The program consists of six regular semesters including summers (two full years of coursework). Institutions which choose to participate agree to implement a standardized program that assures the same high quality mathematical experience for all participating students. The program was launched in 2011 and as recently as May 15, 2013, 405 master's theses have been presented. The motivation for the format of the program is to keep practicing teachers in the classrooms while pursuing their master's degree. Still, given the unreasonable teaching load of many Brazilian teachers, the program includes federal scholarships for participants in the program. The scholarships seek to relieve the participants of some of their teaching responsibilities in order to create some additional time for them to dedicate to their studies in the program. This has proven to be a huge incentive for teachers. It is important to note however that teachers are expected to maintain a minimum teaching load with the public schools throughout their two years of studies.

According to the Brazilian Mathematical Society (SBM, 2013):

> Students claim that the theoretical grounding in content acquired through ProfMat is already having a significant impact on their classroom practice. They claim that previously they did not fully understand the topics they taught; that they knew the content, but not adequate teaching approaches; that they could solve problems, but did not understand the underlying reasons behind their strategies. They affirm that the deep theoretical knowledge acquired has helped them to more adequately answer student questions. One of them states: *Before all I did was teach things, now when students ask, I know how to answer.* (p. 12, authors' translation)

In Figure 1, we present a map of Brazil that illustrates the proliferation of institutions of higher education offering the ProfMat opportunity for

The Network of PROFMAT

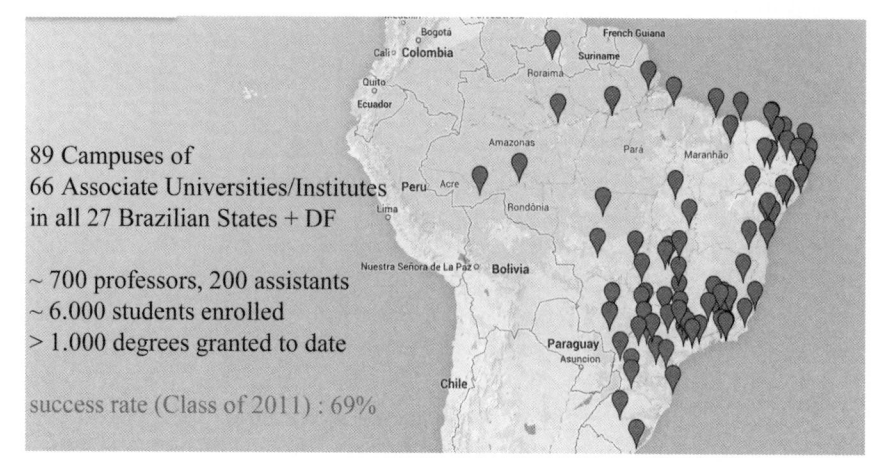

PROFMAT

89 Campuses of
66 Associate Universities/Institutes
in all 27 Brazilian States + DF

~ 700 professors, 200 assistants
~ 6.000 students enrolled
> 1.000 degrees granted to date

success rate (Class of 2011) : 69%

Figure 1. Map showing the distribution of ProfMat opportunities for teachers nationwide. From: SBM, 2013.

teachers. In 27 states, there are 66 institutions offering ProfMat experiences for teachers, with 89 support centers throughout the country. The map shows the results of efforts to distribute the sites across the country and thus reach out to teachers who have typically been underserved by continuing educational opportunities.

Programa Institucional de Bolsa de Iniciação à Docência (Pibid)

In 2007, the Brazilian president signed into law the responsibility of the federal agency CAPES to "... create policies and develop activities to support the professional preparation of teachers for basic and higher education and for the scientific and technological development of the country" (authors' translation, Brazilian Federal Law 11,502 of July 11, 2007).

In response to this responsibility, the *Programa Institucional de Bolsa de Iniciação à Docência* (Pibid) [Institutional Scholarship Program of Initiation to Teaching] was established. It is described thus: "... Pibid is embedded in an educational matrix with three axes: high quality preparation; integration of graduate education, pre-service education, and schools; and knowledge production" (CAPES, 2012, p. 3, authors' translation). In other words, Pibid is an educational program that intends to improve and value the preparation of teachers for basic education, placing future teachers in schools starting very early in their programs of study. The participants receive a scholarship and are placed in public schools as interns. In that capacity, they are expected to collaborate with their host teacher and their university faculty members to develop and implement pedagogical activities. The intent is to approximate university faculty with school faculty and bridge the gap between theory and practice and as a result, improve the quality of education nationally (CAPES, n.d.). Furthermore, graduates who participate in Pibid throughout their teacher education program commit to teach in the public school system upon graduation (MEC, 2013).

There are many events promoted to bring together all participants of Pibid across the nation. The goal of these meetings is to exchange experiences among the institutions that house the Pibid scholars, as well as create mechanisms for evaluating the impact of the program. Participants plan and discuss actions that will improve the program and stimulate student scholars to stay in teaching upon graduation.

The graph in Figure 2 shows the number of student participants in Pibid programs from 2009–2011 by their licensure program areas. The second graph, which we did not include, shows all the program areas with fewer than 772 participants.

As indicated by the graph in Figure 2, the greatest number of participants is in the area of mathematics. This reflects one of the goals of the program that was intended to prioritize and stimulate the improvement of the preparation of teachers for the sciences, which is an area of national growing need.

> Upon its launching in 2007, the priority of Pibid was in the areas of Physics, Chemistry, Biology, and Mathematics for secondary education, given the growing need for teachers in these disciplines. However, with the first positive results, as a policy valuing teaching and the growing demands for teachers, since 2009, the program was extended to support all of Basic Education, including the education of adults, indigenous peoples, rural, and *quilombos* (communities established by runaway slaves). Currently, the determination of the levels to be included and the areas of priorities are to be defined by the participating institutions, according to the

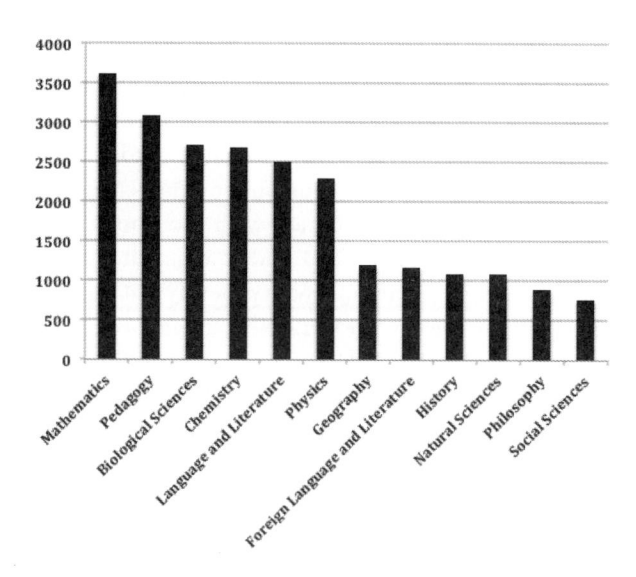

Figure 2. Graph of number of participants in Pibid 2009–2011, by licensure program area. Source: Data from CAPES, 2012, p. 13.

educational and social needs of the location and region. (CAPES, 2012, p. 5, authors' translation)

In the same administrative report of CAPES (2012), Pibid was reported as the second largest scholarship program of that federal funding agency. The latest report of March 2012 indicated that 49,321 scholarships had been granted nationwide. The same report indicates that 195 institutions of higher education are involved with Pibid and 288 projects are developed with roughly 4,000 public schools. With a tendency to grow even more, the intent is that this program becomes a permanent and important component of the preparation of teachers in Brazil. The importance of Pibid for Brazilian education is clear as one notes that during the last National Meeting of Mathematics Education (held in July of 2013), several research studies were presented that discussed the role of Pibid in the improvement of mathematical teaching and learning in the nation.

Universidade Aberta do Brasil (UAB)

In December of 2005, the Ministry of Education launched a program titled *Sistema Universidade Aberta do Brasil* (UAB) [Brazilian Open University

System] with the goal of expanding the offerings of free public higher education throughout the country. UAB is integrated into the existing network of institutions of higher education. With the intent of reaching areas of the country that are remote and distant from existing higher education institutions, the UAB's primary mode of delivery of higher education opportunities is through the use of distance education. The main target of the program is the preparation of teachers, administrators, and other professionals who will work in the Basic Education system. However, other undergraduate programs are also part of the UAB offerings including, for example, the preparation of professionals in the areas of health and agricultural and environmental sciences.

Through the National Plan for the Preparation of Teachers for Basic Education, the federal, state and city governments commit to collaborate on the initial and continued preparation of teachers. Together they identify the regional demands for teachers, administrators and other school professionals, planning and guiding the offerings in the different regions as well as the distribution of face-to-face support centers for the UAB initiative.

According to the UAB (n.d.), by 2010 the UAB aimed to establish 1,000 support centers strategically placed throughout the country. By 2013 the program intended to enroll 800,000 students each year.

The far reach of quality public higher education into areas that are remote and isolated, typically areas of low income and extremely low indexes of educational achievement, will contribute to the national goal of improving the overall quality of teachers. As a result, the quality of education will improve throughout the country and in regions that are most needy. The continued education of practicing school professionals leading to licensure, without the need for travel to large urban centers, is of utmost importance for overcoming the challenges and inequities of the educational system.

Ethnomathematics and Mathematical Modeling

Ethnomathematics and mathematical modeling are two innovations that have stemmed from concerns regarding the ineffectiveness of traditional forms of teaching mathematics to Brazilian children who are not of white middle-class backgrounds. The search for methods of teaching that support learning of all children result in studies that take an ethnomathematical approach, others that look to reality as the motivation for mathematical modeling, and still others that bring together both of these perspectives, choosing cultural dimensions as the source of questions that result in mathematical modeling. The synergy of these two instructional practices is evident in many of the projects developed throughout the nation. Although these two innovations have reached Brazilian schools through different historical trajectories, they have many points of convergence and at times are indistinguishable. For the sake of the discussion here, we have chosen to address each individually, but remind the reader that there is much overlap in their enactments in schools through educational projects.

Ethnomathematics

Brazil has a long tradition of efforts in ethnomathematics. Given the large number of children who are disenfranchised by the Brazilian educational system, including those with non-European roots, it is not surprising that many scholars have sought to understand their cultural heritage and tie the learning of mathematics to their traditions. These scholars believe that through an ethnomathematical approach to the subject children can feel more motivated to study mathematics. Leaders in the field include D'Ambrosio (1990), Ferreira (1993), Knijnik (1996), Clareto (2003), to mention just a few, and many of their students. For examples of studies of indigenous groups to uncover the mathematics embedded in tribal cultural practices we refer readers to the works of Ferreira (2004, 2005), Domite (2010), Silva (2013), Severino-Filho (2010), Scandiuzzi (2009).

Work in ethnomathematics explores the cultural dignity and self-respect resulting from the study of one's roots and the related cultural practices. It proposes a focus on the teaching of mathematics for social justice. The study of mathematics grounded in a cultural perspective puts solidarity, respect and cooperation with others at the center of the educational enterprise (D'Ambrosio & D'Ambrosio, 2013).

The exclusion of children from schools is often tied to the gap between the needs and interests of these children and the extremely traditional curriculum typical of contemporary schools. Brazilian children of *quilombos* (communities established by runaway slaves), urban slums, native or indigenous peoples, and those from the poor rural areas, tend to feel disenchanted and highly unsuccessful in a school system that presents a white middle-class view of the world. As such, the efforts toward building a knowledge base of ethnomathematics with the intent of creating a more culturally relevant curriculum that will result in greater success of the currently excluded children is the goal of many Brazilian scholars.

Mathematical modeling

Also of long tradition in Brazilian scholarship is the use of mathematical modeling as a pedagogical strategy. With a similar genesis as ethnomathematics, interest in mathematical modeling in Brazil emerges from the need to understand the world and to build mathematical models of real situations, often social and cultural. D'Ambrosio (1986, 2009) with a social-cultural focus, and Bassanezi (1994) approaching modeling from the perspective of applied mathematics, lead the development of the field of mathematical modeling in Brazil. Several scholars have emerged over the years and their studies have shaped the field and had great impact on curricular innovations throughout the country. In particular, the works of Barbosa (2006) and Biembengut & Hein (2005) are but a few examples of this ample field of study. In fact, in 2013, the 16th Annual Conference on the Teaching of Mathematical Modeling and its Applications was held in Brazil, with a large number of Brazilian scholars disseminating their work to the international community. A unique feature of the Brazilian scholars' contribution to the conference was in the focus of the papers they presented. Nearly all the presentations were reports of attempts to work with teachers or future teachers on mathematical modeling experiences that would shape their views of the importance of mathematical modeling in learning mathematics and would help them formulate a practice where modeling is in the foreground. Several of the studies presented highlighted how using mathematical modeling students can experience mathematics as a tool to analyze and understand reality.

In the paragraphs above we have dealt with innovations in Brazilian education separately, but in truth these initiatives are all connected. In spite of much scholarship in both ethnomathematics and mathematical modeling amongst Brazilian mathematics education scholars, the trajectory that takes the successes of the studies into the actual classrooms is still a difficult one. The innovations in mathematics teacher education aim to increase the number of qualified teachers, provide continuing education to practicing teachers in remote areas, and bridge the gap between teacher preparation and the practices in schools. These initiatives constitute powerful mechanisms for schooling that provide meaningful mathematical experiences for all children with teachers who understand their responsibility to include all children in the learning of a mathematics that can empower them to participate in the social and political lives of the country. According to these proposals teachers will be prepared to approach school mathematics in a very different way than has been done traditionally.

Final Considerations

This chapter contains two main parts. First we develop a brief history of mathematics education in Brazil, contextualized within the socio-political and educational events in the country. Then we describe some of the concerns with equity and teacher preparation facing the nation, highlight some of the government initiatives targeting these concerns, and describe unique aspects of Brazilian research and development in mathematics education.

Our discussion shows that Brazilian education has been characterized by a series of movements toward the provision of quality education for the entire population. Its history, traced back to the first Portuguese colonies, begins with instruction provided by the Jesuit priests. However, it was only in the first half of the twentieth century that we can document the first concerns with the teaching and learning of mathematics, beginning with the initiatives of the professors of the *Colégio Pedro II* of Rio de Janeiro, in particular with the leadership of Euclides Roxo, considered the first mathematics educator in the country.

Yet, we consider mathematics education to appear, as a field of study, only at the end of the 1980s, when there was a significant growth in the number of graduate programs, national and regional associations, journals, organization of events such as seminars, and congresses specific to mathematics education. In particular, the founding in 1988 of the *Sociedade Brasileira de Educação Matemática* (SBEM) [Brazilian Society of Mathematics Education] was an important marker in the establishment of the field.

Since its origins, research in mathematics education in Brazil has been characterized by a focus on the social and cultural dimensions of society. Ethnomathematics provides the theoretical framework for the robust and vibrant scholarship bridging concerns of teaching and learning with respect for the variety of cultural practices throughout the country. Research in mathematics education has matured to include studies of many different dimensions of teaching and learning. We highlighted the work in ethnomathematics and mathematical modeling, seeing that these are the most focused on honoring and celebrating the diversity of the Brazilian people.

We drew attention to the fact that Brazilian mathematics education emerged within a scenario of official policies striving for democratic education, with a goal of achieving quality education for all Brazilians. As such, many initiatives exist for enhanced teacher preparation, equal opportunities for schooling, and the progress of scientific research and development in the country. As the initiatives potentially impact the teaching and learning of mathematics, they become the focus of research by mathematics educators. Our focus was on these areas of synergy between policies, funding, and research. Of course, there are many challenges to be overcome and it will require the collaboration of all the stakeholders (administrators, teachers, community members, government officials, and students) in order to achieve the ambitious goal of an equitable and high-quality educational system.

There is evidence that Brazilian mathematics education has counted on the participation of all those sectors of the community which have an interest and concern with the teaching and learning of mathematics. At the last *Encontro Nacional de Educação Matemática* (ENEM) [National Meeting of Mathematics Education], the largest national meeting bringing together mathematics teachers, mathematics education researchers, along with undergraduate and graduate students, the unifying theme was "retrospectives and perspectives". In addressing the theme, there was a clear movement of the research community to reach out to the participating teachers in an effort to bridge the gap between these two groups, proposing joint efforts to work on the improvement of mathematics teaching and learning nationwide. The event, commemorative of the 25th anniversary of the SBEM, was referred to as the "new ENEM" due to this collaborative spirit among the professional communities.

As authors, we hope that this chapter not only captures the essence of Brazilian mathematics education, but that it will also lead to a reflection and discussion of creative means of achieving equity in education. Brazil is in need of pursuing educational possibilities that are yet to be uncovered and explored since it is a country whose society is characterized by a tapestry of unique peoples, with their unique cultures, aspirations, and dreams.

Bibliography

Barbosa, J. C. (2006). Mathematical modeling in the classroom: A socio-critical and discursive perspective. *The International Journal on Mathematics Education, 38*(3), 293–301. Retrieved July 25, 2013 from http://subs.emis.de/journals/ZDM/zdm063a8.pdf

Bassanezi, R. C. (1994). Modelling as a teaching-learning strategy. *For the Learning of Mathematics. 14*(2), 31–35.

Biembengut, M. S., & Hein, N. (2005). *Modelagem Matemática no Ensino*. São Paulo, SP: Editora Contexto.

Brazilian Federal Law 11,502 of July 11, 2007. Retrieved June 23, 2013 from http://www.planalto.gov.br/ccivil_03/_Ato2007-2010/2007/Lei/L1502.htm

CAPES (2012). Diretoria de Educação Básica Presencial DNE, PIBID. Relatório de Gestão 2009–2011. Retrieved June 23, 2013 from http://www.capes.gov.br/images/stories/download/bolsas/DEB_Pibid_Relatorio-2009_2011.pdf

CAPES (n.d.). Retrieved June, 23, 2013 from http://www.capes.gov.br/edu cacao-basica/capespibid

Carvalho, J. B. P., & Dassie, B. A. (2012). The history of mathematics education in Brazil. *ZDM Mathematics Education, 44*, 499–511.

Clareto, S. (2003). Terceiras margens: Um estudo etnomatemático de espacialidades em Laranjal do Jari (Amapá). UNESP–Rio Claro, SP. Doctoral thesis.

D'Ambrosio, B. S. (1991). The Modern Mathematics Reform Movement in Brazil and its consequences for Brazilian mathematics education. *Educational Studies in Mathematics, 22*(1), 69–85.

D'Ambrosio, U. (1986). *Da realidade à ação: reflexões sobre educação e matemática.* Campinas, SP: Editora da Universidade Estadual de Campinas.

D'Ambrosio, U. (1990). *Etnomatemática: Arte ou técnica de explicar ou conhecer.* São Paulo: Ática.

D'Ambrosio, U. (2009). Mathematical modeling: Cognitive, pedagogical, historical and political dimensions. *Journal of Mathematical Modeling and Applications, 1*(1), 89–98.

D'Ambrosio, U. (2011). *Uma história concisa da matemática no Brasil.* Petrópolis, RJ: Vozes.

D'Ambrosio, U., & D'Ambrosio, B. S. (2013). The role of ethnomathematics in curricular leadership in mathematics education. *Journal of Mathematics Education at Teachers College, 4*(Spring–Summer), 10–16.

Dassie, B. A. (2008). *Euclides Roxo e a constituição da educação matemática no Brasil.* Rio de Janeiro: Pontifícia Universidade Católica do Rio de Janeiro. Doctoral thesis.

Domite, M. C. S. (2010). The encounter of non-indigenous teacher educator and indigenous teacher: The invisibility of the challenges. *ZDM 42*(3–4), 305–313.

Fischer, M. C. B.; da Silva, M. C. L.; de Oliveira, M. C., & Pinto, N. B. (2007). História do movimento da matemática moderna no Brasil: arquivos e fontes. In E. R. Pacheco & W. R. Valente (Orgs.). *Coleção História da Matemática para Professores.* Guarapuava: SBHMat.

Ferreira, E. S. (1993). Cidadania e educação matemática. *A Educação Matemática em Revista, 1*(1), 12–18.

Ghiraldelli Jr., P. (1991). *História da educação.* São Paulo, SP: Cortez.

INEP (Instituto Nacional de Pesquisas Educacionais Anísio Teixeira) (2013). *Censo escolar da educação básica 2012: Resumo técnico.* Brasília: Ministério de Educação e Cultura. Retrieved July 27, 2013 from http://download.inep.gov.br/educacao_basica/censo_escolar/resumos_tecnicos/resumo_tecnico_censo_educacao_basica_2012.pdf

Knijnik, G. (1996). *Exclusão e resistência: Educação matemática e legitimidade cultural.* Porto Alegre, RS: Artes Médicas.
ProfMat (Mestrado profissional em Matemática em Rede Nacional) (n.d.). Retrieved June 23, 2013 from: http://www.profmat-sbm.org.br/
MEC (Ministério da Educação) (2013). Portal do Ministério da Educação. Pibid. Retrieved June 23, 2013 from http://portal.mec.gov.br/index.php? Itemid=467&id=233&option=com_content&view=article
Miorim, M. A. (1998). *Introdução à história da educação matemática.* São Paulo, SP: Atual.
Niskier, A. (1989). *Educação brasileira: 500 anos de história, 1500-2000.* São Paulo, SP: Melhoramentos.
Plano Nacional de Educação (2010). Federal Law PL-8035/2010. Retrieved July 31, 2013 from http://www.camara.gov.br/sileg/integras/831421.pdf
SBM (Sociedade Brasileira de Matemática) (2013). *Mestrado profissional em Rede Nacional.* Plano Nacional de Pós-Graduação (PNPG), Brasília.
Scandiuzzi, P. P. (2009). *Educação Indígena x Educação Escolar Indígena: uma relação etnocida em uma pesquisa etnomatemática.* 1st edition. São Paulo, SP: UNESP.
Sebastiani Ferreira, E. (2004). Os Índios Waimiri-Atroari e a etnomatemática. In G. Knijnik, F. Wanderer, C. J. de Oliveira (Org.), *Etnomatemática. Currículo e formação de professores. EDUNISIC*, 1, 70-88.
Sebastiani Ferreira, E. (2005). Racionalidade dos índios brasileiros. *Scientific American*, São Paulo, August 8, 90-93.
Severino-Filho, J. (2010). *Marcadores de Tempo Indígenas: Educação Ambiental e Etnomatemática.* Rio Claro, SP: UNESP-RC. Master's thesis.
Schubring, G. (2004). O primeiro movimento internacional de reforma curricular em matemática e o papel da Alemanha. In W. R. Valente (Ed.), *Euclides Roxo e a modernização do ensino de Matemática no Brasil*, 11-43. Brasília: Editora Universidade de Brasília.
Silva, A. A. (2013). *Os Artefatos e Mentefatos nos Ritos e Cerimônias do Danhono* — *Por Dentro do Octógono Sociocultural A'uwẽ/Xavante.* Rio Claro, SP: UNESP. Doctoral thesis.
Silva, C. P. da. (2003). *A matemática no Brasil: história de seu desenvolvimento.* São Paulo, SP: Editora Edgar Blücher Ltda.
UNICEF (2012). *Iniciativa global pelas crianças fora da escola.* Brasília: UNICEF. Retrieved July 28, 2013 from http://www.unicef.org/education/files/Brazil_OOSCI_Study_-_Portuguese_-_Full_report_.pdf
UAB (Universidade Aberta do Brasil) (n.d.). Retrieved June 23, 2013 from http://www.uab.capes.gov.br/
Valente, W. R. (1997). *Uma história da matemática escolar no Brasil (1730-1930).* São Paulo, SP: Universidade de São Paulo. Doctoral thesis.
Valente, W. R. (2004). Euclides Roxo e o movimento internacional de modernização da matemática escolar. In W. R. Valente (Ed.), *Euclides Roxo e a modernização do ensino da matemática no Brasil*, pp. 45-83. Brasília: Editora Universidade de Brasília.

Valente, W. R. (2005). Euclides Roxo e a História da Educação Matemática no Brasil. *Revista Iberoamericana de Educación Matemática*, *1*, 89–94.

About the Authors

Beatriz S. D'Ambrosio is a professor of mathematics education in the Department of Mathematics at Miami University in Oxford, Ohio, USA. Her research interests include the study of the complexities in the preparation of in-service K-12 mathematics teachers, examining socio-cultural issues in mathematics education, and supporting teachers to engage in action research that enhances the children's construction of mathematical knowledge. She began her career as a mathematics teacher in Brazil moving to the USA in 1989. In 2013 she spent a semester as a visiting professor at the Universidade Estadual Paulista "Júlio de Mesquita Filho" (UNESP – Rio Claro, SP Brazil) in the Graduate Program in Mathematics Education.

Juliana Martins is a graduate student in mathematics education at the Universidade Estadual Paulista "Júlio de Mesquita Filho" (UNESP), Rio Claro Campus in São Paulo, Brazil. She has participated in educational projects related to teaching and mathematics education in the early years of Brazilian basic education. Currently she is pursuing her interests in the study of the history of mathematics and mathematics education. She is a member of the Research Group in the History of Mathematics (GPHM), created in 1995, certified by UNESP and credentialed in the directory of research groups in Brazil.

Viviane de Oliveira Santos is an assistant professor in the Mathematics Institute of the Universidade Federal de Alagoas, Brazil. She is also a graduate student in the Graduate Program in Mathematics Education of the Universidade Estadual Paulista "Júlio de Mesquita Filho" (UNESP), Rio Claro Campus in São Paulo, Brazil. Her research interests are in the areas of the history of mathematics, mathematics education and differential geometry She teaches in the Professional Master's Program of the National Network (ProfMat) and has been the Coordinator of the distance education teacher preparation program in mathematics of the Universidade Federal de Alagoas.

Chapter 4

CHILE: The Context and Pedagogy of Mathematics Teaching and Learning

Eliana D. Rojas and Fidel Oteiza

Abstract: Chile was one of the first countries in Latin America to develop formal institutions for teacher preparation and to establish graduate programs in mathematics and mathematics education. It is now going through a difficult stage of reevaluating the neoliberal economic policies that have led to a significant experiment in the privatization of education. Currently groups of mathematicians, mathematics educators, and other stakeholders are involved in far-reaching discussions and collaborations that hold the promise of leading to more equitable high-quality educational outcomes for the Chilean people.

Keywords: Mathematics education; curriculum development; history of mathematics education in Chile.

Introduction

The field of mathematics education, particularly its pedagogy, is going through transformative times in Chile and Latin America. Chile is nowadays not only performing important efforts to radically change and improve the processes of teaching and learning mathematics, but also ensure that all Chilean children are exposed to a rich and rigorous mathematics curriculum. Currently there is a common understanding that it is the quality and the content of instruction, as well as the rigor of the curriculum, which result in the relatively more successful showing of many countries that have demonstrated student success in mathematics learning. Those successful countries also show increases in student high school graduation, access, and retention rates at all levels. For the first time, standards with progress maps are being developed and new strands such as statistics, probability, and mathematical reasoning are being introduced in the national mathematical curriculum. Also, several new approaches, such as mathematical modeling,[1] and new topics including vector geometry are being incorporated. At the same time through private and public funds, national organizations are engaged in a variety of projects looking at developing and implementing innovative teaching materials, software and online multiplayer games to teach algebra, statistics, geometry and mathematical reasoning. New teaching services for students through internet-based interactive tutorials and e-learning programs for teacher training are broadly offered.

The Chilean educational system is organized in eight years of elementary education, four years of secondary education, plus a tertiary level at universities and technical schools. At the moment, the system is moving towards a structure of six elementary and six secondary grades. School attendance is close to 98% in elementary, up to 80% in secondary, and above 55% in the tertiary level (OECD, 2012). School attendance is mandatory up to twelfth grade. The country has around 12,000 schools and, since 1981, private universities have significantly increased university education opportunities.

Professional Development

Professional development on the job using technology has become a widely used approach for improving teaching quality and decentralizing professional development activities. A digital internet resource widely spread at the national level, *Enlaces*, has been in place for the last 20 years. The program was designed to facilitate the use of computers and technology as well as the dissemination of best practices across the country and encourage communication among school districts, as well as to connect teachers, school professionals, and students, including those in the most isolated schools. As a result, schools are in better communication, and teachers have had the opportunity to visit and learn from peers in Chile and abroad.

By 2006 more than 6000 teachers had traveled abroad with *pasantías* (grants) from the Foreign Scholarships Program[2] (MINEDUC, 2006),

---

[1]Representations composed of relationships and variables. Examples include dynamical systems, statistical models, differential equations, or game theory models.

[2]*Programa de Becas en el Exterior* (PBE).

including elementary and secondary mathematics teachers. Teachers were chosen through a rigorous national selection process led by the Ministry of Education (MINEDUC). The hosting institution was required to commit to a program of studies that was submitted to MINEDUC and approved by a designated committee. Many teachers and school administrators took advantage of the experience and returned to Chile with a series of new skills and knowledge that they were supposed to bring back to their school community. MINEDUC intended to embed a significant number of teachers and school professionals in different socio-cultural environments in national and international school communities so as to expose them to a wide range of mathematics curriculum, mathematics teaching and pedagogical discourses. Teachers and school professionals, including higher education faculty and administrators, were exposed to successful research-based mathematics classroom designs and pedagogical strategies. They had the chance to observe and discuss curricular, teaching and assessment practices, teacher evaluation, school management, and parent involvement with mathematics. Some experiences, still not necessarily well developed in Chile, included exposure to differentiated instruction, teaching mathematics to children with learning disabilities as well as gifted and talented children, and practices for teaching mathematics in multilingual multicultural settings.

Chile has a history of a centralized educational system, with the majority of policy-related decisions made by the Ministry of Education, which operates through a network of provincial and municipal offices.[3] The system is comprised of both public (including subsidized) and private schools, with the former currently serving about 90% of the students. The municipal offices control public schools and historically there has been minimal regulation for private schools. A privatization reform was implemented in 1981, under Pinochet's dictatorship and with the input of economic advisors from the United States. The government's support for education was drastically scaled back in favor of a privatized, decentralized system introducing full parental choice through a voucher system. The state pays a fixed amount to both public and private schools (indirect vouchers) on the basis of student enrollment.

Since 1990, over 90% of Chilean students have attended public-municipal (*público-municipales*) or subsidized special schools, and less than

---

[3]*Departamentos Provinciales de Educación y Dirección de Administración de Educación Municipal.*

10% private schools that do not receive government subsidies. Despite the overall stability in enrollment, the educational system has generated an expansion of attendance at subsidized private schools (*particular-subvencionada*) from 32% of the total in 1990 to 45% in 2005 at the national level. This has resulted in an increase in the total number of this type of institution in the country from 2,425 in 1990 to 3,343 in 2005. At the same time, public municipal schools have suffered a decline in enrollment from 61% in 1990 to 48% in 2005 at the national level, which has resulted in a decrease in the total number of such institutions from 6,000 in 1990 to 5,572 in 2005. These changes have occurred mainly in urban areas in cities of more than 100,000 inhabitants (Lara & Repetto, 2009).

Following Pinochet's military coup in 1973, Chile started an educational experiment that some have considered "the most extreme neoliberal educational experiment in the world" (Waissbluth, 2013). Designed by Chilean economists[4] who followed the teachings of Milton Friedman, it was applied systematically from 1973 until its replacement by a center-left coalition in 1990. These educational stratifications over the past 20 years have demonstrated a persistent inequality of educational opportunity across communities (Torche, 2015).

Although the educational reform that began in the early nineties with the return to democracy have focused on improving educational quality and securing equity — and during the last fifteen years the country's economy has been growing — Chilean students' performance on national and international testing (TIMSS, PISA, etc.) has been far from satisfactory. In 2006, a national student movement referred to as the "penguin revolution" rose from the schools and neighborhoods. In their black and white school uniforms, they filled the streets, calling for educational reforms. Students occupied school buildings demanding the end of school segregation and immediate changes to a system that promoted privatization and abuse, including a serious investigation of all public and private organisms involved.

History of Mathematics Education in Chile

The history of mathematics education in Chile (Harding, n.d.) is, indeed, closely related to the history of its socio-cultural and socio-political transformative history. Its beginnings date to the Spanish Conquest of Chile

---

[4]Known as the "Chicago Boys".

(1541–1598). During the pre-Columbus period, our indigenous communities, contrary to other pre-Columbus cultures, did not have a written language. About eighty years before the arrival of the Spanish, our communities received the intellectual influence of the Incas, who also did not have a written language. The Incas did spread the decimal system and mental conceptual abstractions of hundreds, thousands, and multiples of ten. They also developed a counting tool called the *quipu* — a kind of abacus — that is still preserved by the Mapuche community under the name of *pron*.

Chilean mathematics education was institutionalized from the sixteenth to the eighteenth centuries. It started in 1591 with courses offered by the Dominicans following the *quadrivium* — the classical curriculum comprising the four liberal arts of number, geometry, music, and astronomy taught during the Renaissance. Later, a position in the arts was created at the Pontifical University, which had been founded in 1619. The first position in mathematics was created in 1758 at the Royal University of San Felipe. The lectures were delivered in Latin, and the study plan included Euclidean geometry and geography. However, for over one hundred years a degree in mathematics was not offered. In 1798, courses and seminars taught in Spanish were offered at the Royal Academy of San Luis for the first time; the courses were taught by Spanish engineers. In 1813 the National Institute, the most emblematic of pre-university institutions in Chile, was inaugurated. For the first time the teaching of mathematics as systematic courses in Spanish became a reality.

In 1826 Andrés Antonio de Gorbea, a Spaniard expatriated in London, arrived in Chile. He became a mathematics teacher at the *Instituto Nacional*, one of the emblematic high schools in Chile, and conceptualized the scientific perspective of mathematics teaching and learning in Chile. De Gorbea headed the mathematical reform proposed by the French engineer Carlos Ambrosio Lozier.

During the seventeenth and eighteenth centuries, Spain was the center of cultural and scientific influence in Chile. During the nineteenth century, Chile turned to France. Later, during the twentieth century, Chile turned toward the United States.

La Escuela Normal José Abelardo Núñez

La Escuela Normal José Abelardo Núñez (José Abelardo Núñez Normal School), formerly *Escuela de Preceptores de Santiago* (Santiago School for Private Tutors), was founded in 1842. It was the first teacher preparation institution in Chile and in Latin America. The institution focused only on the preparation of elementary school teachers. Its first headmaster was the writer and Argentine politician Domingo Faustino Sarmiento. For over a century, the institution educated elementary school teachers. In 1974, during the dictatorship, it was ordered to close its doors. A long tradition and history of teacher preparation and educational renewal in Chile and Latin America had come to an end.

The Pedagogical Institute of Chile

In 1889, President José Manuel Balmaceda signed the project that created the Pedagogical Institute, the first secondary teacher education institution in Latin America. Teacher education in secondary mathematics was one of the central fields of concentration. The institution was created by the government as part of its 1889 secondary educational reform modeling the *concentric* approach to education. Since then, the concentric system has been the model framing the Chilean educational system. In a concentric approach, learning is conceived to occur in phases from basic principles to the more complex. The same cycles are repeated two or more times, or the curriculum is revised again every few years, which enables students to gain deeper knowledge and understanding of the topic.

To staff the Pedagogical Institute, the government recruited thirty scholars with doctoral degrees from Germany; fifteen of them were mathematicians. The first two mathematics professors at the Pedagogical Institute were Reinaldo Von Lilienthal — who after a couple of months returned to Germany — and Augusto Tafelmacher, who retired in 1907. The way mathematics teachers were prepared in Chile was mostly determined by this German team. In a way, the operational definition of mathematics curricula in the country was tailored in textbooks developed by some of them. Algebra and geometry were taught until the 1960s on the basis of books written by Pröshle and Pöenisch.

With the arrival of the German professors in 1889 to start the Pedagogical Institute, mathematics in Chile acquired a new dimension. Tafelmacher and Pöenisch, both with doctorates in exact sciences, gave mathematics a new dimension as a cultural and autonomous discipline, defining it as a *"corpus of knowledge"* at the service of teaching, a characteristic that they had in Europe from the beginning of the century. Tafelmacher and Pöenisch were passionate and devoted to teaching, and they influenced and prepared a legion of *"discípulos"* who spread that passion and devotion all over the country. Another German mathematician, Carlos Grandjot, founder of the *Instituto de Chile* (1930) also shared with pride his services as a professor of mathematics in the institutions of teacher education in the Republic. His courses included *matemáticas elementales y superiores*, philosophy and physics at the Pedagogical Institute. He also participated in the foundation and development of institutions that cultivated mathematics as a science at the national level. Grandjot was the founder and first president of the *Sociedad Matemática de Chile* (Mathematic Society of Chile, 1953). In 1957 he also founded the *Instituto de Investigaciones Matemáticas* (Institute for Mathematical Research), based at the *Universidad de Chile*, during the presidency (*Rectoría*) of Juan Gomez Millas.

In addition to going in-depth in his lessons on calculus, Grandjot introduced in 1936 in the Pedagogical Institute program, a course on differential geometry. The course, *Geometria Differencial*, gave way to one on *Analysis Vectorial*, and *Fundamentos de la Matematica o Axiomática* became an elective course in the dual degree program. The program graduated teacher candidates with specialization in both Mathematics and Physics *"Profesor de Matemáticas y Física"*. Later with the reform the degree was extended to nine semesters (Gutierrez & Gutierrez, 2004).

Among Grandjot's most outstanding disciples in Chile were Dr. Rolando Chuaqui and Professor Cesar Abuauad. Dr. Chuaqui studied medicine and later earned his Ph.D. in *Lógica y Metodología de la Ciencia* at California-Berkeley in USA (1965). After his return from Berkeley, Dr. Chuaqui became a professor at the *Facultad de Ciencias — Universidad de Chile*. In 1969 he accepted a position as Dean of the Department of *"Ciencias Exactas"* (Mathematics, Physics and Chemistry), at the *Pontificia Universidad Católica de Chile*. Soon after, he developed the first *Programa de Post-grado en Ciencias Exactas* (PEPCE) in Chile, the first Mathematics Doctoral program in the country. His contribution as a researcher in the field of logic is internationally known (Gutierrez & Gutierrez, 2004).[5]

Cesar Abuauad taught the first official courses of this "new science" at the Pedagogical Institute of the University of Chile in 1956: *Algebra abstracta* and *Lógica y Fundamentos de las Matemáticas* (modern mathematics), disciplines that Grandjot had introduced in Chile (Gutierrez & Gutierrez, 2004).[5]

---

[5]Paraphrased and translated paragraphs from Gutierrez & Gutierrez (2004).

Two familiar maxims continue to describe main principles guiding educational reforms around the world: (1) "Without good teachers we can't have good teaching," from the then-Minister of Public Instruction, Federico Puga Borne, and (2) one used by the exemplary educator Valentín Letelier that essentially says that teaching for memorization does not need any pedagogical training.

In 1892, the Pedagogical Institute[6] graduated the first three secondary mathematics teachers in Chile. Until 1981, the next generations of secondary mathematics school teachers graduated from the Pedagogical Institute,[7] which had been integrated into the Faculty of Philosophy and Humanities of the University of Chile. Later, under the military dictatorship, it was separated from the University of Chile and redesigned as the *Academia Superior de Ciencias Pedagógicas* (Higher Academy of Pedagogical Sciences). This aftermath had profound consequences for the ways in which teachers have been recruited and prepared in the country. In 1985, it became the *Universidad Metropolitana de las Ciencias de la Educación* (Metropolitan University of Educational Sciences).[8]

Mathematics as the Basic Science and the Language of Sciences

For the last two centuries in Chile, as in other parts of the world, mathematics has been considered as one of the "basic sciences" in the new world of scientific and technological research. One could argue that, while perhaps the most relevant discipline, it has often been contentious when associated with national and international educational outcomes and educational decisions. In Chile, it took some years before the implementation of formal academic mathematics programs. The first bachelor's degree in mathematics was established in 1963 at the University of Concepción and had its first two graduates in 1968. In 1965, the Faculty of Sciences at the University of Chile launched its first bachelor's degree program in mathematics.

---

[6]http://www.archivonacional.cl/Vistas_Publicas/publicContenido/contenido PublicDetalle.aspx?folio=7187&pagina=1

[7]The 1973 Chilean coup d'état put an abrupt end to the careers of faculty and students from the *Instituto Pedagógico* mathematics department.

[8]http://en.wikipedia.org/wiki/Metropolitan_University_of_Educational_Sciences

The main twentieth century reforms started in 1962 with the establishment of the *Comisión para el Planeamiento Integral de la Educación Chilena*, which started a complete revision of the institutional definition of the entire public education. As a result of this initiative, a profound reform was in place by 1965. It was the first time that a specific notion of curriculum was at the base of a reform in Chile. Schools were built all over the country, education became compulsory up to the eighth grade, a center for on-the-job-training for teachers was established,[9] and the New Math was adopted. One of the most significant effects of the 1965 reform was reaching an 80% elementary school attendance rate.

Licenciatura Académica en Matemática (LAM)

A formal mathematics graduate program started in Chile in 1962 with the inauguration of the *Centro Interamericano de Enseñanza de la Estadística* (Inter-American Center for the Teaching of Statistics) offering a master's degree program in statistics. In 1966, the Mathematics Institute at the Catholic University of Chile opened its first bachelor's degree program in mathematics and by 1979 had its first graduate with a doctorate in logic. In 1967 the former State Technical University (UTE), today the University of Santiago, opened the first graduate program in mathematics in Chile: a master's degree called the Academic Degree in Mathematics (*Licenciatura Académica en Matemática* — LAM). The mastermind of LAM was Jaime Michelow (1929–2013), who in 1962 became the first Chilean mathematician to acquire a doctorate in mathematics. In 1968, LAM had its first two graduates. By 1979, there were 169 graduates of LAM who had positions in universities in Chile and throughout Latin America. Following the military coup in 1973, some students from LAM left the country as exiles. Some completed doctoral programs in the United States, France, Germany, England, Brazil, México, and later returned to the country.

By the late 1970s Jaime Michelow, in collaboration with Bruce Vogeli of Teachers College Columbia University, created a bachelor's degree in mathematics education and computing sciences. A master's degree in mathematics education soon followed.

---

[9] *Centro de Perfeccionamiento, Experimentación e Investigaciones Pedagógicas*, http://www.cpeip.cl

In one of his last interviews (June, 2011), Dr. Michelow discusses the important role of mathematics in the socio-economic development of Chile. He also highlights the role of mathematics in the development of critical thinking skills, and questions the way the system is eliminating *"mathematical proof"* for computations using calculators. He stated, *"How can we eliminate proof if in order to demonstrate knowledge — to know mathematics — you need to prove it?"*

Influential Presidential Terms

Juan Luis San Fuentes (1915–1920)

During the presidency of Juan Luis San Fuentes, **Pedro Aguirre Cerda** was assigned to the *Ministerio de Justicia e Instrucción Pública* and introduced a new significant legislature, the new *Ley de Instrucción Primaria Obligatoria* (Law for Compulsory Primary Education) that increased the 4 years of mandatory schooling to 6 years for all Chilean children in the territory, facing strong opposition from agrarian and church factions of the oligarchy.

Eduardo Frei Montalva (1964–1970)[10]

During the government of President Eduardo Frei Montalva, a significant effort to reform Chilean education began that can be characterized by a deep commitment to literacy-alphabetization across socio-culturally differentiated communities. As in other parts of the world the decade of the 1960s was a time marked by change in Chile. Revolutionary ideas emerged and sectors traditionally forgotten, such as youth and women, acquired great importance, and traditional cultural models were questioned. Analogous to President John F. Kennedy in the USA, education and an incipient agrarian reform highlighted the government of Eduardo Frei Montalva. Some of the important changes relevant to the socio-cultural development of Chilean society relate to his presidential efforts of educational reform. A lack of schools, teachers and enough infrastructures to support demand had delayed educational opportunities for many Chileans across regions.

---

[10] http://historiaeducacional.bligoo.com/ (translated by author).

The reform of 1965 achieved a significant quantitative leap by reducing the school day to two sessions (*jornadas*) — morning and afternoon — consequently doubling the number of students attending schools. However, Chile was not prepared to cover the number of teaching positions generated as a result of the "*doble jornada*". This challenge was minimized by the establishment of a controversial alternate fast route for certification, locally known as "*títulos marmicoc*".[11] Overall, this period is notorious for interventions such as enforcing compulsory primary education for the first eight years of schooling, two types of high schools, strong investment in school infrastructure, and the origin of the *Centro de Perfeccionamiento, Experimentación e Investigaciones Pedagógicas* (CPEIP), an agency from the Ministry of Education that until today supports the level.

Salvador Allende Gossen (1970–1973)

It was during President Salvador Allende's unfinished term that the National Unified School (*Escuela Nacional Unificada*, ENU) concept was developed (see next section).

La Escuela Nacional Unificada (ENU)

The concept of a National Unified School (*Escuela Nacional Unificada*, ENU) was conceived under the presidency of Salvador Allende (1970–1973) and represents the first national movement toward a radical transformation of the Chilean educational system. It fostered the broad social participation of teachers, students, parents and other social organizations in educational decisions. It also supported the principle of the right of all members of the Chilean society to have quality education, starting at preschool age and for the rest of their lives. In this way, it defined education as the fundamental right of every citizen and the avenue for democracy by promoting a pluralistic and participative society. The concept included the intention of providing mechanisms for ample participation on internal decisions of all members of the educational system, particularly teachers, with recognition to equal rights and opportunities for participation. It postulated the end of

---

[11] "*Marmicoc*" means "mocking", referring to a fast route for obtaining a teaching degree or certification.

the curricular and structural differences between the traditional humanistic and technical educational systems at the pre-university level, and access to primary, secondary and post-secondary education to all Chileans and nations in the territory.[12]

During the next decades, the Chilean military regime (1973–1989) set the foundation for what became the current Chilean educational organization and the reason for its deterioration. Starting in 1974, the administration of the educational system fell under judiciary military control. The Ministry of Education took control of the pedagogical and techno-administrative processes, and the *Comando de Institutos Militares* (Military Institutes Command) was in charge of controlling ideological, disciplinary, and national security aspects. At the national level the Ministry of Education, local and regional superintendents, university officials, and other hierarchical positions were assigned to high ranking military officers. Institutions of higher education were assigned to *rectores-delegados* (delegated presidents) (Reyes, 2005). By the end of 1982 there were thirteen "military university presidents", including the presidents of the University of Chile and the University of Santiago, two of the most important national universities. The military control persisted until 1990.

A third transformational period, which occurred during the dictatorship, was characterized by the municipal control of public education in Chile. This period is important for a profound ideological discussion and was characterized by the many debates that took over the media, including educational experts, scientists, politicians and the general public; controversies that revealed the persistent ideological, economic and political differences in Chilean society. The privatization of Chilean education emerged as a promise for better educational opportunities to Chilean families; the period was characterized by a new funding policy, and a new policy targeting the administration of resources. For many authors, these measures relate to national security and not necessarily to a well-intended educational project. In 1977 the Technical-Professional Schools were transferred to private corporations; this process reached its highest point in 1980. In 1986 the process toward national *municipalization* of schools formally ended, leaving Chile with a decentralized schools administration system mostly controlled by the local municipalities.

---

[12] Nations in the Chilean territory: http://www.unpo.org

In their interpretation of the Chilean modernization process, Espinoza and González (2011) remark that the national changes occurred through "the interaction of a clearly intended de-concentration of power and a decentralization of mechanisms of control, while keeping its vestiges of centralism still rooted in the Chilean system".

During the dictatorship, the message of the government was constantly an attempt to "disarticulate the well-recognized teacher unions' organization through an ideological and political depuration, although at the same time intending to dignify the teacher profession" (PIIE, 1984).

Two related and significant policy decisions were made in 1981. The first declared that a certificate in education was not equivalent to a university degree, and the second that public universities were not to offer teacher preparation programs. In addition to the extinction of the normal schools these decisions were to have, and have today, a profound and negative effect on teacher preparation throughout the country.

As stated above, the Chilean educational system is comprised of eight years of compulsory elementary education; four years of secondary schooling; and three types of tertiary education: universities, professional institutes, and technical institutes. There is no academic-ability grouping at any level, and tracking begins at the secondary level, where students have to choose between vocational and academic high schools.

During the years of the military regime two independent organizations that did ground work that would be vital for the educational reforms during and after the 1990s were the Center for Research and Development in Education (CIDE) and the Interdisciplinary Program for Research in Education (PIIE). More than sixty researchers, mainly with international funding, created and tested different strategies intended to introduce innovation in schools.

Mathematics educators became formally organized in 1982 with the establishment of the Chilean Society for Mathematics Education (SOCHIEM). SOCHIEM helped to organize the ninth Inter-American Conference on Mathematics Education (IACME IX) that was held in Santiago in 1995. Other regional mathematics education conferences that have been hosted in Chile include the seventeenth Latin American Meeting on Educational Mathematics (XVII RELME), held in Santiago in 2003, and the sixth Iberoamerican Congress on Mathematics Education (VI CIBEM) in Puerto Montt in 2009.

Following a universal trend, Chile experienced significant educational expansion in the second half of the twentieth century. Between 1990 and

1999, public spending on education increased 2.5 times (MINEDUC, 2000a), and grew from 2.7% of the gross national product (GNP) in 1991 to 3.6% in 1997 (UNESCO, 1999, p. II 500; Ramírez, 2006). Access to education was widely ensured during the 1990s, with the vast majority of the population finishing secondary school. Yet despite wide access to education and the reform efforts, at the end of the 1990s and by the second decade of the twentieth century there was no evidence that students were learning more. The results of the national assessment system and international exams showed that a considerable gap existed between the goals set by the government and the tested achievement of Chilean students (Ramírez, 2006).

Organic Constitutional Law for Teaching

The Constitution for the Republic (1980) clarifies and refers to decrees ruling the principles associated with the rights to education and the Organic Constitutional Law for Teaching (*Ley Orgánica Constitucional de Enseñanza*, LOCE) establishes the juridical and philosophical principles that define the Chilean educational system. Both still rule educational processes in Chile. They are the main reasons for the numerous student protests still carried out in Chile and are the roots for the demands for radical changes in the constitution. The state is subsidiary in both cases, but does not promote nor guarantee the right to quality education.

Education is operated not as social rights but as a commodity, the *sostenedores* (school administrators) who receive state subsidies can benefit financially from them. The private sector, which also receives state subsidies, can select students thus segmenting the school system with the subsequent impediment of access for culturally and socially vulnerable students. Researchers will argue this makes it impossible to have access to the data that could help them evaluate the quality of the educational system at the national level.

The schools management system gives total responsibility to the municipal as well as to the *sostenedores* of the school systems they support. They take charge of main decisions including curricular adaptations, hiring and firing of all personnel (principals, teachers, schools professionals, maintenance, and others).

The municipal *sostenedores* take charge of the education of the large majority of children in poverty; many, also culturally and linguistically diverse. These *sostenedores* receive the same subvention as the private schools, although these children and their families have recognized many more challenges with the subsequent impact on students' performance. The present *sistema de sostenedores* was introduced during the dictatorship at the beginning of the 1980s. The system was thought to be able to help decentralize the administration of schools. By contrast, Chile today demonstrates an educational system that is segmented and segregated by social class.

Program for Strengthening Initial Teacher Education

The first initiative in relation to teacher education and pre-service teachers, at the national level, had its beginnings in 1997 with the Program for Strengthening Initial Teacher Education (*Programa de fortalecimiento de la formación inicial docente*, FFID) and continued in 2003. Since then, the program has been constantly in discussion. Unfortunately the educational reforms in Chile have not been necessarily connected to pre-service teacher preparation programs (OECD, 2004). As a consequence the teacher competence gap has been growing, exposing students to a large number of teachers with inappropriate mathematics pedagogical skills or not enough mathematical content knowledge (OECD, 2004).

For years, Chilean mathematicians, teacher education specialists in pedagogical knowledge, and educators who specialize in mathematics pedagogical knowledge had been resistant to the need for organizing communities or collaborating teams that recognize the fundamental relation between mathematical knowledge and pedagogical knowledge. During the last 10 years, the international mathematics and sciences research communities have been working diligently with the educational research community, including all disciplines (Teachers for a New ERA, 2006) to identify the best research questions and practices that could promote and impact the mathematics learning and scientific attitudes and dispositions of all students. The Chilean mathematics community through the support of national scientific organizations and funds from the government (Ministry of Education) is now engaged in a transformational approach to improve teacher education programs in the nation (FIAC, 2012–2013).[13]

In 2005 faculty and administrators from Chilean universities and professional institutes in charge of teacher education programs, originated a

---

[13] http://www.mineduc.cl/index2.php?id_seccion=3599&id_portal=59&id_contenido=14964

signed document that confirms their commitment to improve the quality of teacher education projects in Chile including planning for responsible accreditation standards measures. In 2006, the presidential advisory council in charge of quality of education contributed to a formal documented diagnostic assessment, and offered recommendations and revisions. Both documents argue the essential need for national collaborative efforts to elaborate a national policy looking at the professional growth of the pedagogical profession, particularly in the field of mathematics instruction.

Mathematics Content or the Pedagogy of its Content

In terms of mathematics and its pedagogy an important and controversial dialogue has spearheaded the Chilean discussions with wide participation on the level of importance that pedagogical knowledge versus mathematical knowledge should have in the process of teaching and learning mathematics for both teachers and students. Only recently has pedagogical mathematics content knowledge (Shulman, 1986, 1987) become an issue for discussion among the mathematics education faculty, mathematicians and the Chilean scientific community. As in other countries in Latin America, decisions in terms of what to teach (curriculum) and how to teach it (pedagogy) have historically been made by mathematicians, school professionals and policy makers, not all necessarily with strong combined mathematical and pedagogical background knowledge. Mathematics education and didactics of mathematics as a field are a new phenomenon in Chile.

In Chile, mathematics education is usually associated with the United States in reference to mathematics educators such as Jeremy Kilpatrick, Bruce Vogeli, Walter Secada, and Alan Schoenfeld. The teaching of mathematics is influenced by the French "*Didactique des Mathématiques*" approach espoused by Yves Chevallard, Guy Brousseau, and Michèle Artigue. Locally, the seminal work of Ismenia Guzmán and Fidel Oteiza should be highlighted since both have contributed to the development of the teaching, learning, and assessment of mathematics in Chile and Latin America. Each represents both lines of thought, the French, European, and the North American, and a combined effort to strengthen the quantity and quality of school mathematics Chilean students are exposed to.

Just in the last 15 years, mathematics education has grown to be a field supported independently as an area of specialization. Traditionally, schools of education had little collaboration with mathematics departments. There was, and in many cases still is, a disconnection between the mathematical content and the pedagogy of that mathematical content. Faculty in the mathematics departments may or may not necessarily have had pedagogical training, and faculty in the education department may or may not have had a strong mathematics or mathematics pedagogical background. Students, on the other hand, were responsible for covering the mathematics coursework during the first two years of university.[14] The final two or three years of their programs often seemed detached from that early work with mathematics content. For the past 10 years, the desire to improve mathematics achievement at the pre-university level has strengthened the commitment of mathematicians and mathematics educators to work collaboratively in an effort to identify best practices in mathematics teacher education.

Mathematics teacher education processes, in particular identifying teachers' mathematical content knowledge, mathematical pedagogical knowledge, and teachers' pedagogical knowledge, are now considered in most mathematics teacher education programs in Chile. National and international research has been strongly committed to identifying the information needed to provide pre-service mathematics teachers with a solid mathematics education preparation program and best practices to mediate professional development practices for in-service teachers in order to improve the processes of teaching and learning school mathematics.

In Chile, as in other places in the world, the international and national educational research and scientific community reaffirm the need to expand the context in which the problem of successful teaching and learning mathematics is being embedded and to look at the challenge as embedded in a macro system. That is, teacher education must be concerned not only with the curricular content and the pedagogical knowledge of that content, but must look also at the context in which teaching and learning take place. The new trends in mathematics education identified by researchers as crucial include areas such as beliefs and understandings about the nature and learning of mathematics; mathematics achievement; preparedness for teaching mathematics; program effectiveness; opportunities to learn in depth the mathematics that one will be teaching and the related pedagogy. There must be knowledge of the history of mathematics; mathematical language; mathematical discourse; and general knowledge for

---

[14]In Chile, a Teacher Education degree takes four to five years of course work. In the last ten years a combination of a bachelor's degree with a master's degree is becoming increasingly common.

teaching. Future teachers must also understand and learn to teach for diversity, become reflective, critical and responsive to their practice, have or seek opportunities to learn in schools through practicums; and be part of a coherent teacher education program (Rojas, 2010).

Mathematics Curriculum

The mathematics curriculum in Chilean schools is increasingly approaching international standards. There have been important influences from Australia (achievement maps), Japan (mainly on teaching methods), Singapore (with some schools using their textbooks, translated into Spanish), and Germany (mainly in technology education as well as at the elementary level of the national curricula).

At present, a "Common Core" standards proposal is being considered.

(1) **Numbers and operations** will go from a naive conception of number to using the different number systems, including complex numbers.
(2) **Algebra** experiments and a new emphasis on the notion and applications of **functions** have been introduced. Some of the latest proposals include asking for recursive models to be used in problem-solving as they relate to human behavioral sciences.
(3) **Geometry** has been the least treated aspect of the curriculum, but it is now being emphasized. It starts with informal Euclidean geometry and moves, at the high school level, to Cartesian and vector geometry.
(4) **Probability and Statistics** were introduced in 1992, but only for the last three years of secondary education. It now starts in the first year of elementary school and is present every year. This includes probabilities in different contexts, random walks, and probability distributions, such as normal and binomial distributions. Statistics aims to introduce the first steps of inferential statistics.

Achievements and Challenges

There is a recognized slow, but continuous improvement in mathematics for Chilean student national and international tests. Nationally, there has been a 10-point average improvement in the 10th grade, between 2010 and 2012. In PISA and TIMSS, the trend is positive. Chile ranked first among Latin American countries in PISA 2012 and is also among the countries with a positive slope. However, no statistically significant changes are observed in the average of the evaluated subjects (literacy in reading, mathematics, and science) when compared to 2009 (OECD, 2012).

Education in Chile still faces some pivotal challenges. One is related to the impact that schooling, the way it is structured, is having on the education of the most disadvantaged Chilean children. This is perhaps the most crucial and essential challenge, and the one with the greatest resistance, due to social and political implications. Other important challenges include:

(1) The quality of teacher professional and personal life and its implication in teacher recruitment, performance and their willingness to pursue possibilities for continued learning.
(2) The lack of obligatory standards for accreditation for all teacher education programs.
(3) The negative impact of national assessments and other control measures that end up being the "observed curriculum", leading to reductions of high-level learning objectives, and a dampening of the creativity of both teachers and schools.
(4) One-test-for-all means one curriculum for all and a lack of differentiated instruction in Chile, while in reality there are great and valuable differences in human capabilities and talents.
(5) The lack of a curriculum development tradition: textbooks, teaching resources, and now computer-based resources need to be created in a research and development environment.
(6) Differentiation at the 11th and 12th grades.
(7) Technical education as part of the secondary school.
(8) The use of digital technologies, although this is a challenge that has been attended to for more than 15 years.
(9) Research and development in mathematics education.
(10) Organizations and professionalization of teachers.
(11) The need for highly-prepared teaching faculty in teacher education programs.
(12) In the classrooms, teachers continue to "tell the story of mathematics", so the main challenge is to have students do mathematics.

Significant National Initiatives: 1995–2013

Comenius: A research and development center

The notions of curriculum development, as a result of research and development efforts, were introduced by a center operating at the University of Santiago de Chile: the Comenius Research and Development Center

(1996–2010). Comenius was influential in generating programs using field-tested learning resources, introducing innovative practices in the mathematics classroom, creating learning spaces specially designed to encourage students to find their own solutions, and guessing. The results observed by Comenius became the guideline for decisions impacting curricular design and characterization of pedagogical practices at the national and local educational levels (Oteiza & Miranda, 2010).

At the *Universidad Santiago* (Santiago University) in Chile, the Comenius Center conducted a series of studies and research projects with emphasis on mathematical education and the use of digital technologies in education. Some of the most significant achievements were innovative practices in the teaching and learning of mathematics. The basic philosophy of the resulting programs was congruent to the one developed by Robert Morris at the University of Syracuse in New York, USA, that can be summarized in Robert Morris' words: "Guess–try–watch what happens — learn what to do next." This was done creating specially-designed learning spaces to encourage students to guess and find their own solutions.

A specific notion of curriculum development, with emphasis on experimentation, field tests and improvement on the basis of repeated phases of design, development, and new versions, was introduced.

In the field of digital technologies in education, Comenius pioneered the introduction of learning digital objects in the country, as applied to mathematical learning, and the design, development and implementation of the first job training seminar for teachers via the Internet. Thousands of teachers, from all over the country, attended courses in geometry, algebra, probabilities, and statistics.

Finally, the results observed by Comenius were influential in defining the present-day mathematical curricula in Chile.

The INICIA Test

The INICIA[15] Test is an obligatory exam for pre-service teachers in order to be able to be hired as teachers in all pre-university municipal and subsidized private schools. The exam was designed to measure the pre-service graduates' pedagogical content knowledge. A similar requirement has been in place in many teacher certification programs in the United States. It was

---

[15] http://www.mineduc.cl/index.php?id_portal=79

implemented in 2008 for recently graduated teachers, and the development of national standards for teacher education was commissioned in 2008 and finalized in 2011. Since 2010 the government has offered scholarships for candidates with high university entrance test scores who apply for initial teacher training. During the last few years, the Ministry of Education and members of the educational commission from the Chilean parliament have been working on a proposal for legislation to create a Law for the Professional Teaching Career (*Proyecto de Ley de la Carrera Profesional Docente*) which proposes to transform the implementation of the teaching career. The Chilean parliament is currently discussing two complementary laws that will establish that passing the exit exam must be mandatory for anyone who wants to work in a state-subsidized school and links teacher test scores to entry salaries (10% of the graduates who demonstrate the best results in the INICIA Test will receive a 30% increment to their salaries). Teachers' salaries have also been promised to be linked to the kinds of schools in which teachers work. The latest document promises to double the salary of those teachers who choose to work in low-performing schools. The document also proposes to increase the minimum requirements for admissions and reduce the number of class-hours.

The Center for Mathematical Modeling

The Center for Mathematical Modeling,[16] as defined by the institution, has as its main objective the development of New Mathematics and the dissemination of the use of theoretical as well as applied contributions to Chile's industrial sector by developing a scientific research environment and promoting the participation and exchange of ideas among faculty, students, and engineers. The center has taken a radical approach by hosting a cutting-edge project called Resources for the Initial Preparation of Mathematics Teachers[17] designed to contribute to the improvement of the mathematics teaching skills and methods of future elementary school teachers in Chile.

The project targets the need to provide students from elementary teacher education programs with the tools that will help them acquire the mathematical knowledge they need to learn and do mathematics. These

---

[16] *Centro de Modelamiento Matemático* (CMM): *Facultad de Ciencias Físicas y Matemáticas AKA Ingeniería, Universidad de Chile.*
[17] *Recursos para la formación inicial de profesores de matemática* (ReFIP), FONDEF D09I–1023.

tools will enable their future students to acquire the mathematical knowledge they need to learn mathematics. Also, collaborations that integrate the many participants in the mathematical development processes of future elementary school teachers are encouraged, including units of local capacity in the different teacher education programs and the promotion of these methodologies used in other disciplines.

The project has produced a collection of texts focusing on:

(a) The mathematical knowledge of elementary teachers in number, geometry, algebra, and data and random data.
(b) Texts for higher education faculty from teacher education programs.
(c) Multimedia materials to support initial teacher training in mathematics.

The key elements that explain the design and the development of the project include:

(1) Resources developed for the initial training of teachers of elementary education in line with the Guiding Standards for teaching careers in elementary education.
(2) Interdisciplinary approaches to the content, which adds the views of specialists in mathematics and mathematics education.
(3) A scheme of work based on the collaborative authorship of texts, which are developed with a shared focus and language.
(4) Three key elements in the development of texts: integration of disciplinary and pedagogical knowledge, emphasis on the mathematical knowledge to be taught, and a rapprochement between theory and practice.
(5) Strong links with the intended users (teacher education students and teacher educators) in a highly participatory and iterative process.
(6) Early validation of the texts through a broad process of pilot testing in 16 universities in the country to adjust the material on the basis of the evidence submitted from its use in university courses.
(7) Project experience records to facilitate the transfer of the work model and its dissemination to other similar initiatives.

Academic Innovation Fund

Academic Innovation Fund (*Fondo de Innovación Académica*, FIAC) is a highly competitive grant that seeks to encourage individual higher education institutions with teacher education programs to redesign their teacher education projects with innovative actions and activities that promote academic innovations. The required parameters have enough flexibility to allow applicant institutions to specify areas of priority and actions in the context of each institution and individual program characteristics. The institutional proposals must be consistent with the institutional strategic planning and the design must include a follow-up plan.

The evaluation of submitted proposals is reviewed by an international and national panel that works diligently under a set of criteria that emphasize rigor and feasibility. The evaluation rubrics are previously known and discussed by the evaluation team. The resources allocated to the best projects are transferred to the beneficiary institution.

FIAC is supported by four strategic indicators:

(1) Identifying the institutional commitment to the improvement of the quality of the students' curricular activities and individual educational experiences.
(2) Opportunities and experiences offered through each program involved in the project.
(3) A commitment to generating greater capacities for research, development, and innovation in the fields of teacher education.
(4) The institutional actions to be developed with a high level of technical support, as well as opportunities for training undergraduate and graduate students, particularly at the doctoral level.

At least ten universities have been awarded these funds. The awardee institutions of higher education (IHEs) have engaged in developing projects that include building strong relations across disciplines and departments within the individual institutions, as well as across awardee IHEs. At the same time, all the awardees contemplated building strong national and international connections with universities, schools, school officials, local governments, and external communities.

Final Remarks

Chile was one of the first countries in Latin America to develop formal institutions for preparing teachers and to develop graduate programs in mathematics and mathematics education. It is now going through a difficult stage of reevaluating the neoliberal economic policies that have led to a significant experiment in the privatization of education. It now appears that groups of mathematicians, mathematics educators, and other stakeholders are involved in far-reaching discussions and collaborations that hold the promise of leading to more equitable high-quality educational outcomes for the Chilean people.

Bibliography

Espinoza, O., & González, L. E. (2011). La Crisis del Sistema de Educación Superior Chileno y el Ocaso del Modelo Neoliberal. In *Barómetro de Política y Equidad, Nuevos actores, nuevas banderas*, Vol. 3, October, pp. 94–133. Santiago, Fundación Equitas-Fundación Friedrich Ebert.

Gutierrez, C., & Gutierrez, F. (2004). Carlos Grandjot, tres décadas de matemáticas en Chile: 1930–1960. *Boletín de la Asociacion Matemática Venezolana*, $XI(1)$, 55–84.

Harding, I. (1978). *Synthesis of development of mathematic in Chile*. Santiago. Unpublished Notes.

Lara, B., Mizala, A. & Repetto, A. (2011). *The Effectiveness of Private Voucher Education: Evidence from Structural School Switches*. CSP study. Educational Evaluation and Policy Analysis. Vol. 33, No. 2, pp. 119–137. DOI: 10.3102/0162237711402990. Available at http://eepa.aera.net

MINEDUC (2000a). Gasto en educación 1990–1999: Nota técnica No. 3 [Expenditures in education 1990–1999]. Available at http://biblioteca.mineduc.cl/

MINEDUC (2004). Estudio explicativo sobre resultados TIMMS-R. Santiago: MINEDUC.

MINEDUC (2005). Comisión sobre Formación Inicial Docente (2005). *Informe de la Comisión sobre Formación Inicial Docente*. Encuentro Nacional: Propuestas de Políticas para la Formación Docente en Chile. Serie Bicentenario. Santiago.

MINEDUC (2006). Serie Bicentenario MINEDUC. Consejo Asesor Presidencial para la Calidad en la Educación (2006). *Informe Final*. Santiago.

OECD (2004). *Revisión de Políticas Nacionales de Educación. Chile*. Paris: Organization for Economic Cooperation and Development.

OECD (2012). *Education at a Glance: Indicators: Organization for Economic Cooperation and Development*. Available at http://www.oecd.org/edu/

Oteiza, F., & Miranda, H. (2010). Tecnología y educación. Condiciones para el uso educativo de las tecnologías digitales. En: El Libro Abierto de Informática Educativa, Enlaces. Centro de Educación y Tecnologías del Ministerio de Educación, Santiago-Chile, Capítulo 9.

PIIE (1984). *El magisterio y la política educacional*, Transformaciones educacionales bajo el régimen militar, v. 1, p. 153. Santiago.

Ramírez, M. J. (2006). Understanding the low mathematics achievement of Chilean students: A cross-national analysis using TIMSS data. *International Journal of Educational Research and Evaluation*, *45*, 102–16.

Reyes, L. (2005). Movimientos de educadores y construcción de política educacional en Chile (1921–1932 y 1977–1994), Tesis para optar al grado de Doctora en Historia, Universidad de Chile, Santiago, 2005, capítulo 3.

Rojas, E. (2010). Using mathematics as a equalizer for gifted Latino/a adolescent learners, in Castellano, J. A. & Frazier, A. D. (Eds.), *Special Populations in Gifted Education. Understanding Our Most Able Students from Diverse Backgrounds*, pp. 353–382. National Association for Gifted Children. Waco, Texas: Prufrock Press Inc.

Rojas, E., & Reyes, A. (n.d.). La contextualización y pedagogía crítica, elementos esenciales en prácticas de formación y desarrollo profesional docente. *NABE Journal of Research and Practice*. Forthcoming.

Shulman, L. (1986). Those who understand: Knowledge growth in teaching. *Educational Researcher*, 15(2), 4–14.

Shulman, L. (1987). Knowledge and teaching: Foundations of the new reform. *Harvard Educational Review*, 57, 1–22.

Teacher for a New ERA (2006). Transforming Teacher Education. Carnegie project: Carnegie Corporation of New York (2001).

Torche, F. (2015). Privatization reform and inequality of educational opportunity: The case of Chile. *Sociology of Education*, 78, 316–43.

UNESCO-ICSU (1999). Declaración sobre la Ciencia y el uso del saber científico. *Ciencia para el siglo XXI. Un nuevo compromiso*. Budapest: UNESCO.

UNESCO (1999). *United Nations Educational, Scientific, and Cultural Organization Statistical Yearbook*. Lanham, MD: UNESCO and Bernan Press.

Waissbluth, M. (2013). Diane Ravitch's Interview (www.mariowaissbluth.com).

About the Authors

Eliana D. Rojas is a professor in the Neag School of Education at the University of Connecticut. She holds degrees from the University of Chile, University of Minnesota, and the University of Connecticut, where she earned her Ph.D. in educational studies. Her research and practice concentrate on the ecology of the processes involved in teaching and learning a content discipline within socio-culturally and linguistically diverse contexts, particularly in the discipline of mathematics.

Fidel Oteiza is a consultant for the Ministry of Education in Chile. He is a former professor in the Mathematics and Computer Science Department at the University of Santiago de Chile. He holds degrees in mathematics and physics from the Catholic University of Chile and a Ph.D. in curriculum and instruction from Pennsylvania State University.

Chapter 5

COLOMBIA: The Role of Mathematics in the Making of a Nation

Hernando J. Echeverri and Angela M. Restrepo

Abstract: This abbreviated history of Colombia's development as a modern nation highlights the role played by mathematics. From the Botanical Expedition in colonial times, to the present with its flourishing universities contrasting against an inefficient basic education system, it narrates the country's realizations while struggling against religious fanaticism and political violence.

Keywords: History of mathematics; mathematics in Colombia; history of mathematics education; mathematics education; nation building.

Mathematics has always been central in the development of any nation, especially in modern times. It goes hand in hand with language skills in the education system. It is also the foundation of technological and scientific progress. In Colombian history, it is remarkable to see how its advances and declines presage or mirror equivalent processes in society at large. It is, after all, a key to enlightenment.

The Colonial Period

Although the south of Colombia was under Inca rule before the Spanish landed in America, nothing worth mentioning has been found regarding the use of mathematics in its territory during pre-Columbian times. Little use of "interesting" mathematics continued during early colonial times. Spain with its Inquisition and its unification undertakings had shut itself to influences from Arabs and Jews, the most industrious sectors of its population, and to Protestants and the New Science that were wakening northern Europe.

As a consequence, it was a latecomer to the Industrial Revolution. Gabriel Poveda (2012) affirms that not even two-column accounting, invented in the late Middle Ages, was used in colonial Colombia. The few incidents of applied mathematics that he mentions were in navigation, commerce, and construction, especially to build the fortifications in Cartagena under the direction of Spanish military engineers.

Education was controlled by the Catholic Church. The elementary schools, called *colegios menores*, which the sons of native dignitaries could attend, were attached to the dioceses. They taught reading, writing, and arithmetic, while their main concern was in Christian indoctrination, civility, and moral values. To pursue further studies, students had to rely on religious seminaries in the larger cities (Poveda, 2012).

In Santa Fe de Bogotá, the capital of the Viceroyalty of New Granada which encompassed what are now Colombia, Venezuela, Ecuador, and Panamá, two main religious orders founded the first universities next to their seminaries. The Dominicans established the University of Saint Thomas Aquinas in 1580, and the Jesuits, the Pontifical Xavierian University in 1623. The latter was an extension of the *Colegio Mayor de San Bartolomé*, a high school or college run by the Jesuits, founded in 1604. In 1653, the archbishop founded the *Colegio Mayor del Rosario*, independent of either order, which was later chartered by Charles III in 1768 to be an American extension of the University of Salamanca tradition. Education in these colleges was directed only to pure-blooded descendants of Spaniards to study medicine, law, and theology with a scholastic methodology. Even in Spain, mathematical instruction in the universities went up to elementary algebra and Euclidean geometry at the most (Poveda, 2012).

Charles III was an enlightened king of the Bourbon dynasty who ruled from 1759 to 1788 and tried to provide a more efficient government, efficiency which meant greater control of the Spanish colonies. During his reign, José Celestino Mutis came to New Granada, originally to be the viceroy's doctor, but ultimately to explore the geography, flora, and fauna with his Botanical Expedition. One of his first initiatives was to establish a higher mathematics course in the *Colegio Mayor del Rosario* in which he taught the subject up to an introduction to infinitesimal calculus (Poveda, 2012). In his courses and lectures, Mutis also taught Newtonian physics and explained the Copernican system for the first time in the region.

The Royal Botanical Expedition of the New Kingdom of Granada (*La Expedición Botánica*) was initiated in 1783 and lasted 33 years. In contact with Carl Linnaeus, the expedition collected and classified 20,000 plant species and 7,000 animal species. Its research included mineralogy and metallurgy, geography, climatology, and astronomy. One of its legacies was the astronomical observatory in Bogotá, one of the first of its kind in South America. After Mutis' death, his most promising student, Francisco José de Caldas, replaced him in his math courses and as director of the observatory. The Botanical Expedition was also a very special introduction to science for a group of young men some of whom would actively participate in the movement that led to independence in 1810 (Bushnell, 1993).

The 19th Century Republican Period

The Spanish colonies' independence was precipitated by Napoleon's invasion of Spain to place his brother Joseph Bonaparte in its throne in 1808. Although there remained some loyalist bastions, most Hispanic–American colonies, except for Perú, took advantage of the situation and declared their independence by the end of 1810, sending viceroys, governors, and administrators back across the ocean with hardly any bloodshed. The Spaniards — troubled at home — could not immediately stop this, but they would return with a vengeance. In the meantime, several monarchist enclaves remained in Colombia (Bushnell, 1993).

To train personnel to combat them, the governor of the province of Antioquia requested Francisco José de Caldas to open an Academy of Military Engineers in the town of Río Negro, which he did with the help of Coronel Manuel Serviez, a French soldier who had studied in the *École Polytechnique* in Paris. This was the first engineering program organized in Colombia. However, it lasted only two years until 1814 when the Spanish returned to re-conquer their territory. More than 300 revolutionaries were executed, including Caldas, and Camilo Torres who had just been president of the nascent United Provinces of New Granada (Poveda, 2012).

The war for independence continued with an army composed of Colombians and Venezuelans under the command of Venezuelan General Simón Bolívar, until 1819 when the Spaniards were ousted from most of New Granada and a provisional government could be formed under the name of the Republic of Colombia. A constitution for a centralized republican government was voted for in 1821 as a trial that would be revaluated at the end of the decade. From the beginning, this Republic of Colombia — which has been known as *La Gran Colombia* to distinguish it from the smaller Colombia of today — encompassed the entire territory of the Viceroyalty of New Granada, including what are now Venezuela, Ecuador, and Panamá. Bolívar was named president and Colombian General Francisco de Paula Santander, vice-president. The latter ruled while Bolívar continued to liberate the territory of the Viceroyalty as well as Bolivia and Perú (Bushnell, 1993).

True to the Liberal ideas of the times, Santander decreed the establishment of elementary schools for boys in all towns of more than 100 families. Since no centralized funding was appropriated, each locality had to fend for itself and tuition was charged, but it was the first effort to institutionalize non-denominational public education. By 1835, 8.7% of the primary age population was attending school (Helg, 1987). Central authorities assumed the direction of secondary education and partially funded it, building new schools in cities outside of Bogotá. A more modern curriculum was developed for these schools with more emphasis on mathematics and science, although laboratories were lacking. French and English languages gained priority over Latin as the Church's influence fell (Poveda, 2012).

Three more universities were established where mathematics and science courses were taught: *Universidad Central* in Bogotá, *Universidad de Cartagena*, and *Universidad de Popayán*. A mission was brought from France to start a mining school and a natural history museum in which the main scientist was French mineralogist and chemist Jean-Baptiste Boussingault. Although the school was never created, courses in advanced mathematics, astronomy, and metallurgy were offered at the museum to professors from the *colegios mayores* and universities in Bogotá (Poveda, 2012).

Cartography and land surveying were also developed, with the first technical map and census made of the territory of *Gran Colombia* yielding 3,064,000 km^2 and 1,233,000 inhabitants. Mining engineers came in from Europe on their own or sent by European banks to exploit and extract enough precious metals to settle Colombian loans from the independence wars. These experts also taught their employees on the job. In this way higher mathematics began to be known in Colombia mainly by its applications in engineering (Poveda, 2012).

The *Gran Colombia* soon began to tear apart at its seams. When Bolívar returned to the presidency from his final liberating battles, he was induced by his followers to become a dictator in 1828. Invested with dictatorial powers he tried urgently to patch troubles and, deeming it more prudent to listen to the Church's complaints, reversed most of Santander's educational reforms. The Church, which continued to be the most popular institution, especially in the rural areas, had spread discontent because it resented the educational reforms that took away its controlling role (Bushnell, 1993). Nevertheless, despite Bolívar's efforts to prevent it, Venezuela and Ecuador both seceded just before his death in December 1830.

Throughout the rest of the nineteenth century, Colombia would have to come to grips with separating itself from its traditional Spanish colonial culture. There were three main issues. First, in a geographically challenging territory with three sparsely populated mountain chains divided by deep river valleys, an important question to resolve was whether to have a centrist government to unify its people or a loose confederation in which each region would mind its own problems. Second, the commercial predicament was whether to protect its artisanal production or approach North Atlantic nations with a more laissez faire attitude importing manufactured goods with relatively low tariffs while trying to export its agricultural and mining commodities. And third, the cultural dilemma, should it continue with Catholic values or embrace the Liberal principles of the times, following Jeremy Bentham's ideas of inalienable rights and freedoms as well as separation of church and state, which dealt directly with education. These issues were politically polarizing, resulting in a state of intermittent civil war that practically lasted throughout the century (Bushnell, 1993).

After the unrest that followed Bolívar's death, a centralist government continued over a much diminished territory under the name of *República de la Nueva Granada*, with Santander as its first president. Under his administration, primary school attendance rose to 8.7% of the age bracket. Although very low, this was higher than in Venezuela which had ended up much richer after independence (Helg, 1987).

With the *Nueva Granada* government leaning toward classical liberalism, slaves were emancipated in 1852. This concluded a process that had started in the early years of independence with the abolition of slave trade and manumission at birth. The economy started to flourish with free trade. The state-owned tobacco monopoly was privatized and the crop became the second export commodity after gold, totaling 25% of the exports. Trade also rose in quinine and coffee.

In the political aspect, the Colombian two-party system of Conservatives and Liberals was started at this time. The main difference between the parties was that the Liberals embraced modern European ideals to free the country from the Church's domination whereas the Conservatives sided with the Church hierarchy. While the dominating educated class discussed the advantages of each system, the poor classes adhered to the parties without a clear reason, as violent soccer fans. Consequently, animosity was great and civil war intermittent (Bushnell, 1993).

General Tomás Cipriano de Mosquera, who came from an aristocratic family of Popayán, started the Liberal hegemony beginning just before the middle of the century. He was an engineer at heart and as such embarked on many engineering projects during his several terms as president, one of which was the *Colegio Militar de Ingeniería* which later became the *Academia Militar y Escuela Politécnica*, imaged after the *École Polytechnique* in Paris. For his undertakings he brought a mission from France to teach at the Academy and to build roads so necessary and so challenging in the Colombian terrain. One road connected Bogotá to the Magdalena River where he reinitiated steam navigation down to the Caribbean Sea. He also signed the contract with US engineers for the construction of the Panamá Isthmus railroad that was completed in 1855. General Mosquera founded the Colombian Science Academy in 1847 and initiated the Chorographic Commission to make maps and describe the different regions of Colombia showing their possible economic resources. For this project he brought in Italian engineer Agustin Codazzi as director (Poveda, 2012).

In the succeeding years of the Liberal hegemony, which climaxed with the Constitution of the United States of Colombia in 1863 and lasted up to 1885, the main population centers were linked by telegraph, and railroad lines were built by bits and pieces, connecting to the Magdalena River system. However, Colombia was sliding more towards a loose confederacy of "sovereign states" weakening the central government's power to coordinate a centralized transportation system, and the terrain made it unavoidably expensive. By 1885 a meager 225 km of tracks had been laid, insignificant if compared to the networks constructed in México (5730 km) and Argentina (2700 km) (Bushnell, 1993, pp. 134–135). Nevertheless, among the upper classes there was a feeling of complacency because they believed they were similar to the civilized European nations with these technical enhancements. There was also an early tendency toward urbanization. While the country's population between 1851 and 1871 rose 32%, from 2,243,054 to 2,951,111, Bogotá's population rose 37%, from 29,649 to 40,833, and Medellín increased an impressive 116%, from 13,755 to 29,765, to become the second largest city. New cities gained, such as Barranquilla on the mouth of the Magdalena River, as colonial cities, Cartagena and Popayán, lagged behind (Bushnell, 1993, pp. 286–287).

In 1867 the *Universidad Nacional de Colombia* was founded; the *Academia Militar* became a part of it, renamed as the School of Engineering, and was later to be called the School of Mathematics and Engineering from 1886 to 1935 (Poveda, 2012). It was to graduate civil engineers, professors of mathematics, as well as professors of similar subjects, such as surveying and hydraulics. The *Universidad Nacional* was to be the most important university in the country, forming professionals in almost all disciplines but placing special importance on science and technology to support the era of the railroad and mechanization (Bushnell, 1993). In mathematics, it would graduate most of the professors of the next generation.

Medellín soon followed Bogotá's example with an Engineering School ascribed to the *Universidad de Antioquia* in 1874, which closed in 1878 due to financial problems, and the School of Mines in 1888, which was fashioned after Berkeley's (Colombia, Presidencia de la República, 2007, Tomás Cipriano de Mosquera).

Aimé Bergeron was probably the most influential mathematics professor in the *Academia Militar* and later the *Universidad Nacional*. Most probably his courses followed French textbooks that he had studied in Paris such as Carnot's *Calcul infinitésimal*, Cauchy's *Résumé des leçons de calcul infinitésimal*, and Lacroix's *Mathématiques* and *Calcul différentiel*. He also wrote books of his own which were later used as texts at the University, such as his *Curso de cálculo diferencial*. His style was probably imitated by other professors who preferred mathematics courses of an abstract nature such as those taught in France. There is evidence that some professors and students voiced their inconformity, demanding a more applied approach to mathematics, but abstract courses for engineering students at the *Universidad Nacional* continued to be the tendency almost to the end of the twentieth century (Poveda, 2012).

The Colombian Society of Engineers was established in 1887 to protect the interests of the fledgling professionals; with it, the first scholarly journal, *Anales de ingeniería*. As an example of its articles, one can mention "An Elementary Introduction to the Calculus of Quaternions" published in the *Anales* in a series of issues from 1890 to 1891. It was written by Pedro J. Sosa, born in Panamá City, who had graduated as an engineer from Troy Polytechnic Institute (now Rensselaer Polytechnic Institute) in 1874, and was a member of the commission that determined the Reclus–Wyse–Sosa route of the Panamá Canal for de Lesseps in 1876 (Poveda, 2012).

The most profuse author was Julio Garavito (Bogotá, 1865–1920). He graduated from the School of Mathematics and Engineering, almost simultaneously obtaining degrees as Professor of Mathematical Sciences and as Civil Engineer in 1891. That same year he was named professor of the

School, where he taught differential and integral calculus, rational mechanics, and astronomy. Two years later he became the director of the Astronomical and Meteorological Observatory in Bogotá. His articles dominated the *Anales* until his death, covering many aspects of celestial mechanics, but he also wrote about actuary and economics. To this day he is considered one of the most important Colombian mathematicians and scientists.

The Liberal regime also made an effort to develop primary education. In 1870, by decree it declared primary education free, mandatory, and non-denominational — imitating what president Domingo Sarmiento was doing in Argentina. The credit in Colombia goes to President Eustorgio Salgar and his interior minister Felipe Zapata (Colombia, Presidencia de la República, 2007, Eustorgio Salgar).

They also started the first normal schools to prepare teachers for their jobs, bringing a mission of German educators who taught pedagogy, based on the reforms of Swiss pedagogue Johann Pestalozzi. Still, the responsibility for education rested on both the central government and the regions, neither of which had sufficient funding. Nevertheless, in four years the number of students in elementary schools rose 40% from 60,000 to 84,000, approximately. In 1875, 18% of the age bracket was enrolled (Helg, 1987, p. 25).

The religious neutrality of the proposed school system, and the fact that some of the German educators were Protestant, however, infuriated the clergy and their Conservative allies. Their alliance finally toppled the Liberal hegemony. The laissez faire economic strategy failed as commodity prices fell and the excessive slackness of the confederate government caused untenable political instability. Finally, Rafael Núñez, a lawyer from Cartagena who had militated in the Liberal party, was elected president in 1885 and autocratically changed the constitution the next year.

The new Constitution of the Republic of Colombia, sanctioned in 1886, called for a centralist government. The "sovereign states" were demoted to "departments" and governors and mayors were no longer elected but designated by the executive branch led by a strengthened president whose term was extended to six years and who could be re-elected to succeed himself. Although Núñez had been a Liberal, he was a pragmatist who saw his political power in the joint backing of the Conservatives together with his loyal Liberal supporters. Moreover, although not a devout Catholic, he used the restitution of the Catholic Church as the nation's moral guarantor to his advantage (Bushnell, 1993).

Backed by an 1887 concordat with the Vatican, the new constitution gave ample provisions for education to be conducted by the Roman Catholic

Church, which would approve curricula, texts, and even teacher appointments. As a result, instead of international missions, Catholic religious orders were invited to enrich the educational situation. As such, the Jesuits were authorized to return. They had first been expelled from the Spanish Empire by Carlos III in 1767 due to their meddling in politics. Tentatively invited to return to Colombia in 1842, they were expelled again in 1850 amid a wave of expulsions that befell them across Europe and Latin America in the middle of the nineteenth century.

Among several other orders, the Christian Brothers were invited in 1890 to teach at the Normal School. Their French education especially sponsored rigorous mathematics courses and they soon established the *Instituto Técnico Central* in Bogotá that first provided intermediate technical education and later formed civil and electrical engineers at the university level (Poveda, 2012, p. 116). Today they run private elementary and secondary boys schools around the country, as well as several universities, all of which are well accredited.

Although the Constitution of 1886 reverted many tax revenues back to the central government, these funds continued to be low, and the transportation system remained almost the same as during the Liberal hegemony. Only some urban conveniences appeared in several cities, such as telephones, electric lighting, and modern aqueducts with metal pipes, but these only served government buildings and the homes of the very wealthy, emphasizing class differences. However, what most concerned the Liberal party was that all executive positions — regional and central — went to the Conservatives, who by coercion and corruption also managed to control almost one hundred percent of the seats in congress. Civil strife erupted first in 1895 and finally went all out in the Thousand Days War which lasted from 1899 to 1902 (Bushnell, 1993).

The First Half of the 20th Century

Besides the enormous cost in lives and resources, the war had another detrimental consequence, the separation of Panamá. In 1903 a chauvinistic Colombian Senate unanimously rejected the Hay–Herrán Treaty — already signed with the United States — that would have leased, in perpetuity, a six mile wide strip of land across the isthmus to build the canal. Angry Panamanians revolted and, with the support of the United States, gained their independence.

In 1904, General Rafael Reyes, a Conservative, was elected president, with the challenging task of reconstruction and reunification. He was forced to close an almost unanimously Conservative, difficult Congress, and as a dictator convoked a national assembly, to which he named Conservatives and Liberals to rule with him. The constitutional amendments that were issued during the term of his successor were intended to promote a smooth alternation of power between the two parties and provided for a minimum representation of the opposing party in the government, with a shortened presidential term of four years (Bushnell, 1993).

Imitating Mexican dictator Porfirio Díaz, Reyes immediately set out to modernize the country, creating the Ministry of Public Works to build roads and railways, finally connecting Bogotá and Medellín to the Magdalena River by rail, and Cali to the Pacific Ocean. To generate conditions for an economic recovery he strengthened the peso and maintained its value. He also provided security for foreign investment by transforming the military forces, making them professionals, respectful of the Constitution, and above partisan politics.

The Conservative presidents who followed Reyes up to 1930 were anything but remarkable, with the exception of Pedro Nel Ospina (1922–1926). Ospina had graduated as a mining engineer from University of California, Berkeley, and although he had experimented in politics — his father had been president of Colombia — he preferred his career. He was an important entrepreneur, owning mines and a foundry, then, toward the end of his life, he was elected president. As Reyes had done before, he dedicated the four years of his term to modernizing the economic infrastructure, making use of the indemnity the country had received from the United States for the independence of Panamá, as well as loans from Wall Street banks. Reflecting his interests and enthusiasm, the Public Works Ministry's budget grew from 3.6% of the total in 1911 to 35% at the end of his mandate, while military spending fell from 30% to 8.8%. Thus, the total length of the railways almost doubled, from 1,481 km in 1922 to 2,434 in 1929, and new roads filled in the gaps between the train lines. In the financial aspect, with expert advice from the Kemmerer Mission of the US, he created the Central Bank or *Banco de la República* to regulate money supply and exchange rates (Bushnell, 1993).

In the first third of the century, peasants had spread throughout the mountain ranges, planting coffee which then became the principal export. It was a crop that was financially sustainable even in minor plots, so that many small farmers with their families grew it along with their food crop, providing a large population with better living conditions. The larger exporters established a national trade association, the *Federación de Cafeteros*, to regulate prices and coordinate exports, in which all producers, large and small, were members. This assured credit to the small producers and gave them a sense of belonging, while controlling the quality of the coffee.

Two other export commodities became important: bananas and oil. However they were mostly managed by foreign interests, and although workers received relatively higher wages, the abysmal inequality in living conditions of the Colombian workers vis-à-vis the foreign employees caused serious labor unrest recurrently.

Several industries were also established to provide goods for local consumption. Mining had been important since colonial times, especially for gold and silver that were exported. Coal, on the other hand, was consumed locally, and in the end of the nineteenth century, scrap-iron foundries were started. Among the first consumer goods industries were textiles, foods, and beverages. A German immigrant family successfully established the Bavaria brewery in 1889 — one of the largest Colombian firms today — as well as the bottle factory for its supplies. However, serious industrialization really began during Reyes' term, when tariff policies were set to protect the newly-formed industries. Textile mills were started, especially in Medellín, where Coltejer — also an important firm today — was established in 1907. In the Caribbean ports flour mills were set. In Barranquilla, the most important port — with river and ocean access — foreign and local entrepreneurs made the city prosper with their businesses. All of these firms required better transportation so rail construction thrived, although gradually, due to the challenging terrain. To transcend these difficulties, the first commercial airline in the Americas was established in Barranquilla by German immigrants and Colombian visionaries. No runways were necessary at first, for hydroplane routes linked Barranquilla to a port up the Magdalena River which was connected by rail to Bogotá (Bushnell, 1993).

All these enterprises offered employment to engineers who were graduating from the new engineering schools. Not only were they required for specialized jobs, engineers were soon recognized as excellent managers and entrepreneurs, due to the analytic and problem-solving skills that they acquired through their mathematics courses. As a result, engineering became a popular career for the sons of well-to-do families, especially among those with prominent mathematical aptitudes.

Education for the general population, on the other hand, made little progress. In the rural areas where 82% of the population lived during the first third of the century, there were primary schools with the first three grades. Only 52% of the children in the age bracket attended at one time or another. Schoolhouses had only one classroom in which a female teacher taught all three grades together, with about 50 to 80 students. Physical punishment was the rule and school materials were non-existent or extremely scarce. Since children had to be separated by gender to abide by the Vatican's directions, boys would attend one day, girls, the following day, for six working days a week (Helg, 1987).

Being a primary teacher was the only possible employment for independent women of the underprivileged classes who did not join a nunnery. Their instruction was minimal, elementary schooling or possibly less; more important was their good conduct and submission to the Catholic faith. Their appointment was made by a county official and signed by the parish priest. The convenience of the post's location depended on the teacher's family's influence. If it was far, a family member would live with her as a chaperone. The children's parents had no say and were not interested in what was taught, although they did watch over the teacher's moral conduct. At times the school was so inaccessible that the parish priest could visit it only once or twice a year and the teacher actually became his representative before the village.

In the urban areas, 45% of the primary school teachers were men, 38% of them had received pedagogical instruction in a normal school, as opposed to women of whom 56% had attended a normal school. Depending on the size of the town, the primary schools, separated by genders, had six grades. There could be several teachers in each school, but only women could teach at girls' schools. There were school inspectors who made their rounds and teacher conferences, in which new pedagogies were discussed and physical punishment was criticized. In smaller towns the teachers, still watched over by the parish priest, were part of the elite since they were relatively well paid and could afford good clothes and comfortable living conditions. In the larger towns and cities, however, salaries did not meet the living expenses. Besides due to the localities' fiscal difficulties, payments could be delayed for several months (Helg, 1987).

The curriculum was imposed by the central government, although how or if it was followed is questionable. By a decree of 1904, it consisted of four main courses, catechism, reading, writing, and arithmetic, and a fifth time slot that could be civics — namely urbanity or courteous behavior — geography, history, geometrical drawing, music, and/or physical education. Arithmetic concentrated mainly on the four operations with whole numbers in the first three grades, first numbers from 0 to 50, then from 0 to 1,000 and finally 0 through 10,000 and decimals. Fractions followed in the fourth grade, and proportions, geometry, and elementary accounting in the fifth and sixth grades. In the rural areas, where each class met half the time each week, due to the gender issue, the arithmetic topics of the first four years were crammed into three years (Helg, 1987).

Exams were usually oral in front of parents and school or town authorities and were a special social event in small towns. A student who could not read somewhat fluently at the end of second grade had to repeat it. By this account, literacy was roughly equivalent to very basic numeracy. Learning was generally by memorization since there were no textbooks or notebooks. However, those who completed all six grades had a reasonable training to be an aide in many trades, or a teacher.

There was not much difference between private and public secondary schools. Both charged for tuition, and while most of the secondary schools attended by the children of the wealthy were run by religious orders, many of the public schools were too. Perhaps the most popular secondary school alternatives for poorer classes were the women's or men's normal schools in each province which trained teachers, for primary schools in four years, and for secondary schools, in an extra year. The government provided more scholarships for students entering these schools. Besides, they gave a well-balanced general education which also served to enter a university and with some teaching experience it was even possible to land a university job without further education. The classic high schools or *bachilleratos*, which prepared for the universities, usually run by European religious orders, were preferred by the better off, especially for their daughters (Helg, 1987).

For the Liberals, who shunned religious education, a few private non-denominational schools had been established, of which the best known, the *Gimnasio Moderno*, was founded in Bogotá in 1914 by wealthy Liberal personalities and run by Agustín Nieto Caballero, a convinced pedagogue of the New School, who had studied in the Sorbonne, the *Collège de France*, and Teachers College at Columbia University. Several non-denominational bilingual schools were also established, the first of which were the German schools in Barranquilla (1912) and Bogotá (1922), the French *Lycée* in Bogotá (1934) and the pro-US schools in Barranquilla (Karl C. Parrish School, founded in 1938) and Bogotá (*Colegio Nueva Granada*, founded in 1938). They were also coeducational, which to the Conservatives was a sacrilegious fashion (Helg, 1987).

Towards the 1930s there was a significant change in Colombia's sociocultural and political composition. Internally, it was promoted by urbanization and demographic growth, by industrialization, with its consequent budding proletariat, middle class, and labor unrest, as well as by the excessive liquidity brought about by the Panamá Canal indemnification. Externally it was provoked by post World War I circumstances, such as the United States' growing involvement in Latin American affairs, substituting Great Britain as world power, the expansion of communism and fascism, nationalist movements in México and Perú, the novel student movement in Argentina, and the Great Depression. This general setting, combined with the Conservative party's disunion, brought about — among other consequences — the Liberal party's comeback beginning in 1930, for a sixteen year period that has been called the Liberal Republic (Bushnell, 1993).

After more than forty years out of power, the Liberal party had changed its platform from the classic laissez faire and federalist policies to economic protectionism and social interventionism, while still strongly advocating separation of church and state, especially in education. For the first years, however, to avoid a confrontation with the Church, ministers of education were named from the Conservative party which had already voiced its intention to reform the educational system nationally.

Conservatives and Liberals were convinced that education for the popular segments of the population was of utmost importance to form morally straight, law-abiding, economically productive citizens. For President Alfonso López Pumarejo (1934–1938, 1942–1945), it became the number one priority (Sánchez, 2001, p. 7). The main effort went into the Constitutional Reform of 1936 which tried to implement governmental control of the public educational system. The responsibility for primary education was divided between the department authorities that paid the teachers and the central government that provided the materials and structured the educational programs. Yet neither had sufficient funds and despite the administration's intention to extend public education, it failed to declare that elementary education should be free and compulsory (Helg, 1987). Agustín Nieto Caballero as General Director of Education (1932–1936) pressed for changes to the educational practices that were maintained by the Catholic clergy, promoting the more scientific approach of the New School. Several representatives of these ideas came to Colombia between 1925 and 1940, including Ovide Decroly and Maria Montessori. Nieto Caballero tried to raise rural education standards, although not very successfully. However, he made important changes in teacher education commending the ideas of active learning proposed by the New School. More normal schools were established throughout the country with scholarships for more than half of the future teachers. Their curriculum phased out religious discourse giving more scope to modern pedagogy, psychology, anthropology and pediatrics. Practice teaching was made possible in adjacent schools and became a central part of the programs. An official national ranking system (*escalafón*) was begun in 1936, which categorized teachers according to their degrees, experience, and qualifying examination scores, and served as a basis to assign salaries. Independent courses and conferences were offered to train teachers on the job to make promotions possible. By 1943, 80% of the primary school teachers had been classified (Herrera, 2006).

In all, this was one of the most dynamic periods for the renovation of public education. Although many issues remained in the planning and regulating stages, there was an important effort to establish the main policies and goals, such as a schooling system independent of the church, based on generally accepted pedagogical theories. In practical terms, although teacher standards and school inspections were launched, teachers maintained a low status in society, due to their meager salary level and their lowly social origins. School attendance remained minimal and much of the educational system continued in the hands of religious orders and private educators, especially in the secondary level. Despite all efforts, illiteracy actually rose from 35% in 1937 to 37% in 1947, since population growth outpaced alphabetization (Herrera, 2006).

The achievements of the Liberal Republic at the university level were perhaps the most important. The goal of its far reaching reform was to form an intellectual elite that would propose theories to explain the national reality and design new means of coexistence. The term "national reality" encompassed all that makes Colombia unique. It was believed, naively, that if it could be explained theoretically it would be possible to guide the nation to a more peaceful existence.

Well prepared graduates were also needed to occupy positions that were emerging as a consequence of economic growth, including those in the country's administration. To accomplish its goals, the nation's university system had to become more democratic and scientific, developing research, which was totally inexistent (Herrera, 2006).

The *Universidad Nacional* in Bogotá would be the nucleus of the university system, becoming a model for the rest of the country. It was to group all existing public institutes of higher learning in Bogotá in a central campus, called *la Ciudad Universitaria*. Construction began after its legal consolidation in 1935, absorbing a good portion of the central government's budget for education during the next 10 years. It started to function in 1940 and by 1946 had 3,673 students that represented 50% of the total number of university students (67% of the student body in public universities). However, there was still no actual professorial career in the universities, full time dedication only started to grow in the 1950s and 1960s (Herrera, 2006).

The different schools within the *Universidad Nacional* started offering new programs such as economics, architecture, and statistics, although, in 1943, 80% of the students continued to select the traditional careers of medicine, law, and engineering. The School of Education *(Escuela Normal Superior*, later renamed *Universidad Pedagógica de Colombia)* which was established in 1936 began to impart university level instruction in the fields of education and social sciences with the promise of starting research in these fields. Women, who were already a majority in the education profession, obtaining their credentials in the normal schools, were finally admitted to the university in 1937, but initially only in careers "compatible" with their gender (Herrera, 2006).

The *Universidad Nacional* in Bogotá would also lead and coordinate four satellite universities which already existed, in Medellín (*Universidad de Antioquia*), Pasto (*Universidad de Nariño*), Popayán (*Universidad del Cauca*), and Cartagena (*Universidad de Cartagena*). This restructuring motivated the opening or strengthening of private universities throughout the country, both lay and religious. They were tentatively regulated by the central government which tried to consolidate careers along the same lines set by the *Universidad Nacional*.

Adjacent to the *Universidad Nacional*, learned societies were established which would begin a series of periodical publications meant to disseminate new ideas and discoveries. Among them, the *Academia Colombiana de Ciencias Exactas, Físicas y Naturales* (1936); its journal was published from 1936 to 1947 and from 1950 to 1976, but carried very little on mathematics (Herrera, 2006).

From Mid-Twentieth Century to the Present

The secularization of education during the Liberal Republic was constantly repudiated by the Catholic Church and a sector of the Conservative party with accusations that it was antichristian, Marxist, and immoral. When the Conservative party finally took over the executive in 1946, many of the reforms were dismounted together with the bureaucratic apparatus that had been organized around state education, and there was a return to Christian and Hispanic traditions (Herrera, 2006). However, partisan violence continued to escalate and culminated with a bloody uprising in Bogotá on April 9, 1948, known as the *Bogotazo*. Violence continued through the Conservative administration until 1953 and kept on during the military dictatorship from 1953 to 1957, although at a lesser degree. From 1958 to 1974 Colombia was governed jointly by the two parties — in an effort to end the conflict — and partisan fighting actually ended, but hostility has persisted to this date led by Marxist guerrillas and paramilitary forces (Bushnell, 1993).

Nevertheless, after the middle of the twentieth century, the country became more industrialized. Heavy industry had begun with a tire production plant toward the end of World War II, and then followed with a state-sponsored steel mill funded by taxpayers and two automobile complexes, which the government considered basic requirements of an industrialized nation. With stiff tariff protection backed by government policies of import substitution, consumer industries, such as kitchen appliances and home electronics were also started, leading to growth in communications and other services.

Population growth reached a maximum of 3% annually in the 1950s and 1960s. Urbanization peaked as urban population increased from 39% in 1951, to 52% in 1964; urban growth rate surging from 19.5% to 26.1% in the same period. For the rising middle class, university education became a prerequisite for upward mobility. As the existing universities gained prestige, they were not able to admit the flood of candidates, and what were called "garage universities" sprouted throughout, hoping to make a profit offering the barest minimal conditions (Ramírez & Téllez, 2006).

Although school attendance grew at the fastest pace in the century during the 1950s and 1960s, primary and secondary education had difficulty keeping up. The government's priority during these decades was in the numbers. Educational planning was initiated, funding school construction and teacher education, but the increase was also due to public schools having two sessions — one group of students attending in the morning and another in the afternoon — and even three sessions, with an evening shift in the secondary schools. The consequence of the multiple sessions was, of course, fewer hours of schooling for each student and less learning. Figures 1 and 2 show gross enrollment ratios (GER) for primary (1940–2000) and secondary schooling (1950–2000). Note: GER may register numbers beyond 100%

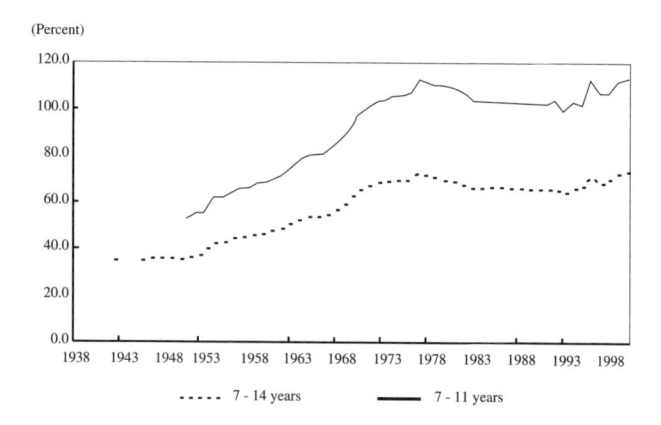

Figure 1. Gross enrolment ratio for primary schools, 1938–2000.
Source: Ramírez, M. T. & Téllez, J. P. (Gráfico 17, p. 489).

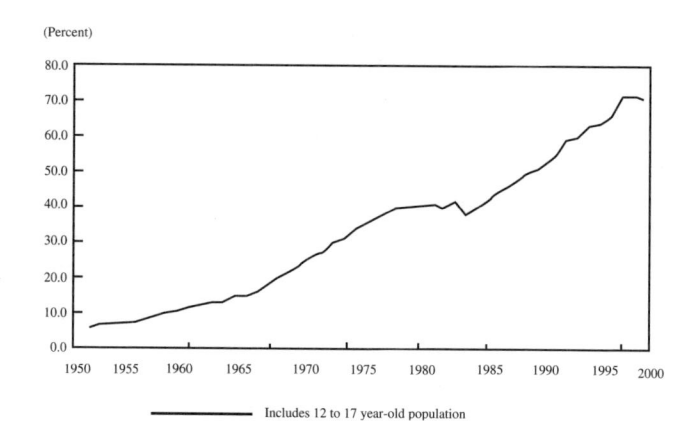

Figure 2. Gross enrollment ratio for secondary schools, 1950–2000.
Source: Ramírez, M. T. & Téllez, J. P. (Gráfico 21, p. 498).

due to overage enrollment, and is therefore not a very accurate indicator (Ramírez & Téllez, 2006).

General policy, however, was hesitant, sometimes to the point of being contradictory, with an effort to centralize educational management up to 1990, followed by decentralization, especially after the 1991 National Constitution. First, in the 1960s the Alliance for Progress Project sponsored teacher education and curricular revision, fostering Instructional Technology and the New Math (Aristizábal, 2012). Then, in 1968 the German Pedagogical Mission (GPM) took over.

The vertiginous increase in school enrollment, the fewer hours of classes due to the multiple sessions in each school, and general teacher incompetence resulted in low quality education and consequent high dropout rates. Therefore, the GPM's focus was in designing classroom guides, materials, and textbooks, as well as in training inspectors and teachers to guarantee an effective use of the materials, for grades 1 through 5. This most ambitious curriculum development project — covering language skills, arithmetic, science, social studies, music, physical education, and art — went on for nearly ten years. During 1974, its most active year, its design committee was made up of thirty-six Colombian and seven German experts (Rojas, 1982).

Another educational reform in 1976 led to a deterioration of the relations between the GPM and the Ministry of Education and to the dissolution of the Mission in 1978. In the end, there was no concluding evaluation of the project. Of the promised materials, the mathematics textbook manuscripts were not given an approval to be printed and, in general, the distribution of what was produced was erratic. However, most public school teachers managed to obtain classroom guides which they used constructively even though many never received the training they were supposed to (Rojas, 1982).

By 1978, there was even more concern with the quality of the educational programs, more people in Colombia and the Ministry of Education were knowledgeable in educational matters, and there was an aspiration to develop a homegrown schooling system. Carlos Vasco, a very scholarly Jesuit with a Ph.D. in Mathematics from Saint Louis University, became the chief consultant at the Ministry. He designed an interesting frame of reference by which he fit several school disciplines in diverse "systems". There were concrete systems that arose from everyday situations from which conceptual or mathematical systems were abstracted, and these, in turn had different symbolical systems (Molano, 2011).

However, this reform also met resistance because it was too formal for primary school teachers, and, although preliminary material was generated, it was never implemented except in an experimental context (Molano, 2011, pp. 179–181). The main problem was the lack of continuity. During the 28 year period from 1962 to 1990 there were 17 ministers of education. Consequently many projects begun by one minister were dismissed by the next. Besides, most ministers were appointed as political favors with no regard to their expertise in education, a practice that occurs in the public education sector, even today, down to the hiring of teachers (Molano, 2011).

As was mentioned above, public education was decentralized as of the 1990s. Although funding for public education is funneled through the Ministry of Education, each school today — public or private — must develop its own "Institutional Educational Project (PEI)", elaborating and explaining its goals and methods which should respond to the necessities of the targeted population. The Ministry however, has a website in which it explains the basic educational standards. For mathematics, these are documented under *"Estándares Básicos en Competencias en Matemáticas"*. Similar in treatment to the NCTM Math Standards and Expectations, they are divided into five types of thought: (1) numerical, (2) spatial or geometrical, (3) metrical and measurement, (4) probability and data systems, (5) variable (functional), algebraic, and analytic. They are also divided into five processes: (1) formulating, treating, and solving problems, (2) modeling, (3) communication, (4) reasoning, and (5) formulating, comparing, and exercising procedures or algorithms. At the end of the document, tables suggest a distribution of specific competencies among grades 1 through 11 (Ministerio de Educación, 2006). In general, there is no 12th grade in the Colombian educational system.

With 11,500,000 students, and 300,000 teachers under the supervision of the Ministry of Education — 84% of which correspond to public education — decentralization has made the task of evaluating schools of utmost importance. In 1968 the ICFES (Colombian Institute for the Promotion of Higher Education) was created to regulate higher education. Universities promptly asked the ICFES to provide a standardized college entrance examination — now called *Saber 11* — along the lines of the American SAT. In 1980, this test, which measured content comprehension and aptitudes, became compulsory for all students finishing 11th grade — the end of the secondary school — to serve as an assessment for this educational stage.

In order to adjust to this second purpose, the test underwent changes in the 1990s to account for the wider target population. Then in 2000, it had another significant change to assess the competencies that were established by the Ministry in its Educational Standards documents.

Tests were also developed to be applied in grades 3, 5, and 9 — *Saber 3*, *Saber 5*, and *Saber 9* — which, after an experimental period, have become compulsory every three years as of 2001. Then in 2002–2003, an end-of-college examination, called *Saber Pro*, was applied for the first time. This exam also became compulsory in 2009, with the purpose of assessing college education. It covers generic competencies, such as quantitative reasoning, analytical reading, written composition, citizenship, and English, as well as competencies specific to major professional areas, such as engineering, health sciences, natural and exact sciences, etc. With an entry and an exit examination, the Ministry hopes to have a good indicator of the universities' performances. Consequently, ICFES has become the chief assessment entity, while the Vice Ministry of Higher Education took on its other responsibilities.

With all the recent changes in the exams, however, ICFES statistics are not very reliable yet. There are no statistically significant results, aside from a very noticeable difference in achievements between public and private schools. Higher scores for the latter can be statistically attributed to socio-economic variables. In the *Saber Pro* for mathematics, the large variation in the number of students and institutions evaluated also makes comparison between results unreliable (ICFES, 2010, 2011).

Of greater interest are the results from the Trends in International Mathematics and Science Study (TIMSS) or the OECD-sponsored Programme for International Student Assessment (PISA). Figures 3 and 4 show TIMSS percentages of students in different levels of achievement in mathematics, grades 4 and 8, in 2007. Only Colombia and El Salvador participated among Latin American countries. Outcomes for science are similar except that Colombia comes before Algeria. These scores also indicate private schools doing better than public ones, urban areas better than rural and boys better than girls. A comparison with 1995 outcomes shows a significant improvement in both math and science.

As for the PISA tests, Figure 5 shows Colombia lagging behind Uruguay, Chile, México, Argentina, and Brazil, but in front of Panamá and Perú, in math in 2009. Figure 6 shows the evolution of math and science averages from 2006 to 2009 for the first six Latin American countries.

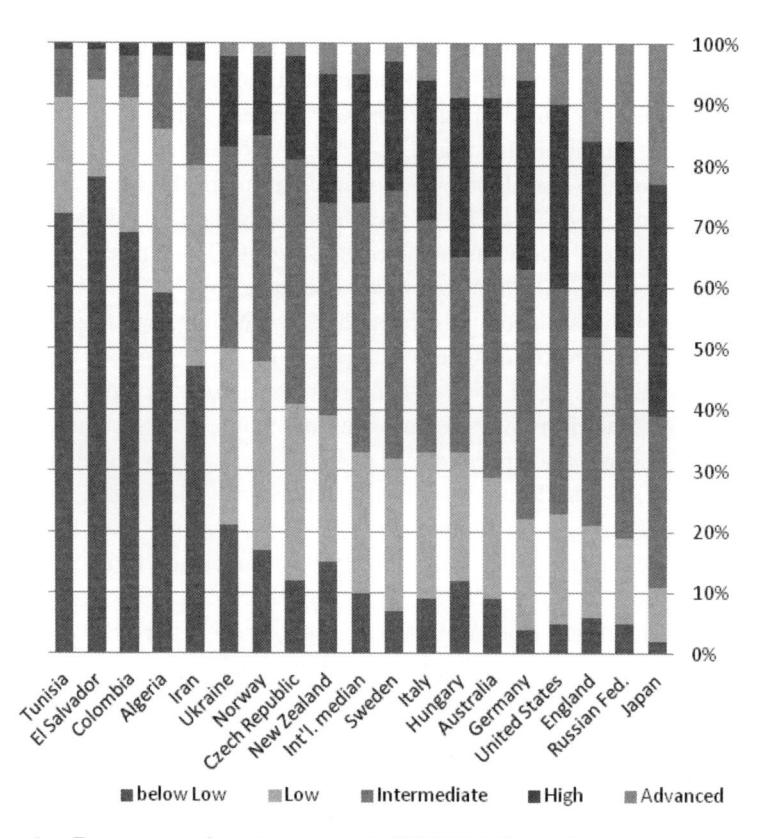

Figure 3. Percentage of students in each TIMMS 2007 mathematics achievement benchmark. Results for 4th Grade in selected countries.

Data Source: IEA's Trends in International Mathematics and Science Study (TIMSS) 2007 (Mullis, Martin, & Foy, 2008, pp. 34, 70).

Another important indicator for top level students is their participation in the Mathematical Olympiads which have been sponsored in Colombia since 1980. Competitions have become very popular, perhaps because they give students the opportunity of traveling as well as promoting their self-esteem. Colombia has won over 200 medals in the international contests, including gold. However, the program is mainly directed at elite private schools.

**Percentage of students in each achievement benchmark
Mathematics 8th Grade, selected countries
TIMMS 2007**

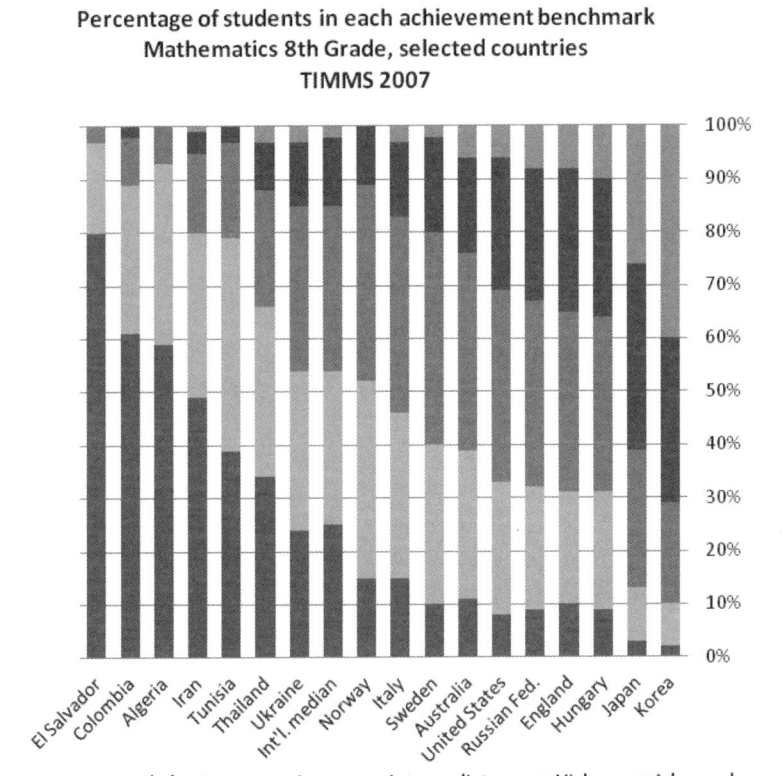

Figure 4. Percentage of students in each TIMMS 2007 mathematics achievement benchmark. Results for 8th Grade in selected countries.

Data Source: IEA's Trends in International Mathematics and Science Study (TIMSS) 2007 (Mullis, Martin, & Foy, 2008, pp. 35, 71).

University Mathematics Today

Higher education also had an important turn towards the middle of the twentieth century, especially in mathematics. Perhaps the first time set theory was explained in Colombia was in 1942 when Francisco Vera, a Spanish professor exiled during the Spanish Civil War, gave a short course

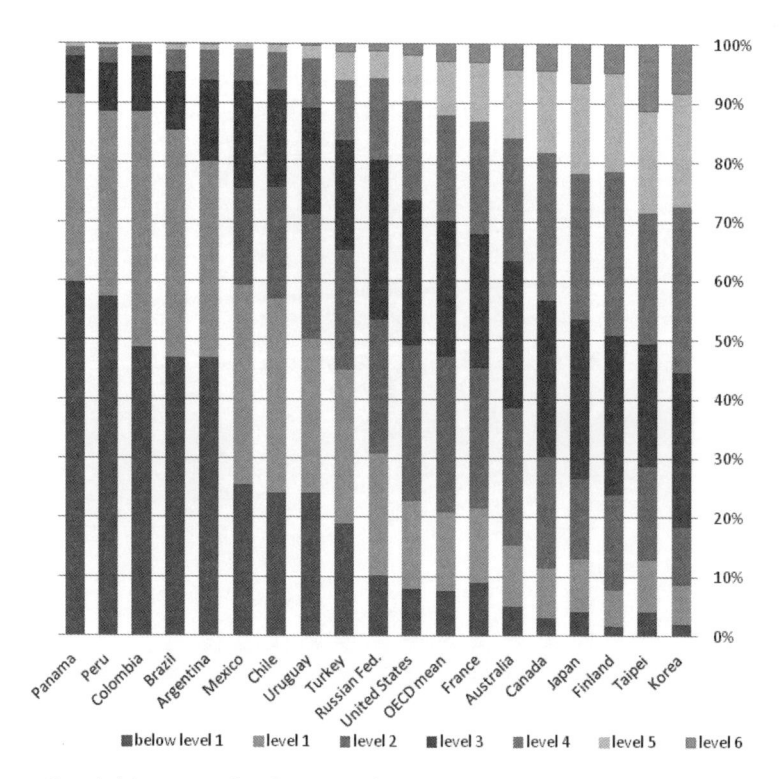

Figure 5. Achievement levels in mathematics. PISA 2009 results for 15 year-olds in selected countries. Based on data from OECD (2010), PISA 2009 Results: What Students Know and Can Do: Student Performance in Reading, Mathematics and Science (Volume I), PISA, OECD Publishing. http://dx.doi.org/10.1787/9789264091450-en

on the subject in the *Sociedad Colombiana de Ingeniería*. In 1943 he offered a course in the *Universidad Nacional* called Introduction to Modern Mathematics of which the last part was on topology. Until then, modern mathematics was not covered in the School of Mathematics and Engineering. In fact, Garavito had dismissed non-Euclidean geometry as a game with no applicability. Except for the article on Quaternions mentioned above, math production in the *Anales* was never above advanced calculus and it was practically inexistent until the 1940s. The School continued to use nineteenth century textbooks such as Sturm's *Cours d'analyse* (Sánchez, 2001).

Then, in the midst of the havoc caused by the *Bogotazo*, Mario Laserna, a young maverick from a provincial well-to-do conservative family,

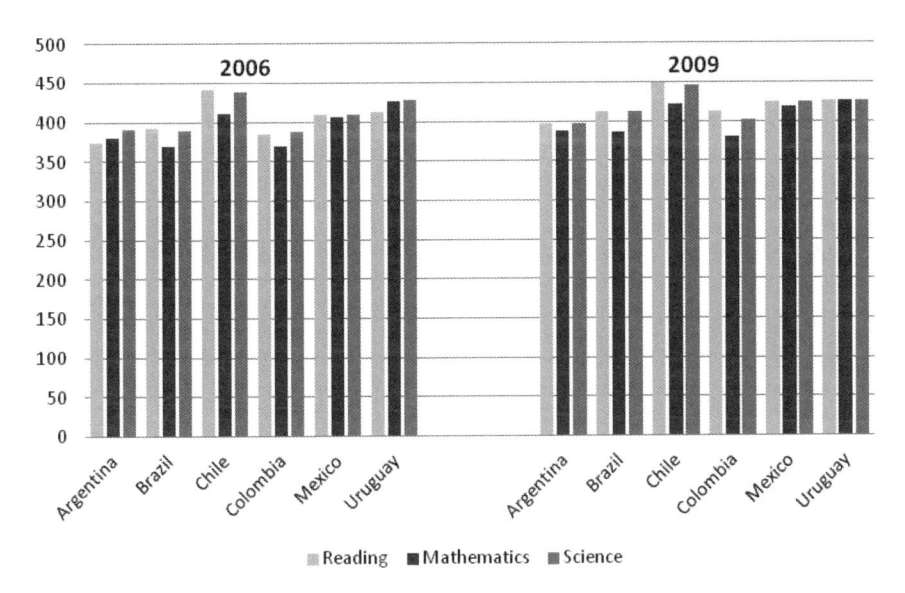

Figure 6. Variation of average scores: PISA 2006 to 2009. Based on data from ICFES (2010, PISA, p. 221, 224) and OECD (2010), PISA 2009 Results: What Students Know and Can Do: Student Performance in Reading, Mathematics and Science (Volume I), PISA, OECD Publishing. http://dx.doi.org/10.1787/9789264091450-en

convinced a group of intellectuals and executives around the country to found a university in Bogotá, the *Universidad de los Andes*, which admitted its first students in March 1949. It made a statement by being politically neutral. Its board of directors was divided evenly between the two parties, crossing two or three generations; in the beginning its presidents would also alternate between Conservatives and Liberals. Although non-denominational, it also conformed to a Catholic society by mentioning the Creator in its principles and asking the Vatican's blessing through its Nuncio for the inauguration. It would also play on the Conservative governments' lack of confidence regarding the *Universidad Nacional's* left-leaning political stance (Bell, Pinzón, Morales, Rojas, 2008).

Feeding to the growing need for engineers that had increased with the gross national product and the new government policy of import replacement, its main focus was on science and technology but its humanistic core curriculum emulating American universities made it different. The Mathematics Department (called faculty or school in the beginning) was to be the keystone of the university, and to make sure of this, Mario

Laserna himself was named chairman until a more appropriate one could be found. He had graduated from Columbia University (1948) with a major in mathematics and physics, and with his overwhelming personality had also convinced a number of professors at Columbia, Princeton and other east coast universities to back his project by participating in the new university's Advisory Committee. This was another distinctive feature of the *Universidad de los Andes*; from the start it had an international outlook (Bell, *et al.*, 2008).

Hungarian-born, János (John or Juan) Horváth was named chairman of the Mathematics Department and arrived in 1951, after visiting his mathematical acquaintances in Princeton, Yale, and Harvard. He had obtained his doctorate in 1947 at the University of Budapest under Lipót Fejér and worked at the *Centre National de la Recherche Scientifique* (CNRS) in Paris as a research attaché under Szolem Mandelbrojt and Jean Leray. He too had important connections in the mathematical world. Working together with Laserna they invited internationally known personalities as visiting professors; among the mathematicians were John von Neumann, Salomon Lefschetz (they came to Colombia before Horváth in 1950), Marc Krasner, Jean Dieudonné (member of the Bourbaki group), Laurent Schwartz (Fields Medal 1950), and his wife, Maria Helena Schwartz (Horváth, 1993). Regular professors were found among some Colombians educated in the *Universidad Nacional* or the *Normal Superior*, and many European immigrants who had arrived towards the end of World War II. They were all required to attend conferences given by the international experts as well as a weekly seminar, whether they understood or not. And although Mario Laserna had written an outline of courses for a mathematics major, the only careers offered initially were architecture, engineering, and economics. For the latter two, students would do their first five semesters in Bogotá and finish their degree in American universities, where there were more resources, so that the math courses taught at *los Andes* were mainly, algebra, trigonometry, and calculus (Schotborgh, 2004). These international programs became very popular, resulting in a rapid expansion of the Engineering School which soon accounted for about half of the student body. Economics also flourished when the school that was started in the *Gimnasio Moderno* merged with *Universidad de los Andes*. With an uncommonly strong groundwork in mathematics, its students excelled in their graduate studies and rapidly occupied key governmental positions (Bell, *et al.*, 2008).

In 1951, the *Universidad Nacional* started a Specialization in Higher Mathematics — which ended with the degree of *Licenciado en*

Matemáticas — organized and directed by Italian born mathematician, Carlo Federici. This was the origin of the first modern mathematics program in Colombia. Ironically, its most knowledgeable professor by far was the chairman of the Mathematics Department at the *Universidad de los Andes*, Horváth, who had just arrived and was more than glad to teach the more challenging courses — Fourier Series, Measure Theory, Hilbert Spaces and von Neumann Algebras — that had little audience at *los Andes*. It also took advantage of the extraordinary set of visiting professors who were contacted by Horváth and funded by the state sponsored *Fondo Universitario*. In 1952, Horváth also helped start the journal that was to become the *Revista Colombiana de Matemáticas*, and the Colombian Mathematical Society in 1955 (Sánchez, 2001), and during the Second Latin American Mathematics Symposium in Argentina, 1955, he made the first contacts with South American mathematicians (Horváth, 1993).

The collaboration that Horváth initiated between the two universities has continued to the present; common seminars are still given each semester and some dissertations are sponsored conjointly. When Horváth left in 1957 to become a professor in the Mathematics Department at the University of Maryland in College Park, several of the students who had graduated from the *Universidad Nacional* became instructors at *los Andes*. Visiting professors from abroad continued to come to both universities for periods of two or more years, including two German professors, Konder and Baessler, who had taught in *los Andes* and *Nacional* in the 1950s and returned to help initiate the mathematics program in the *Universidad de los Andes* in 1965. In particular, Peter Konder established a cooperation agreement with the University of Mainz and the German Academic Exchange Service (DAAD), through which professors continued to come to Colombia and Colombian students were able to continue their studies in Germany. The French Government followed suit in 1972.

One of the main attractions of the mathematics program at *los Andes* was the possibility of following a double major. Many students pursued mathematics simultaneously with Systems Engineering (Computer Science), Industrial Engineering, or Economics. A close integration was begun with the Engineering School, with courses such as logic, computer mathematics, numerical analysis, and operations research initially taught jointly, and many math students continued their graduate studies in these areas or vice versa.

Simultaneously, beginning in 1963, with the publication of the *Atcon Report on Latin American Universities*, the Alliance for Progress and the

Peace Corps, together with the Rockefeller, Ford, and Kelloggs Foundations, directed reforms which included facilities and installations, particularly in the universities of *Valle*, *Antioquia*, and *Industrial de Santander* (UIS). Masters and doctoral studies abroad were also offered, and promoted by the new government agency, ICETEX (*Instituto Colombiano de Crédito Educativo y Estudios en el Exterior*). The so-called Atcon Plan helped position these universities among the best in the country focused in science and technology; oddly, the Caribbean region was missing in this project (Padilla & Hidalgo, 2004).

Soon *Universidad del Valle* (Cali, 1967) and *Universidad de Antioquia* (Medellín, 1969) also started their mathematics programs. Then by 1972, the *Universidad Nacional* opened its Masters in Mathematics. *Universidad de los Andes* started it in 1975 and *Valle* and *Antioquia* followed soon after. In 1994 the first students were admitted to the doctoral program in the *Universidad Nacional* in Bogotá, the Medellín campus was also authorized to initiate it. By 2002 a third doctoral program — in Applied Mathematics — was opened at the EAFIT (*Escuela de Administración, Finanzas y Tecnología*) in Medellín (Sánchez, 2001). At the *Universidad de los Andes*, an excessively cautious faculty delayed its initiation until 2006.

University and program accreditation began in 1998 at the national level. It requires that the program or institution should have graduated students for the past five years at the least, among many other considerations. The number of mathematics programs accredited or in process of accreditation, as of July 2013, was: 11 undergraduate programs in mathematics or statistics, 9 undergraduate programs in mathematics education, 2 masters programs in mathematics and 1 doctoral program in mathematics, compared to 239 undergraduate engineering programs (CNA, 2013). The historically best known programs belong to *Universidad de los Andes* (Bogotá), *Universidad Nacional* (Bogotá and Medellín campuses), *Universidad de Antioquia* (Medellín), and *Universidad del Valle* (Cali). These four universities were the best placed Colombian universities in the 2013 QS ranking of Latin American Universities. *Universidad de los Andes* appeared as fourth, after *São Paulo*, *Católica de Chile*, and *Campinas* (Brazil); *Universidad Nacional* was ninth, *Antioquia* 32nd, and *Valle* 53rd (QS, 2013).

Colombia is a Group I member of the International Mathematical Union (IMU) through the *Sociedad Colombiana de Matemáticas* and it is planning to apply for admission to Group II. The IMU, an international nonprofit organization, was established to promote international cooperation in mathematics. It does so mainly through the International Congress of

Mathematicians, the most important conference for mathematical topics, best known because during its second meeting (Paris, 1900), David Hilbert announced his famous 23 problems, and because the most coveted mathematical prize, the Fields Medal, is awarded during its sessions every four years. There are seventy-seven member countries in the IMU, distributed among five groups according to their activity in mathematics. Dues and voting rights are also assigned conforming to the group's level. Other Latin American members are: Cuba, Perú, and Venezuela in Group I, Argentina, Chile, and México in Group II, and Brazil in Group IV; Ecuador is an associate member.

The letter postulating Colombia in 2007 as a full member of the IMU, cites more than 400 publications in the five preceding years by "mathematicians working at Colombian universities" in indexed journals such as *Advances in Mathematics, Annals of Pure and Applied Logic, Communications in Mathematical Physics, Comptes Rendus de L'Académie des Sciences, Illinois Journal of Mathematics, Journal of Algebra, Journal of Differential Equations, Journal of Number Theory, Journal of Symbolic Logic, Nonlinear Analysis, Proceedings of the AMS, The Quarterly Journal of Mathematics, SIAM Journal of Optimization, Tohoku Mathematical Journal, Topology and Its Applications* (SCM, 2007).

It names the universities most active in research at that moment as: *Universidad del Norte, Universidad Antonio Nariño, Universidad de Antioquia, Universidad del Cauca, Universidad de Córdoba, Universidad de los Andes, Universidad del Valle, Universidad EAFIT, Universidad Industrial de Santander, Universidad Nacional de Colombia*, Bogotá, *Universidad Nacional de Colombia*, Medellín, and *Universidad Tecnológica de Pereira*. It also provides evidence of its multiple mathematical activities in several national and international meetings at all levels, the latest of which was the *XIX Congreso Colombiano de Matemáticas* in July, 2013.

Conclusions

Although mathematics and science had an early start with the Botanical Expedition in the late colonial period, clearly, continuous political strife has slowed Colombian progress in technology as well as generally. Today, political and religious fanaticism have almost disappeared as the population lost interest in these issues. Nevertheless, it has not been possible to completely end violence, a situation that hinders the country's overall development.

Guerrilla warfare and illicit crops have been pushed to the remote outlying regions. But, even if a peace agreement is reached in the near future, there is a large displaced population that has to be resettled before normalcy is achieved. The rural areas, which have always been the poorest, are the worst hit in this respect. As in many other countries, organized crime, drug-trafficking, street gangs, and vandalism are taking their toll in the cities. Violence has bred in poor neighborhoods, families, and schools — especially public schools — where bullying and teacher harassment are common. Under these challenging circumstances, education must step in to restructure social values, fostering toleration and nonviolent means of conflict resolution, as well as honesty and proper work ethics. It also needs to promote social mobility to alleviate social inequality.

The weakening of the guerrillas in the past decade has been accompanied by economic growth and improved general welfare in the cities. However, as former chief economist for Latin America at the World Bank, Guillermo Perry, puts it, this improvement was limited by several problems, the first of which was the meager progress in providing quality education. "The majority of the graduates are not competent enough to contribute to the construction of a developed nation," he says, adding that this restricts economic growth, diversification, and competitiveness, limits social mobility, and prolongs social inequality (Perry, 2013).

Another problem that Perry emphasizes is the lack of incentives to encourage innovation and creativity which is reflected in an unimaginative entrepreneurial sector. Industry and agriculture are consequently going through a phase of stagnation. These shortcomings can be traced back to an educational tradition which rewards conformity. It also has to do with poor instruction in mathematics, technology, and science. Primary school teachers are not sufficiently competent in elementary mathematics, to say the least of their ability to teach it. In secondary school, algebraic notation and syntax are glossed over as are problem-solving and argumentative skills. As a consequence, many students reach college lacking basic mathematical concepts, a liability that diminishes their capacity to understand new fields (Perry, 2013).

The government's focus on education for the past fifty years has been to raise enrollment to nearly 100%. It is aware that it must now work on the quality of education, but it has paid only lip-service to this end. Corrupt political practices continue to burden primary and secondary schooling, with current lack of continuity in educational policy as one of the lesser problems.

The main challenge for schools today is to be more attractive to students. This could be accomplished by transforming them into community centers that can fill the void left by the Church as a consequence of the country's secularization. They should provide the glue that holds the community together. For this new role, schools should not only be better-kept and more appealing, they must offer services to the whole community, such as adult education, cultural events, as well as library, computer, and sports facilities; more interesting incentives can be provided through exciting extracurricular activities. An effort in this direction is being led by cities such as Bogotá, Medellín, and Barranquilla.

Teachers, as well as school administrators and staff need to be educated in this regard. This calls for a modernization of the institutions that train them, to turn out professionals with a more creative and imaginative attitude. They must also be paid accordingly, to give them the social status required to be community leaders, and to make their careers attractive to future generations.

For the higher end of the social spectrum, universities are flourishing and young talented people have found they need not study the classical professions of medicine, law, or engineering, for it is possible to lead a decent life while pursuing their passion for research and the sciences. Young professors, who received their doctorates abroad, had the opportunity of a postdoc position, and are returning with excellent global connections. They have become role models for students who follow in their footsteps, as they direct their doctoral dissertations and help to form them as researchers in local universities. With modern communications, international online discussions are easy and publishing in first-class journals is no longer an impossible ordeal.

However, there is a link that is still underdeveloped. As in earlier times, in which engineering projects motivated the study of mathematics, research in leading-edge information, communication, and biological technology or other high-tech endeavors can be an engine of mathematical innovation today. But it has not really started in Colombia, which lags behind even in traditional engineering projects, such as highways, railroads, and port facilities. High-end research should be jump-started by sponsoring science and technology parks around the top universities, with government and private support, like the Stanford Research Park which was crucial in the creation of Silicon Valley. It would, no doubt, stimulate students as well as research in mathematics and science.

To summarize, in a technologically globalized world, mathematics has become even more important for Colombia's development. Fortunately, at the university level its growth has started to take off. Internationally competent professionals and Ph.D.'s are graduating to enrich the country's faculty and research community in mathematics as well as in the sciences. However, there is still a long way to go.

Work has to be done on two fronts to promote careers in science and technology. The first one is in primary and secondary school education. Motivating students to prepare for these careers is a significant challenge for the country's educators in science and mathematics. It is necessary to broaden this group of students to be able to enhance the country's prospect of growth in the knowledge-producing quaternary sector of the economy.

The second one has to do with the jobs for graduates, which must be challenging as well as economically rewarding. Creativity and ingenuity at this front will not only promote innovation in the entrepreneurial sector, it will also encourage more young people to follow their calling. In return, to keep the engine going, those who enter this science-powered workforce must produce new technology, to further the country's development as has happened in countries like South Korea. How Colombia, as a whole, invests in these undertakings will largely determine its future.

Bibliography

Aristizábal, M. (2012). La irrupción de la teoría curricular a partir de 1960 y su influencia en las reformas educativas en Colombia. *Acción pedagógica*, *21*, 29–37. Retrieved from http://www.saber.ula.ve/bitstream/123456789/36643/1/dossier03.pdf

Bell, G., Pinzón, P., Morales, L., & Rojas, R. (2008). *Historia de la Universidad de los Andes*. V.1. Bogotá, Colombia: Ediciones Uniandes.

Bushnell, D. (1993). *The Making of Modern Colombia*. Berkeley, CA: University of California Press.

Colombia, Presidencia de la República (2007). Presidente de la Nueva Granada, General Tomás Cipriano de Mosquera 1845–1849, 1861–1864, 1866–1867. Bogotá, Colombia: Presidencia de la República. Retrieved from http://web.presidencia.gov.co/asiescolombia/presidentes/09.htm

Colombia, Presidencia de la República (2007). Presidente de la República de Colombia, Pedro Nel Ospina 1922–1926. Bogotá, Colombia: Presidencia de la República. Retrieved from http://web.presidencia.gov.co/asiescolombia/presidentes/45.htm

Colombia, Presidencia de la República (2007). Presidente de los Estados Unidos de Colombia, General Eustorgio Salgar 1870–1972. Bogotá, Colombia: Presidencia de la República. Retrieved from http://web.presidencia.gov.co/asiescolombia/presidentes/23.htm

Consejo Nacional de Acreditación (CNA) (2013). Consultar Programas Acreditados. CNA República de Colombia. Retrieved from www.cna.gov.co

Consejo Nacional de Acreditación (CNA) (2013). Listado de Programas en Proceso de Acreditación. CNA República de Colombia. Retrieved from www.cna.gov.co

Helg, A. (1987). *La educación en Colombia 1918–1957*. Bogotá, Colombia: Fondo Editorial CEREC.

Herrera, M. C. (2006). Historia de la educación en Colombia, La República Liberal y la Modernización de la educación: 1930–1946. Bogotá, Colombia: Universidad Pedagógica de Colombia. Retrieved from http://www.pedagogica.edu.co/storage/rce/articulos/rce26_06ensa.pdf

Horváth, J. (1993). Recuerdo de mis años en Bogotá. *Lecturas Matemáticas*, *14*(1–3), 119–128.

Instituto Colombiano para el Fomento de la Educación Superior (ICFES) (2010). Exámenes de Estado de Calidad de la Educación Superior SABER PRO (ECAES), Análisis de resultados del período 2005–2009. ICFES República de Colombia. Retrieved from http://www.icfes.gov.co/

Instituto Colombiano para el Fomento de la Educación Superior (ICFES) (2010). Resultados de Colombia en TIMSS 2007, Resumen ejecutivo. ICFES República de Colombia. Retrieved from http://www.icfes.gov.co/investigacion/component/docman/doc_view/15-informe-resultados-de-colombia-en-timss-2007-resumen-ejecutivo?Itemid=

Instituto Colombiano para el Fomento de la Educación Superior (ICFES) (2010). Colombia en PISA 2009, Principales resultados. ICFES República de Colombia. Retrieved from http://www.plandecenal.edu.co/html/1726/articles-308346_archivo.pdf

Instituto Colombiano para el Fomento de la Educación Superior (ICFES) (2011). Examen de Estado de la educación media, Resultados del periodo 2005–2010. ICFES República de Colombia. Retrieved from http://www.icfes.gov.co/

Ministerio de Educación, República de Colombia (2006). Estándares Básicos de Competencias en Matemáticas. Bogotá, Colombia: Ministerio de Educación República de Colombia. Retrieved from http://www.mineducacion.gov.co/1621/articles-116042_archivo_pdf2.pdf

Molano, M. (2011). Carlos Eduardo Vasco Uribe, Trayectoria biográfica de un intelectual colombiano: una mirada a las reformas curriculares en el país. *Revista Colombiana de Educación*, *61*, 161–198. Retrieved from http://revistas.pedagogica.edu.co/index.php/RCE/article/view/860

Mullis, I. V. S., Martin, M. O., & Foy, P. (with Olson, J. F., Preuschoff, C., Erberber, E., Arora, A., & Galia, J.) (2008). *TIMSS 2007 International Mathematics Report: Findings from IEA's Trends in International Mathematics and Science Study at the Fourth and Eighth Grades.*

Chestnut Hill, MA: TIMSS & PIRLS International Study Center, Boston College. Available at http://timss.bc.edu/timss2007/PDF/TIMSS2007_InternationalMathematicsReport.pdf

OECD (2010). *PISA 2009 Results: What Students Know and Can Do: Student Performance in Reading, Mathematics and Science (Volume I)*. PISA, OECD Publishing. DOI: 10.1787/9789264091450-en. Available at http://www.oecd.org/pisa/pisaproducts/48852548.pdf

Padilla, A., & Hidalgo, H. (2004). Implementación de las reformas educativas en la educación superior en Colombia 1948–1980. Universidad Nacional de Córdoba, Argentina. Retrieved from www.reformadel18.unc.edu.ar/privates/amalfi.doc

Perry, G. (2013). Los próximos 20 años. *El Tiempo*. 28 Sept 2013. Retrieved from http://www.eltiempo.com/opinion/columnistas/guillermoperry/ARTICULO-WEB-NEW_NOTA_INTERIOR-13089601.html

Poveda, G. (2012). *Historia de las matemáticas en Colombia*. Medellín, Colombia: Ediciones UNAULA.

QS Quacquarelli Symonds Limited (2013). *QS Latin American University Rankings 2013*. QS Top Universities. Retrieved from http://www.topuniversities.com/university-rankings/latin-american-university-rankings/2013

Ramírez, M. T., & Téllez, J. P. (2007). La educación primaria y secundaria en Colombia en el siglo XX. *Economía colombiana del siglo XX, Un análisis cuantitativo*. Robinson, J. & Urrutia M. (Eds.). Bogotá, Fondo de Cultura Económica, pp. 459–515.

Rojas, M. C. (1982). Análisis de una experiencia: la Misión Pedagógica Alemana. Universidad Pedagógica Nacional. Retrieved from http://www.pedagogica.edu.co/storage/rce/articulos/10_05ensa.pdf

Sánchez, C. H. (2001). 50 años de Matemáticas Modernas en Colombia. *Boletín de Matemáticas*, Nueva Serie *8*(2), 3–28.

Schotborgh, A. (2004). *El Departamento de Matemáticas de la Universidad de los Andes 1949–2003*. Bogotá, Colombia: Universidad de los Andes.

Sociedad Colombiana de Matemáticas (SCM) (2007). Letter to the International Mathematical Union. Retrieved from http://www.mathunion.org/fileadmin/IMU/AO/ColombiaEvidence.pdf

About the Authors

Hernando Echeverri Dávila is an associate professor in the Department of Mathematics at *Universidad de los Andes* and holds a Ph.D. in mathematics education from Columbia University. At *Universidad de los Andes*, where he has taught for more than 30 years, he pioneered in computer-assisted instruction in mathematics and promoted the development of the first end-of-college national assessments. Today he is chief editor of *Hipótesis*, a popular science journal published by the School of Sciences at his institution.

Ángela María Restrepo is an assistant professor of mathematics education in the *Centro de Investigación y Formación en Educación* (CIFE) at *Universidad de los Andes* in Bogotá, Colombia. She holds a Ph.D. in mathematics education from the *Université Joseph Fourier* in Grenoble, France. Her main interest is the teaching and learning of mathematics and STEM integration. She has worked as a consultant in the Colombian Institute for the Promotion of Higher Education (ICFES).

Chapter 6

COSTA RICA: History and Perspectives on Mathematics and Mathematics Education

Ángel Ruiz

Abstract: A description of mathematics and its teaching in Costa Rica beginning in the nineteenth century is presented. It includes curricula, researchers, and practitioners (with special mention of foreign influences and individuals), educational institutions, reforms, and research in mathematics education. Perspectives on a recent ambitious school mathematics reform are elaborated.

Keywords: Costa Rica; mathematics education; mathematics; curriculum reform; history of mathematics; research in mathematics education.

Mathematics and its teaching in Costa Rica have been conditioned by institutions, factors, and significant historical events. Some of them correspond to variables within the National Education System, others are a result of very broad social and national processes (including political), and others are due to domestic circumstances within higher education. These factors have been:

(1) The creation of the *Casa de Enseñanza* (Teaching House) and later the *Universidad de Santo Tomás* (University of Saint Thomas): between 1814 and 1888 it constituted the institutional space of reference for national education.
(2) The Educational Reform by Mauro Fernández from 1985 to 1988: grasping the Liberal Ideology it closed the *Universidad de Santo Tomás*, created the Physical and Geographical Institute, the National Museum and National Library, created and supported secondary schools, and attracted foreign teachers.

(3) The establishment of the *Escuela Normal de Costa Rica* (Normal School of Costa Rica): from 1914 to 1974 it was the institution responsible for preparing teachers and — in the absence of a university until 1940 — was the main intellectual center of the country.
(4) The creation of the *Universidad de Costa Rica* in 1940: it rebuilt all the possibilities of educational, scientific, and cultural development of the country.
(5) The University Reform by Rector Rodrigo Facio: it allowed the creation of the Department of Physics and Mathematics in 1957 and opened the way for higher mathematics development and enhancement of its teaching.
(6) The creation of the Department of Mathematics in 1971–1972: this was a quantum leap in higher mathematics in the country and created the foundation for the development of this scientific discipline to the present.
(7) The creation of new public universities between 1971 and 1979: this diversified and expanded the initial preparation of teachers of mathematics.
(8) The emergence of dozens of private universities in the 1990s: it has rebuilt, although not in positive terms, the role, the teachers' professional status, and some conditions of mathematics education in the country.
(9) A new mathematics curriculum for all pre-university education was approved in 2012: its design and implementation currently influences all aspects of the teaching and learning of mathematics in Costa Rica.

The Nineteenth Century

The history of Costa Rica cannot be separated from its membership in the *Captaincy General of Guatemala* within the Spanish Empire. Guatemala was always the center of economic and cultural welfare of the region. It was also the place where the political processes, which would act on the rest of Central America, had the strongest impact. On the contrary, Costa Rica was the more distant and poorest province, formed by a few quite isolated villages that always wanted to impose their influence over the others, even if no blood was shed except on two occasions.

Before the nineteenth century, there was an element related to mathematics that needs to be mentioned. The former *Reyno of Guatemala* had an expert in exact sciences and philosophy: Costa Rican-born José Antonio Liendo y Goicoechea. He lived and worked in Guatemala and not in Costa Rica. In Guatemala, Goicoechea taught philosophy, physics and mathematics and, furthermore, directed the reorganization of studies in the *Universidad de San Carlos* (University of Saint Charles) within the tendency of the "New Science" (Liendo y Goicoechea was an "encyclopedist", pupil of Escoto and Feijóo); in 1869 he imparted a course of philosophy where he introduced experimental physics, confronted with the Thomist–Aristotelian instruction. From 1765 until 1767 he studied in Spain; when he came back to Central America he brought with him some machines and instruments for experimental physics, books, globes, armillary spheres, planetary systems, maps and hydrographic cards, tables of longitudes and latitudes; "he studied mathematics and taught this subject privately ..." (González-Flores, 1945, pp. 65–67; Láscaris, 1982b; Rodríguez & Ruiz, 1995).

The 19th century represents the end and the beginning of two eras in Costa Rican history: it is the end of colonial life and the beginning of republican life. The first half of the century was destined mainly to establish the basic political and social institutions, and to build up the nation. The second half of the century was determined by the conflict between Liberals and Conservatives, especially the decade of the 1870s (González-Flores, 1945). The same conflict took place in almost the entire Central American region.

Coffee production and its exportation was an element that added a new universe to the national life. Even though it was introduced in the country at the beginning of the century, it did not develop intensively before the 1840s. In 1843 the first shipment of Costa Rican coffee to London took place. The new activity stimulated the creation of new dominant groups and a social stratification that moved away from a precarious agricultural economy and a shared level of poverty and (more or less) equality of opportunities.

Costa Rican leaders have always given priority to educational development. Already in 1849, under President Dr. José María Castro Madriz, an important administrative reform took place in education, which impelled the creation of a Normal School, a Grammar School for girls, and a more efficient coordination and inspection of primary education. In 1858 and in 1862, it was decided that formal education should be mandatory for all social classes and for children of both sexes. A few years later, in 1869, the Costa Rican Political Constitution incorporated free and mandatory primary education, financially supported by the State (González-Flores, 1945).

One of the most interesting features of Republican life in the Costa Rica of the 19th century was the presence of foreign intellectuals and teachers, a phenomenon which took place in a permanent way and which defined some tendencies in the process of shaping the local educational institutions. Mexicans, Cubans, Guatemalans, Nicaraguans, Colombians, Spaniards, Swiss, Germans, British, and Italians taught in Costa Rica, some of them strongly influenced by liberal ideas in politics, education, and culture (Láscaris, 1982b). A certain level of regional interaction was also characteristic, above all in the first half of the century, as intellectuals moved from one country to another.

Casa de Enseñanza and *Universidad de Santo Tomás*

The *Casa de Enseñanza de Santo Tomás*, created in 1814, imparted lessons of first reading, as they were called, in which the children learned how to read and write, to count and, besides these elemental subjects we could find lessons of philosophy and Castilian and Latin Grammar. It cannot be said that this institution was an institution of primary or secondary education, but rather a mixture of both. After several reforms and reorganizations, in 1824 the institution really acquired pre-university characteristics, so that they established the *Bachillerato* degree (a Baccalaureate, for accomplishing secondary education) according to the constitution of the *Universidad de San Carlos* in Guatemala. We also know, according to Rodríguez and Ruiz (1995), that the *Casa de Enseñanza* taught arithmetic and geometry.

In 1843, President José María Alfaro signed the decree by which the *Casa de Enseñanza de Santo Tomás* was transformed into a university. It offered a mixture of different levels of education. According to the foundation statute, this institution would include minor studies (Castilian and Latin Grammar, philosophy and mathematics) for obtaining the Bachelor's degree in philosophy, and major studies in medicine, theology and law. Before 1848 we could not find but failed attempts to establish a *"cátedra"* (chair) of mathematics, and in 1861 this chair obtained university status (González, 1989; Rodríguez & Ruiz, 1995).

Careers in civil engineering, architecture, and land surveying in Costa Rica started with the Mexican engineer Angel Miguel Velázquez, who arrived in Costa Rica in 1862 (he had studied Science in New York and graduated as an architect in Rome). He wrote an elementary work, which was improved to be used as a textbook at the *Universidad de Santo Tomás*:

Tratado Elemental de Matemáticas, Primera Parte. With the opening of these careers, the mathematics chair was closed (González-Flores, 1921, pp. 291–292; González, 1989). Rodríguez and Ruiz (1995) found no evidence of a graduated civil engineer or architect in the *Universidad de Santo Tomás*, only a statement about graduates of land surveying who were called by the peculiar title of "*licenciados geómetras*" (licensed geometers) who "had to know arithmetic, geometry, and furthermore, trigonometry". This program was closed in 1866.

In 1874, a secondary school was created within the university: the *Instituto Nacional* (National Institute). In December 1879, for economic reasons, the *Instituto* became a private institution. In 1883 it was closed and reopened as *Instituto Universitario* (University Institute) in 1884. It had three sections: Preparatory, Secondary Education and Special Lessons. In the *Instituto Universitario* there existed a first course of arithmetic and algebra, a second course of plane geometry, rectilinear trigonometry and astronomical geometry; the third course was on solid geometry and spherical trigonometry and the fourth course on differential calculus, with which the Baccalaureate's degree was completed. Among the special math courses taught were first-year arithmetic and algebra for the *Mercantile Technicians*; and arithmetic and algebra, geometry and trigonometry (one course each year) for the *Land Surveyors* and the *Construction Builders*. The program for the fourth year was entirely on algebra and differential calculus (Ruiz & Rodríguez, 2002, pp. 107–108). There is no evidence this syllabus was effectively implemented.

In 1888, National Congress decided to close the *Universidad de Santo Tomás*, and with this a period of higher education in Costa Rica ended (Fischel, 1991; Quesada, 1991).

The first book published in Costa Rica

In 1830, Rafael Osejo (who was hired in Nicaragua and was the most influential person in the *Casa de Enseñanza*) circulated the first book ever published in Costa Rica: *Brebes Lecciones de Aritmética*. In 1838, a second edition was included in the *Revista de los Archivos Nacionales* (National Archives Journal). As Zelaya (1973) states the book was designed as "catechism", in other words: by questions and answers. The first part is "About Arithmetic", and starts with the question: "To which of the human sciences does it correspond?" Afterwards: "... it continues explaining the operations of adding, subtracting, multiplying and dividing, without showing neither proofs nor justifications of these operations." The second part is about "The fractions", and mentions how to express them. In the third part, it talks about the "Theory of the decimals", defined as follows: "This is how we call the fractions which have as a denominator a number with one, two, three or more zeros; this means that the denominator is multiplied by ten, according to the numbering system, and for this reason they are called decimals." The fourth part is about the "dominions or potentialities" and says "thus we call the product of a number multiplied by itself for a certain number of times" (Zelaya, 1973, pp. 109–112). The work in mathematics performed by Osejo shows the state of mathematics in those years.

Primary and secondary education

For many decades, the efforts in primary education were about building schools; nevertheless, the absence of qualified teachers retarded the academic concerns. Mathematics instruction was reduced essentially to arithmetic.

Based on a few textbooks it is possible to know the main contents of arithmetic which were taught in the Costa Rican primary schools for a large part of the nineteenth century: elementary operations with integers and fractions, roots and powers of numbers, ratios and proportions, the rule of 3, simple and compound interest, resolution of first degree equations, measuring tables, and the decimal metric system (from the 1970s). The mandatory introduction of the decimal metric system was an important issue (many of these textbooks were known thanks to a book published by L. Dobles-Segreda in 1929).

Secondary education was born in Costa Rica in 1869, the year the *Colegio San Luis Gonzaga* was founded in the city of Cartago. In 1886, secondary education existed in four institutions: the *Colegio San Luis Gonzaga*, the *Seminario* in San José, the *Colegio San Agustín* in Heredia, and the *Instituto Universitario* of the *Universidad de Santo Tomás* in San José (Azofeifa, 1973, p. 18). At the *Colegio San Luis Gonzaga*, in 1873, the program was for five years (and one preparatory for university), in which many mathematics courses were included: arithmetic and geometry (preparatory), arithmetic and geometry (first year), arithmetic and commercial calculation, and algebra (second year), geometry, algebra and trigonometry (third year).

Normally, the primary and secondary schools offered courses in arithmetic, geometry, algebra, and trigonometry; in a few cases, geometry was rectilinear, in others plane or objective. Sometimes commercial applications were included, and land surveying. The programs of the *Instituto Universitario* were the most ambitious of the whole century (Rodríguez & Ruiz, 1995).

The Reform by Mauro Fernández

During the 1880s there was a deep Liberal Educational Reform led by Mauro Fernández, Secretary of Public Instruction and at the same time of the Treasury Department. The reform pretended to transfer to the State the control of education policies, and also promoted the laicization of education (Fischel, 1987). Fernández was influenced by Herbert Spencer, whom he knew personally, by the North American Horace Mann and the educational practice of Jules Ferry, and also, as in many countries of Latin America, by the ideas of Andrés Bello and Domingo Faustino Sarmiento (Fischel, 1987; González-Flores, 1921, pp. 99–103). Besides the educational reform, the National Museum, the National Library, and the National Physical-Geographical Institute were created (Ruiz & Barrantes, 2000).

It is interesting to remark that the intellectual ideas that influenced Costa Rican liberals were from the British-styled economic liberals, the ideas of the French Revolution about the national State and the importance of education, and positivism. Spanish Krausism was relevant in Costa Rican education. Krausism was the most important philosophical and pedagogical process in 19th-century Spain. Krause's pantheistic point of view was adapted by Sanz del Río and his pupils to become a "harmonistic" religion but based on Rationalism and Liberal Humanism. In education they rejected boarding schools and punishment, and advocated for the exemplary conduct of the teacher, the philological sense of culture, and self-responsibility of the student (Láscaris, 1982a, p. 156). Krausism was the philosophical base of the secondary education syllabus in the Costa Rica of those years. The Spanish brothers Valeriano and Juan Fernández-Ferraz, who were pillars of the Costa Rican educational development during those years, carried out this influence (Láscaris, 1982, pp. 155–176; Ruiz, 2000, pp. 673–674; Chacón, 1984, pp. 128–150; Ledesma-Reyes, 1995).

Mathematics and science instruction were taken into consideration. From the years in which the reform began the syllabus of mathematics in primary as well as in secondary school went through some changes, especially an increase in the number of subjects to study. Two important secondary schools were created: *Liceo de Costa Rica* in 1887 (Liceo de Costa Rica, 1930) and *Colegio Superior de Señoritas* in 1888 (Colegio Superior de

Señoritas, 1939). It also established conditions for hiring foreign teachers in science and mathematics.

When the *Liceo de Costa Rica* began, it was organized as follows: the Elementary Division (including arithmetic), the Inferior Division (including arithmetic and elements of geometry), and the Superior Division (including arithmetic, algebra, geometry, and trigonometry). The Elementary Division had 25 weekly hours of mathematics for five years (Azofeifa, 1973).

The *Universidad de Santo Tomás* was closed during this reform in order to diminish conservative influence of this institution. Instead, a "Napoleonic style" Polytechnic Institute was desired, but it never occurred; only the *Liceo* resembled the French model (Ruiz & Barrantes, 2000). But the reform succeeded in attracting foreign teachers for secondary schools, although not to the extent the liberals wanted. Besides the ideological bias, there existed also an economic feebleness in the country. There was not enough money to support a higher education institution. After this closure, independent faculties offered some higher education: Law, Agronomy, Fine Arts, and Pharmacy (Fischel, 1991; Quesada, 1991; Ruiz & Barrantes, 2000).

Mathematics curriculum

After the liberal reform began, primary as well as secondary education in Costa Rica changed, especially because of the increase in the number of subjects to be studied.

The reform of Mauro Fernández intended to give great support to primary education, as well as restructuring the secondary school on more solid foundations. However, the growth of the secondary school from the reform and until 1940 was much smaller than the primary. Barrantes and Ruiz (1995c) point out that in 1886 there were only three official institutions of secondary education: the *Instituto de Heredia*, the *Instituto Universitario* and the *San Luis Gonzaga*. With the reform of 1886 the *Instituto de Heredia* was removed, the *Instituto Universitario* became the *Liceo de Costa Rica* and the *Colegio Superior de Señoritas* and the *Instituto de Alajuela* (1889) were created.

For the new secondary schools Fernández intended a unified syllabus, which was issued for all institutions of secondary education of the country. This syllabus had much less content than the one used before in the *Instituto Universitario*, but was much more realistic and had more content than in the other secondary schools.

According to Barrantes and Ruiz (1995c), the mathematics syllabus of 1892 established the study of one of the mathematical branches every year. So, in the preparatory course, arithmetic and geometry were taught. The arithmetic part studied different systems of numbering, fractions, divisibility (greatest common divisor and least common multiple), the decimal metric system and other measures used in Costa Rica, second and third powers, square and cubic roots, logarithms, the rule of three and progressions. Moreover, the resolution of application problems was proposed. Concerning the part on geometry, it was constituted by a review, with a major grade of intensity of what was studied in primary school (including some aspects as congruency, similitude, perpendicularity, etc.).[1]

Barrantes and Ruiz (1995c) inform: as for primary education, the programs were divided into two parts: arithmetic and geometry. In arithmetic, natural numbers (operations and basic elements of the theory of numbers like divisibility and calculation of the greatest common divisor), fractions (operation, mixed numbers) and measuring units (length, capacity, weight and time) were studied. As an application, the study of the simple rule of three was proposed as well as calculation of interests and discounts, and relations to other disciplines in the curriculum. In geometry, straight lines and curves, plain and space geometrical figures; calculation of perimeters, areas and volumes, all this from the intuitive point of view and without the

---

[1] During the first year, only arithmetic was taught, from a more rigorous point of view, establishing definitions and theorems. Some elements of the theory of numbers (theorems about divisibility, greatest common divisor, least common multiple, prime numbers, the fundamental theorem of arithmetic), fractions, potencies, radicals; proportions, the decimal metric system, the simple and composed rule of three, arithmetical and geometrical progressions, logarithms, and some elements of financial mathematics. In the second year algebra was studied: powers, radicals, algebraic expressions, integers, polynomials (operations, factorization, maximal common divisor, minimal common multiple of polynomials), algebraic fractions (operations, simplification), first and second grade equations, first grade equation systems with two or more unknowns, elements of combinatorial analysis, inequalities, logarithms, basic notions of determinants, and continuous fractions. The program for the third year included a study of geometry from a more rigorous point of view. In the fourth year, trigonometry was studied: trigonometric ratios for acute angles, basic formulas, tables, resolution of triangles, some application problems and element of spherical trigonometry. Finally, the program for the fifth year was totally related to the study of projective geometry (Barrantes & Ruiz, 1995c, pp. 71–72).

use of many formulas were studied. In general, the first years were about recognizing lines and figures and the next years, about performing a few elementary calculations. Moreover, the relation of geometry to the remaining disciplines and professions was proposed (use of the plumb, the watermark and drawing triangle).

Some Foreigners in Costa Rican Science and Mathematics

Europeans, Mexicans, Central Americans, Colombians, and Cubans had an influence on cultural life in Costa Rica. Chile had an influence on pedagogy and Argentina on legislation. Costa Rica was of interest to naturalist and geological studies, especially because of its extraordinary biodiversity. Nevertheless, most of the time, these scientists and researchers did not get involved in national life nor played any important role in the institutionalization of the sciences, education, or mathematics.

According to González-Flores (1921), after the Spaniards, the most important influence on Costa Rican legislation, culture and education was the French (p. 251). Books, texts, the ideas of the French Revolution or later ideas by thinkers like Comte, were the main means through which French influence was felt. The national flag of Costa Rica was designed having a direct French influence. However, the presence of French teachers or scholars who stayed or remained in the country was very low. The United States also exerted another important influence through its Constitution, which was key to this country's own constitution. Moreover, the school organization model and even some pedagogical ideas were introduced due to the United States' political and economic hegemony in the region. During the nineteenth century, a very large collection of US explorers, biologists, botanists, and geologists visited Costa Rica and wrote many papers and books about this country; however, almost none stayed. Something different occurred with some particular European nationalities: German, Swiss and Italian teachers or scientists taught Science and Mathematics in Costa Rica; some of them had a direct deeper impact in national culture (Ruiz, 2000, pp. 661–688). Two conditions attracted immigrants or visitors to Costa Rica during the nineteenth century: a relatively peaceful political atmosphere compared with the convulsed and unstable ones in countries of the region, and a certain good economical development due to the coffee-based economy and, later, during the 1880s and 1890s, a proactive policy of the liberal reformers to attract European scholars or teachers.

Many German scholars and engineers visited or lived in Costa Rica or wrote about this country. One particular wave of Germans came right after the German revolution of 1848–1849. Some were associated to the coffee production or commerce, to the national public administration or infrastructure development (Herrera, 1988). Others came as a consequence of the interest in its natural history or because of Costa Rica's geological and seismological characteristics. González-Flores (1921) mentions some names: the naturalist Alexander von Frantzius (from Danzig, since 1854 stayed 15 years in Costa Rica); the physician and botanist Carl Hoffman (Stettin, Prussia, arrived in 1854 and died in Costa Rica in 1859); the naturalist Wilhelm Nanne (born in Hannover in 1828); Franz Rohrmoser (member of a huge family from Prussia who came to Costa Rica; he developed some meteorological studies); the engineers Franz Kurtze (Hamburg) and Ludwig von Chamier (Prussia); Friederick Maison (from Bavaria, who taught mathematics in the USA and came to Costa Rica in 1862. He made meteorological studies and was the director of the Office for National Statistics until his death in 1881); Hellmuth Polakowsky (born in Berlin in 1847) who was hired by the *Instituto Nacional* in 1874 and developed important biological and botanical studies (he was very much influenced by Alexander von Humboldt); and the chemist Carl Beutel (born in Baden in 1869) who was one of the founders of the School of Pharmacy and lived in the country until his death in 1913. Moreover, as Herrera (1988) points out, two names are connected to the selling of books until today: Antonio Lehmann (from Bonn) and Carl Federspiel; Lehmann founded a relevant printing business (p. 109). One of the most influential persons was the priest Bernhard August Thiel (born in Elberfeld in 1850, a Jesuit who lived in Costa Rica in the years 1877–1884 and 1886–1901 and was appointed Bishop in Costa Rica). Some of these surnames are very familiar to Costa Ricans.

Within the liberal reform, the *Colegio Superior de Señoritas* was organized by the German teachers Laura and Elisabeth Hinrrichs, Francisca Schardinger and Ana Ferrier (González-Flores, 1921, p. 66).

In connection to mathematics, between 1889 and 1894, Otto Littmann lived in Costa Rica. He held a doctoral degree in Philosophy from the University of Bratislava, and engaged as a teacher for the *Liceo de Costa Rica*. During his stay in Costa Rica he wrote two textbooks, one of them for teaching arithmetic that was inspired by the methods of the groups of Grubbe. González-Flores (1921) affirms that he had much influence on pedagogy (p. 67).

Although, he did not live in Costa Rica, August Tafelmacher (University of Göttingen) exerted influence on Costa Rican mathematics through students to whom he taught at the *Instituto Pedagógico* in Santiago, Chile (organized by German professors in 1888). Some of his textbooks were used in Costa Rica (González-Flores, 1921, p. 74).

What is really interesting is that, because of the influence of the liberal reform, in 1886 several Swiss scientists arrived in Costa Rica and exerted a very important influence on the country's education (González-Flores, 1921; Ruiz, 2000). Among them: Paul Biolley arrived in 1886, Henri Pittier and Johann Sulliger in 1887, Gustav Michaud, Jean Rudin and Adolphe Tonduz in 1889.

Sullinger was a student and engineer of the Polytechnic Academy of Zurich (however, he stayed in Costa Rica just one year); he taught Mathematics at the *Liceo de Costa Rica*.

Biolley (born in 1862, Neuchatel) had a degree in literature and was professor in The Netherlands and candidate to be lecturer at the University of Bonn. He was invited by Mauro Fernández, and worked at the *Colegio Superior de Señoritas* at the *Liceo de Costa Rica* and *Colegio San Luis Gonzaga* as a science teacher. González-Flores (1921) affirms that he helped to develop the national sciences through his research on topics such as mollusks of the Coco Island and invertebrate animals of Costa Rica. He also published *Elements of Botany* and *Elements of Greek Grammar*. He established himself in Costa Rica until his death in 1908.

Pittier (born in Bex in 1857) had been a teacher of physical geography at the Academy of Lausanne. He was one of the most important naturalists in the New World. In Costa Rica, he was a teacher at the *Liceo de Costa Rica* and the *Colegio Superior de Señoritas*, and was Director of the *Instituto Físico-Geográfico* for 15 years. He made very important scientific contributions through his research, much of it on the national history of Costa Rica and also on meteorology. The *Instituto Físico-Geográfico* became what is today the *Instituto Geográfico Nacional*. At this *Instituto*, Pittier prepared the first official map of Costa Rica. He also gave the first detailed description of river basins and mountain formations in Costa Rica and the first serious and continuous weather measurements. In the field of botany, Pittier published *Primitiae Flora Costaricensis* in 1895 (*Anales del Instituto Geográfico-Nacional*) and also *Ensayo sobre plantas usuales de Costa Rica* in 1908 (Washington D.C., USA). In zoology he did the first inventory of Costa Rican mollusks (with Biolley). Pittier organized many expeditions through all the national territory including Cocos Island. He built the National Herbarium, with more than 5,000 different species and more than 18,000 collections of plants, some of which bear his name in the international scientific notation (Valerio-Zamora, 2000; Conejo, 1975). In 1904, the United States Secretary of Agriculture hired Pittier (for 14 years), and in 1919 he established his residence in Venezuela until his death in 1950 (Rosero, 2011, pp. 2–3).

Due to Pittier's influence, the Swiss naturalist and botanist Adolphe Tonduz (born in Pully, 1862) came to Costa Rica and stayed in this land from 1889 to 1921 (Ossembach-Sauter, 2008, p. 13).

This "golden age" of naturalists in Costa Rica included the French Carl Wercklé (born in Alsace in 1860), who made an important contribution to Costa Rican botany. His main work was *La subregión Fitogeográfica Costarricense* (1909) with a wide description of the flora of different regions of Costa Rica. He died in San José in 1924 (Gómez, 1978). British naturalist Charles Lankester (born in Southampton, 1879) should also be mentioned. He worked with Pittier, Tonduz, Wercklé and some Costa Rican naturalists, and died in Costa Rica in 1969. Nowadays, in the lands he bought there is a famous botanical garden run by the *Universidad de of Costa Rica* that bears his name (Urbina-Vargas, 2005).

González-Flores (1921) mentions that Michaud was born in 1860 in Geneva. He obtained a Bachelor's degree in Physical and Natural Sciences and a Doctorate in Physical Sciences from the University of Geneva; he was known in Switzerland for having written two scientific works for publication: *Chémie populaire* y *Terre, l'eau, l'air et la feu*. In 1886 he became a doctor in Physical Sciences at the University of Geneva. He was a teacher in his home country and in the United States. In Costa Rica he worked at several secondary schools and other institutions. It is interesting to note that Michaud had been engaged in establishing and directing a "Physical-Mathematical School". When, for political reasons, Mauro Fernández left the Secretary of Public Instruction in the hands of Ricardo Jiménez, the contract of Michaud was modified; he then taught Science at the *Liceo de Costa Rica*. Between 1895 and 1905 he was a professor at the American International College in Springfield, Massachusetts. In 1905 he returned to Costa Rica, but carried on being a frequent collaborator of several journals in the United States, Switzerland, and France (González-Flores, 1921, pp. 298–299).

Rudin (born in 1849, in Muttenz, near Basel), also engaged by Mauro Fernández, was dedicated to studies in mathematics, physics, geology, and astronomy. He had worked in Hungary and in his home country, Switzerland, he collaborated with Hermann Kinkelin in statistics; many of his principal activities in Switzerland were related to astronomy. In Costa

Rica, he worked at the *Colegio Superior de Señoritas* (1895–1904), at the *Liceo de Costa Rica* and at the *Colegio San Luis Gonzaga*, being the latter's director when he arrived in Costa Rica. He also collaborated at the *Instituto Físico-Geográfico* (1893–1897). His pedagogical influence on Costa Rican primary schools was remarkable (González-Flores, 1921, pp. 299–301; Rodríguez-Vega, 1988; Ruiz & Barrantes, 2000).

Although the Italian scientists or engineers who came to this country were not many, there is one that deserves special attention: Rodolfo Bertoglio. According to González-Flores (1921), Bertoglio was born in Milano in 1844 and has a Bachelor in Physics and Mathematics from the *Liceo St. Alesandro*. He had attended the University of Naples, where he obtained his degree, and the *Politecnico di Milano* (Polytechnic University of Milan) where he became an engineer. Bertoglio was one of the disciples of the eminent and wise engineer Ignacio Porro. He arrived in Costa Rica in 1875. He was a teacher at the *Instituto Nacional* as well as its Director, Director and Founder of the Engineers' School, and a member of the Council for Public Instruction. Among the works realized in Costa Rica, there are his topographical studies about Santa Clara, the plan for the *Asilo Chapuí*, and the building of the sewage system in San José. He left important unpublished studies about properties of numbers, logarithms, several topics of geometry and the notable *Espiral of Bertoglio*, an ingenious instrument that allows solving complicated problems by the use of the compass. He exerted a great influence on the mathematics of this country. Mauro Fernández said: "The assiduous representative and propagator in our country of the modern instruction methods for mathematics from 1879 until 1886 was the engineer Rodolfo Bertoglio. It can be affirmed that he founded a school" (González-Flores, 1921, p. 265). In January of 1886 Bertoglio left for Nicaragua and a year later, on February 13, 1887, died in León, when Costa Rica was about to make use of his services again.

From the *Escuela Normal de Costa Rica* to the *Universidad de Costa Rica*

In November 1914 the *Escuela Normal de Costa Rica* was founded, with the principal aim of solving serious problems of teacher preparation. This action was the second big moment after the Reform by Mauro Fernández, which had confined this preparation to "Normal Sections" in two high schools. The creation of the *Escuela Normal* started a new stage in national education. According to Barrantes and Ruiz (1995a) it had three sections: General Studies (3 years of secondary school), Normal (3 years of the proper Normal and professional preparation) and Application (which were the 5 years for primary school where future teachers carried out their "practice").

Two decades later (1936) the internal structure of the *Escuela Normal* was changed. A Section of Pedagogy would prepare the teachers for primary and secondary school. At the end of the second course, this section would give the diploma of a Normal Primary School Teacher, and at the end of the fourth that of a Normal Secondary School Teacher. When the *Universidad de Costa Rica* was created, years later, the Section of Pedagogy was turned into the Department of Pedagogy of the new university (Barrantes & Ruiz, 1995a; *Escuela Normal de Costa Rica*, 1940).

Between 1915 and 1940 the *Escuela Normal* also functioned as a secondary school. Until 1940 there were no new official secondary schools, although for some years there were a few private schools like the *Seminario*, *La Esperanza*, *La Escuela Nueva* and *Evans*. It was in these schools where the major part of the little mathematical knowledge available at that moment in this country was taught. Until 1940 secondary schools were not nationwide, since these few existing schools were concentrated in the Central Valley (a high plateau where most of the Costa Ricans lived).

The *Escuela Normal* was influenced by the pedagogical ideas of John Dewey, which differed from the pedagogical vision of Herbart, Ziller, Rein, Dittes, Dilthey, and established the so-called "pedagogical socialism". The Pragmatism, which is the base of Dewey's ideas, had important consequences on the curriculum and methods of Costa Rican Education (*Escuela Normal de Costa Rica*, 1940).

An interesting detail pointed out by Barrantes and Ruiz (1995a): between 1929 and 1940, around the *Liceo de Costa Rica* existed a *Centro de Estudios Newton* (*Newton Study Circle*) which was used as a propaedeutic for young people who afterwards would be professionals of importance in the country. Inspired by the Swiss teacher Charles Borel, this center, motivated the study of science and technology by high school students. Many of the students who belonged to that center afterwards became professors at the *Universidad de Costa Rica*.

The creation of the *Universidad de Costa Rica* in 1940 opened a new stage within the evolution of Costa Rican education and culture and, particularly, a new historical context for the development of mathematics and its teaching in this country.

Mathematics at the *Universidad de Costa Rica*

At the *Universidad de Costa Rica* (UCR) between 1940 and 1972, two stages in the history of mathematics can be considered: (1) from 1940 to 1957, and (2) from 1957 to 1972.

Within the professional faculties

From 1940 to 1957 mathematics was taught by departments and faculties of the *Universidad de Costa Rica* where it was needed: Engineering, Science, and Social and Economic Sciences. The new institution was essentially a federation of professional faculties and not a structured unit with homogeneous academic standards; mathematics besides its little development was subordinate to the existing professional careers. Scientific standards and quality relied on the needs of the programs these faculties offered: Engineering had the highest level, the Faculty of Sciences the lowest, Economics and Social Sciences was in between but had teachers with good preparation such as José Joaquín Trejos Fernández and Bernardo Alfaro Sagot. Alfaro Sagot's initial preparation was in Pharmacy (Solano & Ruiz, 1995c).

Solano and Ruiz (1995a) inform that in those years there existed among the Engineering (and other disciplines) students and faculty an interest in developing higher mathematics. A group was formed by the engineer Luis González, who was the first Costa Rican who generated original mathematical results (Herrera, 1993 and 1995c). To that group belonged José Joaquín Trejos Fernández (who had studied at the University of Chicago, where he received the support of Marshall Stone) and Rodolfo Herrera (who was later repeatedly Dean of the Faculty of Engineering).

The Department of Physics and Mathematics

The creation of the *Departamento de Física y Matemática* (Department of Physics and Mathematics) in this university and its subsequent development occurred in the broader context of a deep reform commanded by its Rector, Rodrigo Facio, in 1957. This reform generated the institutional framework and the university model for almost all-public higher education in Costa Rica. This process with the aim of giving the institution unity and integration, adopted two basic objectives: to invigorate the general humanistic student preparation, and to centralize a wide variety of the university chairs.

Although a University Assembly held in 1955 had approved its existence, the Department of Physics and Mathematics started on March 1, 1957. From that moment, almost all mathematics courses that were taught in different faculties joined the Department (Solano & Ruiz, 1995b).

Solano and Ruiz (1995b) mention that the first Director of the Department was José Joaquín Trejos Fernández (who was president of Costa Rica in 1966–1970), who was the Dean of the *Facultad de Ciencias y Letras*. Due to the many functions he had at the time, Trejos Fernández asked Bernardo Alfaro Sagot to organize the Department. Alfaro Sagot took office with the informal title of "Coordinator". However, afterwards, Trejos left the position of Director to Alfaro Sagot, who held it for two consecutive terms.[2]

The development of mathematics in an independent Department was essential for defining the particular perspectives of this discipline. Having mathematics instructors in each of the faculties would have prevented their own scientific progress as well as the quality of mathematics received in the same faculties. A bachelor's degree in physics and mathematics in the new Department began to be created in 1957. Also offered was a *"Profesorado"* (a first degree for Teaching) which included courses offered by the Faculty of Education.

Beginning in 1967, the Department offered the titles of Teacher of Physics, Bachelor's degree in Physics, *Licenciado* in Physics, Teacher of Mathematics, Bachelor's degree in Mathematics.

Since then most of the universities of Costa Rica have offered one, two or three of these study programs:

- The University *Bachillerato*: four years of preparation for the teaching of a subject. It issues a Bachelor's degree.
- The *Licenciatura* is a program of studies of around 5 years with a thesis. It issues a *Licenciado* degree.
- The *Profesorado* is a pre-service program of around three years of study for teaching at the secondary school level. It issues a secondary school teacher's degree.

---

[2]It is not known why it was decided to bring together mathematics and physics in a department. Some people claim that it was something incidental, others that it was due to the influence of Alfaro who had fostered a relationship between physics and mathematics from his years as a professor at the *Liceo de Costa Rica* (institution created during the reformation of Mauro Fernández in the 19th century). However, this decision defined the mathematics of the UCR for years.

Olgierd Alf Biberstein, a Polish mathematician, exerted some influence during this period. He had studied in France (with the Nicolás Bourbaki group of mathematicians) and Canada, taught in the United States and Canada, and was in the UCR department from 1959 to 1962 (Herrera, 1995b, pp. 315–331). According to Morales-Luna (2013), after Costa Rica he was in Chile for seven years and finally in México (since 1970) in the *Escuela Superior de Física y Matemáticas* of *Instituto Politécnico Nacional* (National Polytechnic Institute). This mathematician would help define some characteristics of the curricula that would develop the Department.

Though more brief, a visit to Costa Rica by Dirk Struik in 1965 exerted influence on some teachers and students (Herrera, 1995a, pp. 333–345).

The Department and School of Mathematics: a new stage in the mathematics of Costa Rica

In April 1971 the Department of Physics and Mathematics approved its division into two departments: Physics (its first director was Neville Clark) and Mathematics (its first director was Francisco Ramírez). Both were installed in 1972. The name of the Department was changed to School due to global decisions of the *Tercer Congreso Universitario* (Third University Congress) at the *Universidad de Costa Rica*, which had started in 1972. With the name of School of Mathematics this academic unit was installed in March 1974 and offered the programs of Teacher in Mathematics, Bachelor's and *Licenciado*'s degree in Mathematics. According to Ruiz, Barrantes and Campos (2010), in 1974 Bachelor's and *Licenciado*'s degrees for high school teaching of mathematics were also created and in that same year the degrees of Bachelor and *Licenciado* in Computer Science. Under the School of Mathematics, just a few courses of the Computer Science Bachelor's degree differed from the Mathematics degree. Computer Sciences ceased to belong to the School of Mathematics in 1981 with the creation of an independent Informatics and Computing School within the Engineering Faculty.

Ruiz, Barrantes and Campos (2010) inform that between 1972 and 1993 three *Licenciados* graduated in physics and mathematics and 61 in mathematics. Between 1978 and 1981, 25 graduated in Mathematics, a historic "peak". This group of people formed the basis of the School of Mathematics until the beginning of the 21st century. During the 1970s and 1980s, the UCR sent students to foreign universities to get their Master's and Ph.D. degrees (mainly France and United States, but also Canada, Germany,

Romania and the former Soviet Union). This provided that institution with qualified human resources for three decades.

From the end of the 20th century to the first decade of the 21st century, generations of mathematicians who entered in the 1970s and 1980s began to qualify for retirement. While the university continued with the strategy of sending students abroad to study, the number of mathematicians who returned did not correspond to the large number of professors who have retired. This created a deficit of mathematicians in the country, affecting all the institutions because UCR was the only institution that generated them.

In 1980 UCR created a Master's program in mathematics, initially oriented to pure mathematics. In the beginning this graduate program sought to enhance the preparation of professors of the School of Mathematics who did not go abroad to study. Alfaro *et al.* (2013) point out that the program had few graduates, and because of this during a period it was closed by the *Consejo Nacional de Rectores* (Council of Rectors of Public Universities). During the first decade of the 21st century it diversified its offering by creating an emphasis on applied mathematics and in mathematics education (named Educational Mathematics), but with few students and graduates.

At the beginning of the 21st century this school created the Bachelor's and *Licenciado* degrees in actuarial science.

The creation of the Department and then School of Mathematics was the result of academic progress that occurred within the old Department of Physics and Mathematics. Before 1957 mathematics was subordinated to various faculties. Between 1957 and 1971 they became independent, but essentially for training teachers for secondary education. In the 1970s, with the motivation of ideas and teachers of the previous decade, UCR began to build a space for mathematics not intended essentially for the training of teachers but dedicated to higher mathematics.

From 1967 until now, UCR has exclusively assumed (except for a few years) all preparation in pure and applied mathematics in Costa Rica.

Initial Training of Teachers for Pre-university Education

The postwar population boom caused one demand on universities in the seventies of the last century. Already in 1971 the *Instituto Tecnológico de Costa Rica* (ITCR, Technological Institute of Costa Rica) was approved (although opened in 1973). Shortly afterwards, in 1973, the *Universidad Nacional* (UNA, National University) was founded that integrated the *Escuela Normal de Costa Rica*, *Escuela Normal Superior* (which had been created a few years earlier to meet a shortage of high school teachers), and regional Normal Schools in Guanacaste, San Ramón, and Pérez Zeledón. A new School of Mathematics was created in 1974. In 1975 the *Universidad Autónoma de Centro América* (Autonomous University of Central America) was created: the first private university in the country. In 1979, the *Universidad Estatal a Distancia* (UNED, State Open University) was founded, following more or less the model of the British Open University.

In 1974, the School of Mathematics of UNA created Bachelor's and *Licenciado*'s degrees in the Teaching of Mathematics. Although its purpose was also to produce teachers and professors for the early years of higher education, the vast majority was incorporated into secondary institutions. That same year Bachelor's and *Licenciado* degrees in Pure Mathematics were created, and Bachelor's and *Licenciado* degrees in Applied Mathematics followed suit in 1980 (Adolio, González & González, 1995, pp. 219–230). These last two degrees did not last very long; however, the first contributed to the preparation of professors who would form that school, a generation of academicians that would last until the beginning of the 21st century, when most of them retired. Although it cannot be denied that there was academic and "ideological" influence of UCR in UNA, they also inherited approaches from the Normal Schools that gave rise to that University.

Ruiz, Barrantes and Campos (2010) consider that the Bachelor's and *Licenciado* degrees in the Teaching of Mathematics were approved in UCR in 1973 and were installed in 1974 *to replace the Profesorado*. Its goal was the same as the *Profesorado* program, but with greater development on the mathematical aspect as in the pedagogical. Since its inception and until the mid-1970s the Faculty of Education did not have much influence on these programs but between 1976 and 1989 diverse actions taken by the University gave to this Faculty formal administrative control, something that had provoked for a long time institutional frictions with the School of Mathematics. In 2013 this tense situation generated the proposal for a new program in the Teaching of Mathematics with direct administration by the School of Mathematics, that so far has not been accepted by this university.

In 1980 UCR closed its *Profesorado* program. This had a negative impact because it contributed to a shortage in teachers of mathematics in the country (Ruiz, 2008). The national response was the creation of a Teaching Program of Mathematics now offered by several institutions: UCR, UNA and UNED, with the support of the Council of Rectors of Public Universities and the support of a loan from the World Bank. Its curriculum was quite different from the one that had existed in UCR since 1967. This was the opportunity that allowed UNED to start a program in this area. In 1999, the UNED created a Bachelor's degree and in 2014 this university will offer the *Licenciado* degree. There is a section in the *Escuela de Ciencias Exactas y Naturales* (School of Exact and Natural Sciences) that manages the Mathematics Program (Ruiz, Barrantes & Gamboa, 2009).

The Department (later called School) of Mathematics of the ITCR, born in 1980, created in 1996 the Bachelor's degree in the Teaching of Mathematics Assisted by Computer. This new program added to the Pedagogy and Mathematics a component in Computing (Astorga, Meneses & Rodríguez, 1995, p. 241). In 2007 it began offering the *Licenciado* degree, but never had the *Profesorado*.

In the 1990s the offer of pre-service programs was expanded with the rise of dozens of private universities (45 private universities were created between 1986 and 2000). In different ways, several private universities offered this program, however in 2013 apart from the public institutions, the main private universities that graduated students were *Universidad Americana* (UAM), *Universidad de San José* (USJ) and *Universidad San Francisco Labrador* (UISIL).

In 2003 UCR put an emphasis on "*Matemática educativa*" ("Educational Mathematics", a term coined by the Mexican mathematician Eugenio Filloy) in seeking to revitalize their Master's program, but had problems with the curriculum design: it never went from being a curriculum of mathematics of significantly lower level than the one offered in the preparation for undergraduate students in pure mathematics with scarce elements of mathematics education. (Alfaro, et al., 2013). This program has been useful only to persons who teach mathematics in universities and with no solid preparation in pure mathematics. The program does not prepare students for research in mathematics or mathematics education. The country lacks a true Master's program in mathematics education that could impact classroom practice.

Ruiz, Barrantes and Gamboa (2009) provide some comparative elements: between 1997 and 2006 the public universities graduated a total of 635 secondary school teachers (251 from UNA, 182 from UCR, 113 from UNED, and 89 from ITCR). In the same period, the private universities graduated 409 bachelors: UAM 175, *Universidad Católica* 31, *Universidad Latina* 86, *Universidad Adventista* 13, *Universidad de San José* 70, and *Universidad Central* 34. The total number of high school graduates from public and private universities was 1044 (Ruiz, Barrantes & Gamboa, 2009, pp. 159–162).

Up until 2006 the public universities still graduated the majority of secondary school teachers in different programs (*Profesorado, Bachillerato, Licenciatura*): from that year the trend changed in favor of the private universities, and it is expected that in the following years the vast majority of teachers in middle and high schools will come from these private institutions. This is a phenomenon that occurs in all professions.

The initial teacher training offered by all of these universities has had serious weaknesses: a drastic gap between mathematics courses and those of pedagogy, a significant absence of specific pedagogy of mathematics, and mathematics courses not suitable for the mathematics teacher (in general, with a perspective of the pure mathematician but providing less content and depth). These programs are essentially a juxtaposition of the main thematic components (mathematics, pedagogy). In general, they have not incorporated important findings of the research carried out internationally in mathematics education, as in problem solving, modeling, as well as curriculum, assessment, class management and specific didactics of mathematics, or the appropriate role of history or technologies in the classroom (Ruiz, 2008; Ruiz, Barrantes & Gamboa, 2009, pp. 227–234). In addition, these programs have been dissociated from the official national curriculum and real classroom practice (Chaves, 2007).

The reasons why these programs exhibit these deficiencies are multiple. One reason is international: research in mathematics education as an independent science of mathematics and of general pedagogy (and as a separate science-based profession) is relatively recent in the world, around 60 years. Similarly, although Shulman (1986) raised the need to incorporate the pedagogical knowledge of the content in the preparation of teachers, this has not been emphasized not only in Costa Rica but also in other parts of the world.

Another reason has been that until recently the negative attitude and ideology of the Costa Rican mathematicians formed during the 1970s and 1980s: they did not understand nor promote math education, and often even denigrated it. In UCR that resulted in just a few students pursuing graduate studies abroad in mathematics education and in the reluctance of university mathematicians to incorporate mathematics education specialists as faculty (Ruiz, Barrantes & Campos, 2010, pp. 113–114).

Although, as we shall see, in UCR relevant research has been developed in mathematics education that is recognized internationally, the institution has not integrated the findings into the programs for pre-service teacher training. A true contradiction! As Ruiz, Barrantes and Campos (2010) affirm, this was essentially due to a social, professional, and political division that developed at that School in the 1980s, and that only in recent years has been weakened (it essentially ended due to the retirement of dozens of mathematicians who formed that School in the previous decades and the emergence of a set of new young mathematicians).

The model of UCR has influenced other institutions that offer pre-service training for teaching in secondary school. ITCR had an opportunity to renew this model with the addition of technology. However, as Alfaro, *et al.* (2013) consider, it failed to design a different program: the role of the computer was not according to appropriate pedagogy of mathematics (this program only included some programming courses and software) and added rather more levels of separation between its curricular components.

In what refers to private universities: they have copied the model of the public universities, with the aggravating circumstance of offering Bachelor's and *Licenciado* degrees with a lower level of quality. A Bachelor's degree from a public university takes four years of study, in a private institution 2 years and 2 quarters. In the public ones a *Licenciado* degree implies between 5.5 and 6 years of study, in the private less than 4 years. In the past years, the number of *"express"* graduates has increased significantly, further weakening the quality and the role of the pre-service teacher preparation of the country, and at the same time the professional status of teachers. This is a phenomenon that affects not only mathematics.

In terms of pre-service teacher training for the primary school, UCR, UNED, and UNA offer degrees to teach at these levels, as do 18 private universities. Most programs do not offer more than two mathematics courses, usually of poor quality. This affects the standard of mathematics classes in primary schools. *Universidad de San José* (www.universidadsanjosecr.com) is the only one that — at least on paper — offers a specialty in primary education with emphasis in mathematics (8 courses related to mathematics). UAM offered a Master's degree in Teaching Mathematics for Primary School, but it has had very few students and its impact has been minimal.

Professional Development of Teachers of Mathematics

This is one of the main weaknesses of the Costa Rican education system. There is no provision in the working hours for in-service teacher training or to conduct research on the lesson that might advance classroom practice, nor are there permanent integrated national programs for this work. That means that isolated training has dominated, given by the public universities, the *Colegio de Licenciados y Profesores en Letras, Filosofía, Ciencias and Artes* (an obligatory enrollment association for secondary school teachers), or by the Ministry of Public Education.

Participation in symposia, conferences or seminars on Mathematics Education (organized by different entities) has been an important platform for professional development (mostly for secondary education teachers) but usually funded by the teachers themselves. The professional career of teachers employed in the Ministry of Education recognizes these courses or events to improve teacher compensation. In mathematics in Costa Rica, this participation has been developed since the 1980s. One of the first events was the *Congresos Nacionales de Matemáticas* (National Congress of Mathematics); the first one was held in 1983, and the last two editions in 1985 and 1990. Although four public universities participated in its organization, UCR was the strongest partner. According to Alfaro, et al. (2013), the longest tradition in math education events has been the *Simposio Costarricense sobre Matemáticas, Ciencias y Sociedad* (Symposium on Mathematics, Science and Society) with 25 editions between 1991 and 2012. It was always organized by the UCR's *Programa de Investigaciones Metamatemáticas* (Program of Meta-Mathematics Research), although between 2001 and 2008 in close collaboration with the School of Mathematics of UNA. As de Faria (2003) informs, Ángel Ruiz chaired the scientific and organizing committees of most of these 28 events (pp. 323–328).

Another tradition has been the *Congresos Internacionales de Enseñanza de la Matemática asistida por computadora* (International Congress on Teaching Mathematics Computer-Assisted), six editions from 1999 to 2011, which is organized by the School of Mathematics at ITCR. Also *Festivales internacionales de Matemáticas* (International Festivals of Mathematics) organized by the *Fundación para el Centro Nacional de la Ciencia y la Tecnología* (Foundation for the National Centre of Science and Technology, CIENTEC), eight events between 1998 and 2012 in cooperation with some universities (de Faria, 2003, pp. 328–333). Also *Encuentros de Enseñanza de la Matemática* (Meeting on the Teaching of Mathematics) by UNED (four from 2006), and more recently the *Encuentro sobre Didáctica de la*

Estadística, la Probabilidad y el Análisis de Datos (Meeting on the Teaching of Statistics, Probability and Data Analysis, EDEPA) by the School of Mathematics of ITCR (2009, 2011).

Though not in math education, the *Simposios sobre Matemáticas Aplicadas a la Ciencia* (Symposiums on Applied Mathematics to Science) organized by the School of Mathematics of UCR, have had a very important continuity: 18 events between 1978 and 2012.

This significant number of events should be highlighted; although their quality is uneven, they have been until recently the main point for the gathering and in-service preparation of the mathematics education community of Costa Rica.

In the local teachers' professional development the existence of an *Asociación de Matemática Educativa de Costa Rica* (Association of Educational Mathematics of Costa Rica, ASOMED, www.cimm.ucr.ac.cr/asomed) founded in 1998 should be mentioned; it has promoted conferences, round tables, workshops and forums. It is, however, very small. Its active membership does not exceed 30 persons (a great percentage of them retired professors). Their influence is very small. Another professional group is the *Sociedad Costarricense de Matemática Pura y Aplicada* (Costa Rican Society of Pure and Applied Mathematics), created in 2011; it is associated with the *Unión Matemática de América Latina y el Caribe* (Latin American and Caribbean Mathematical Union) which has less activity than ASOMED.

Math School Curriculum

Since the late XIX century the contents of the school mathematics curriculum in primary and secondary education had not changed much until 1964 (Barrantes & Ruiz, 1995d, 1995e). The mathematics curriculum approved at the end of 1963 by the *Consejo Superior de Educación* (Higher Council of Education) and that began to be implemented in 1964 was influenced by the famous Reform of Modern Mathematics (New Math), that occurred in several parts of the world beginning in the 1950s (Ruiz, 1990; Barrantes & Ruiz, 1995e, 1998). Besides the contents, the official commission at that time in charge of writing the curriculum produced a fairly extensive document that consisted of two parts entitled: "A renewed high school math program" and "Characteristics for a good program of mathematics for high school and correlative methodological considerations", which showed the "New Math" approach for the reform (Barrantes & Ruiz, 1995e, pp. 379–392).

To prepare teachers for the secondary schools the Department of Physics and Mathematics of the *Universidad de Costa Rica* gave some courses during the summers of 1964 and 1965 that offered information on teaching the new topics. In that period a studies program was launched that would culminate with a degree of Teacher of Mathematics in 1967. The aim was to train professionals who would later teach this subject in the Secondary Education fully imbued with the ideas of modern mathematics. This is important because it means that the initial training programs for teachers of mathematics were, since their inception, influenced by the New Math.

Between 1964 and 1995 no big changes were made in the country's mathematics programs (Ruiz, 2013, pp. 16–19). In 1995 new programs were designed that would experience some minor modifications in 2001 and 2005. A constructivist intention was declared in the theoretical foundations, a certain amount of contextualization for the contents was proposed, as well as the aim to include intuitive and empirical heuristic strategies when introducing concepts and mathematical methods. These programs played a positive role, representing a move away from behaviorist approaches. As pointed out in MEP (2012, pp. 484–485), these programs failed to accomplish most of their purposes (posed abstractly) and they had serious weaknesses. They exhibited a strong inconsistency between the issues raised and stated in the theoretical foundations (for instance, the constructivist abstract declaration) and the contents and approach present in the syllabus (which were introduced with a behaviorist approach).

The syllabus of this curriculum used "programmed objectives". To each objective, isolated procedures, methodology, and evaluation were assigned. This approach pushed a disconnected treatment of the objectives, and it distorted assessment because each objective had to have an assessment item that was usually disconnected from the others. With this approach lesson planning and classroom development based on problem solving were not favored. Such programs did not allow developing an integrative and constructive approach to deal with content and desired skills or abilities (MEP, 2012, pp. 484–485; Ruiz, 2013, pp. 16–19).

Mathematics Olympiads in Costa Rica

Mathematics Olympiads for Secondary Schools in Costa Rica (OLCOMA) were created in 1989 by an agreement between UNA, UCR, ITCR and

UNED, the Ministry of Public Education and the Ministry of Science and Technology (Buján & Jiménez, 2003, p. 313). The Costa Rican Mathematics Olympiads for Primary Schools (OMCEP) were the result of a project developed as a final requisite for *Licenciado* degree in the Faculty of Education at Universidad de Costa Rica circa 1992 and 1993. Since 1995 *Agrupación de Consultores para la Atención y Promoción del Talento* (Association of Consultants for Talent Care and Promotion, ACAPTA, www.acapta.org) has organized this event (Buján & Jiménez, 2003, p. 314).

In 1996, ACAPTA began to develop and apply the Mathematics Olympiads for school students aged between thirteen and fifteen years old (OMATEC). However, OMATEC and OLCOMA were integrated in 1999 (Adolio, González & Mora, 2003, p. 300).

OLCOMA works in two phases: National Olympiads and the Iberoamerican Olympiads of the Organization of Iberoamerican States. National Olympiads classifies students to participate at the Iberoamerican Olympiads. Costa Rica (host in 1996 and 2011) obtained bronze medals in 1990, 1991, 1994 through 2000, 2002, 2004, 2011 and 2012. It obtained silver medals in 1991, 2000 and 2011 and the gold in 2011 and 2012 (OEI, 2009, 2013; *Universidad Nacional*, 2013).

The Education System in Costa Rica and the *Colegios Científicos Costarricenses* (Costa Rican Scientific High Schools)

The structure of elementary and secondary school education in Costa Rica can be represented as shown in Table 1.

Preschool is also available, although the State provides it in a limited way.

At the end of the diversified education cycle, students must take national tests in Spanish, Math, Civics, either English or French, and a choice of Biology, Chemistry or Physics, which awards the high school *Bachillerato* (Baccalaureate diploma), which is a requirement to enter higher education. These tests were reintroduced in 1988 (they were suspended for several years); they are standardized tests constructed with multiple-choice questions.

The country has a centralized educational system administered by a Ministry of Public Education (MEP). However, there is a *Consejo Superior de Educación* (Higher Council of Education) that defines the general

Table 1. Structure of elementary and secondary school education in Costa Rica.

		Cycles	Ages and years in each cycle
General basic education	Primary education	Cycle I	7 to 9 years (1st, 2nd and 3rd)
		Cycle II	9 to 12 years (4th, 5th and 6th)
	Secondary education	Cycle III	12 to 15 years (7th, 8th and 9th)
Diversified education		Cycle IV	15 to 18 years (10th, 11th, 12th depending on the branch)
Diversified cycle has three branches: the *Academic* is two years (tenth and eleventh), the *Arts* is two years and the *Technique* is three years (tenth, eleventh and twelfth).			

Source: *Ministerio de Educación Pública de Costa Rica* (2013).

educational policy of the whole system, and in particular is responsible for approving changes to curricula.

The *Colegios Científicos Costarricenses* (CCC) were created in 1989, thanks in particular to the work of Víctor Buján under the administration of the minister of education Francisco Antonio Pacheco (1986–1990) (Meoño, 2008, p. 97; Buján & Jiménez, 1991; Buján, 1995). Its purpose was to allow selected students in grades 10 and 11 of secondary education (*Academic* branch within the diversified education cycle), to delve into the areas of mathematics, physics, chemistry, biology and computer science in addition to studying regular official programs.

These institutions generally operate within some public universities: *Universidad Nacional* (UNA), *Universidad de Costa Rica* (UCR), *Instituto Tecnológico de Costa Rica* (ITCR) and the *Universidad Estatal a Distancia* (UNED). Some of the courses the students take can be validated by universities.

In 1989 the first two CCC were opened: in the city of Cartago sponsored by ITCR and the other by UCR in San Pedro de Montes de Oca. In 1993 four more of these high school programs were opened in Liberia and San Ramón (sponsored by UCR), San Carlos (by ITCR) and Pérez Zeledón (by UNA). Three more were created later: Atlántico and Alajuela (by UNED), and Puntarenas (by UCR).

Each institution accepts a maximum of 30 new students per year, bringing the number of students that each CCC has in one year to between 40 and 50. Across the country the total does not exceed 500 students each year.

The selection criteria for students are:

- Be enrolled in the 9th year of the General Basic Education.
- Have an average of above eighty five (scale 0–100) in the seventh, eighth and in the first two quarters of the ninth year.
- Pass approved specific academic aptitude tests.
- Pass an interview that takes place in the company of the student's parents (Buján & Jiménez, 2003; CCCPZ, 2013, p. 1).

If needed, students receive socio-economic support. Also, to remain in the CCC, students must average grades of at least 70 (scale from 0 to 100).

Systematically, the CCC students get the best average scores of all public institutions in the country in the *Bachillerato* national tests, they participate and win awards in various academic olympiads and competitions (in mathematics, physics, chemistry, robotics), they enter public universities with good averages in their entrance exams, and a high percentage of them follow a career in engineering or health sciences (CCCPZ, 2013, p. 2).

Research in Mathematics Education

Research in almost all disciplines has been developed essentially in UCR, due to resources and the historical academic maturity of this institution. There has been no difference in the case of mathematics and mathematics education.

In 1997 two research centers were created within the UCR associated with mathematics: *Centro de Investigaciones Matemáticas y Metamatemáticas* (Center for Mathematics and Meta-Mathematics Research, CIMM, http://cimm.ucr.ac.cr) and *Centro de Matemáticas Puras y Aplicadas* (Center for Pure and Applied Mathematics, CIMPA, www.cimpa.ucr.ac.cr). The first was built based upon two research programs that had existed since 1990: *Programa de Investigaciones Metamatemáticas* (Program for Meta-Mathematics Research) and *Programa de Modelación Matemática en las Ciencias Físicas* (Program for Mathematical Modeling in the Physical Sciences).

Programa de Investigaciones Metamatemáticas has been the only program in Costa Rica where there has been research in history, philosophy, sociology and mathematics education ("Meta-mathematics" was reconceptualized to mean multiple studies about mathematics). Researchers who had done a lot of work during the 1980s especially on topics of history and philosophy of mathematics and science and their relation to education founded this program. Alfaro, *et al.* (2013) point out that this group participated in the *Asociación costarricense de Historia y Filosofía de la Ciencia* (Costa Rican Association of History and Philosophy of Science) which was active between 1983 and 1995 and organized five *Congresos Centroamericanos y de El Caribe de Historia de la Ciencia y la Tecnología* (Central and the Caribbean Congresses of History of Science and Technology) once every two years between 1985 and 1993. This group, led by Ángel Ruiz, was responsible for the translation into Spanish of *Disquisitiones Arithmeticae* by Gauss (Gauss, 1995); an online version can be seen at http://centroedumatematica.com/da. They also published *Historia del Comité Interamericano de Educación Matemática* (The History of the Inter-American Committee on Mathematics Education); both books were published by the *Academia Colombiana de Ciencias Exactas, Físicas y Naturales* (Colombian Academy of Exact, Physical and Natural Sciences), respectively in 1995 and 1998. They also produced *Matemáticas en Costa Rica* (Mathematics in Costa Rica) in 1990, and in 1995, *Historia de las Matemáticas en Costa Rica: Una introducción* (History of Mathematics in Costa Rica: An Introduction), by the publishing offices of UCR and UNA.

Between the years 2001 and 2009, taking advantage of a job association of Ángel Ruiz with the School of Mathematics of UNA, this group of researchers established a working partnership with some professors and talented students from that institution. According to Alfaro, *et al.* (2013), this allowed him and his colleagues to generate research in mathematics education, which up until then had been virtually non-existent; and more than that, this process made it possible to support the academic reconstruction of the School because at that time there was a threat that it would collapse with the imminent retirement of the majority of their professors.

In 2006, based on the results of the experience developed within UCR and UNA, a *Programa Interinstitucional de Investigación y Formación en Educación Matemática* (Inter Institutional Program for Research and Training in Mathematics Education) was created, which brought together researchers from these two institutions and also from UNED. In 2011 this program became the *Centro de Investigación y Formación en Educación Matemática* (Center for Research and Teacher Training in Mathematics Education, CIFEMAT, www.cifemat.com).

There have been other research activities in Costa Rica in math education: associated with the Faculty of Education at UCR, especially during the 1980s and 1990s under the *Instituto para el Mejoramiento de la Educación Costarricense* (Institute for Improvement of Costa Rican Education) that is now called *Instituto de Investigación en Educación* (Institute of Research in Education), at the *Centro de Investigación y Docencia en Educación* (Center for Research and Teaching in Education) of UNA, in the *Escuela de Matemática* at ITCR, at the *Escuela de Ciencias Exactas y Naturales* (School of Exact and Natural Sciences) of UNED, at the Schools of Mathematics at UCR and UNA independently of CIFEMAT, and finally at the Ministry of Public Education.

However, CIFEMAT has been the main core of research in mathematics education in the country and the Central American region: Edison de Faria, Hugo Barrantes and Ángel Ruiz (from UCR) and Edwin Chaves (from UNA) have published more than 150 articles and 30 books in this discipline. The influence of their work has transcended the country, a fact that is expressed in their positions in boards of international societies: de Faria as an executive member of the *Comité Latinoamericano de Matemática Educativa* (Latin American Committee for Educational Mathematics, www.clame.org.mx), Barrantes as executive member of the *Comité Interamericano de Educación Matemática* (CIAEM), and Ruiz as two-time President of the CIAEM and also two-time Vice-President of the International Commission on Mathematical Instruction, ICMI, www.mathunion.org/icmi/home. Various themes of mathematics education within the international community have been addressed by this group: Problem Solving, Modeling, Teaching of Geometry and Algebra, Teaching of Statistics and Probability, Curriculum, Beliefs about Mathematics and its Teaching, Mathematical Competencies and Abilities, History of Mathematics and its Teaching, Use of Technologies, Philosophy of Mathematics and its Teaching.

Since 2006, this group has nurtured the *Cuadernos de Investigación y Formación en Educación Matemática* (Journal of Research and Training in Mathematics Education, http://revistas.ucr.ac.cr/index.php/cifem); this periodical published by UCR is the leading journal of research in this discipline in Central America and has a great international impact. The Journal in 2011 established a close partnership with the CIAEM. It has included articles derived from the conferences this Committee organizes since 1961 (for example, numbers 7, 8, 9 and 10).

Nowadays, *Cuadernos* also has a close connection to the *Red de Educación Matemática de América Central y El Caribe* (Mathematics Education Network of Central America and the Caribbean, www.redumate.org) that was founded in 2012 on the initiative of ICMI through its Capacity and Networking Project (CANP), one of the most important programs that ICMI developed. Indeed, CIFEMAT organized the Second CANP in Costa Rica in August 2012, the first one was held in Mali in 2011, and in Cambodia in 2013 (http://www.mathunion.org/icmi/activities/outreach-to-developing-countries/canp-project-2012-central-america-and-the-caribbean/).

In Costa Rica, in addition to this publication, other journals publish articles on mathematics education:

- *Las Matemáticas y su enseñanza* (Mathematics and its Teaching), to provide educational resources (with support from four state universities UCR, UNA, UNED, ITCR; it existed from 1989 to 1996).
- *Ciencias Matemáticas* (Mathematical Sciences), focuses on various mathematical topics including teaching (published by School of Mathematics of UCR; it existed from 1990 to 1992).
- *Revista digital Matemática, Educación e Internet* (Digital Journal of Mathematics, Education and the Internet), which accepts contributions in mathematics education, published by the School of Mathematics of ITCR since 2000 (http://www.tec-digital.itcr.ac.cr/revistamatematica).
- *Revista Educación* (Journal for Education) edited by the Faculty of Education at UCR (http://www.revistas.ucr.ac.cr/index.php/educacion).
- *Uniciencia*, published by the Faculty of Exact and Natural Sciences of UNA (http://www.revistas.una.ac.cr/index.php/uniciencia). Two separate special issues of Volume 20 of this journal, dated 2003 but came to light in 2004, were devoted to mathematics education and were prepared and edited by members of the Program of Meta-Mathematics Research and some authors from UNA.

Reform in Mathematics Education of Costa Rica

In 2010 the Minister of Public Education of Costa Rica requested that Ángel Ruiz design a new curriculum for primary and secondary schools; he accomplished this with a team composed of researchers from the CIFEMAT and in-service teachers of primary and secondary education. On May 21, 2012, this curriculum (MEP, 2012) was approved by the Costa Rican Higher Council of Education and was installed progressively in Costa Rican classrooms from 2013. Now that it has been approved, the majority of national institutions and individuals support more or less the implementation of this curriculum. However, as is fully described in Ruiz (2013), in the beginning there was resistance from teachers' unions, officers of the Ministry of Public Education, and some departments or schools at three of the public universities: ITCR, UNA, UNED, which were joined in their opposition by UCR's Faculty of Education. On the other hand, the new proposal was fully endorsed by the School of Mathematics of UCR (although with suggestions for improving it). This official curriculum was a second version where the authors decided to include some recommendations by universities and individuals in order to improve the text, to reach a more inclusive proposal and to obtain the necessary political support to get it approved.

One of the reasons for creating a new curriculum were indicators such as performance of Costa Rican students in the PISA 2009 Plus (carried out by OECD). The Costa Rican results were not among the worst in the Latin America countries that participated. In reading, only Chileans surpassed Costa Ricans, but in mathematics they ended behind Chile, Uruguay, México and Trinidad and Tobago, and — what is relevant — they were positioned 87 points below the average obtained by OECD countries (MEP, 2012, pp. 480–481). The most useful result for educational policy: 23.6% of Costa Rican students did not reach even level 1 of performance in those tests. 56.7% did not pass the first level, 84.5% did not pass level 2. The issue was especially serious in mathematics, because in reading and science, Costa Rican students who did not reach level 1 were only 1.3% and 9.6% respectively (Australian Council for Educational Research, 2011; OECD, 2010). This is just one of several indicators that showed that a radical strategy needed to be adopted.

The same group of people who developed the new curriculum (reinforced with other professionals in the areas of informatics and virtual education) has designed a crucial instrument to lead its implementation: the *Proyecto Reforma de la Educación Matemática en Costa Rica* (Project Reform in Mathematics Education of Costa Rica, www.reformamatematica.net). Its aims must be accomplished before 2015. A private *Fundación para la Cooperación Costa Rica Estados Unidos* (Foundation for Cooperation Costa Rica United States, www.crusa.cr) is funding this project. The Ministry of Public Education provides an important counterpart. The project has written gradual transition plans for each year of curriculum implementation, documents for teaching support, has designed and conducted very original pilot plans to gauge the rhythm of the reform, designed and conducted blended courses with the highest international standards (face-to-face sessions, independent distance work and assessment virtual sessions using the platform Moodle), designed online courses using cutting-edge technological strategies, and designed and administered a powerful virtual community of mathematics education. All of these actions, which are developed in an integrated and synergic manner, are accomplishing a pioneering experience in the region (Ruiz, 2013).

This new scenario has led to two results in the evolution of research in the country around CIFEMAT. First, their research has shifted to specific topics for the curriculum and its implementation within the context of a developing country. Indeed, these authors have synthesized theoretically an original curricular approach they call the *Praxis Perspective in Mathematics Education*. Second, it has empowered the group's place in the national educational life.

Below are some aspects of the new curriculum:

- The main focus of the new curriculum is a pedagogical strategy that the authors called "Problem Solving with an Emphasis in Real Contexts" which essentially proposes the organization of the lessons to build learning through carefully selected problems (promoting real context problems or that can be perceived as such). This emphasis is not adding or subtracting content. It is a departure from the teaching style of teaching of mathematics through lessons initiated with theory, followed by examples and routine practice, and maybe a final challenging or contextualized problem (MEP, 2012, pp. 13–14).
- The new paradigm is reinforced with curricular emphases that are fully operationalized around the syllabus: one is the explicit promotion of positive attitudes and beliefs about mathematics, another is the intense use of digital technologies (albeit gradual and adequate), and finally the use of the history of mathematics (MEP, 2012, pp. 35–40).
- It proposes to build higher cognitive abilities in students (transversal competencies) and mastery of specific abilities associated with mathematics content through an appropriate *teaching mediation* that involves mathematical tasks of increasing complexity and transversal activities (which are designated "mathematical processes") (MEP, 2012, pp. 24–34).

- It is an integrated curriculum from the first year of primary to the last year of secondary and finally organized into five areas of mathematics: numbers, geometry, measurement, relations and algebra, and statistics and probability.

As MEP (2012) mentions, the curriculum incorporates original research findings and international experience in mathematics education, a "biased" reading of the Japanese experience in problem solving, some ideas of the National Council of Teachers of Mathematics (USA), Realistic Mathematics Education (Freudenthal and the Netherlands) and the OECD's PISA framework, and some theoretical results of the French School of Mathematics Education (*Didactique des Mathématiques*). This *Praxis Perspective* assumes that a curriculum cannot be designed "*in vitro*" and then implemented, but implementation should illuminate the design: a good curriculum is one that — taking into account the international findings and experiences — can be implemented in a local reality. The syllabus contains about two thousand indications and problem examples for conducting the teaching these authors propose. Ángel Ruiz has directed both curriculum design and implementation.

The mathematics education scenario in Costa Rica has been substantially modified by the action of this group of researchers and educators within one equation combined with important government support. Not only has Costa Rica adopted a high quality curriculum with international standards but adjusted to the national context. In addition some other elements should be highlighted:

- There are blended courses conducted each year. These allow training of thousands of primary and secondary school teachers, something that was not usual in the country (Ruiz, 2013, pp. 67–79).
- These blended courses seek to provide preparation in specific mathematics pedagogy, trying somehow to correct some of the weaknesses produced by pre-service preparation.
- In this training strategy there are "two times": one for leaders, and one for large groups of teachers (the latter conducted by the leaders). A "quality control" of the massive courses is done through the testing within the virtual platform. Through this approach a set of teachers can be prepared and articulated as a social base of the desired Educational Reform.
- In 2013, a strategy to use entirely virtual training in diverse technologies (Geogebra, spreadsheets, etc.) deeply connected to traditional online

materials (PDFs) was conducted; this very new approach in Latin America seeks to avoid some of the possible local difficulties (no computer availability within schools or teachers' time limitations).
- There are other courses conducted entirely virtually using the methodology of the Massive Open Online Courses (MOOCS) that strengthen the implementation and construction of this educational leadership (cursos.reformamatematica.net).
- There is a virtual community in mathematics education that constitutes a reference to the reformers and brings up an efficient tool for dealing with the concerns and actions of teachers in the country (www.reformamatematica.net).

These actions are changing the way that in-service preparation is being done in the country. First of all, they have offered continuity. They support educational leadership preparation. They have promoted quality in courses for in-service teachers offered on a wide scale. They have incorporated intense and cutting edge use of the technologies of communication with and into the courses.

Similarly, this scenario pushes universities that prepare teachers to adjust their pre-service training to be in line with the reform perspectives.

In this period, this project has constituted the main meeting and reference center for the mathematics education community of Costa Rica.

This period of reform of mathematics education has opened a new stage for the teaching and learning of mathematics in Costa Rica.

Acknowledgements

The author wishes to thank:

- Hugo Barrantes, Danilo Solano, Rodolfo Herrera, Norma Adolio, Carmen González, Fabio González, Alcides Astorga, Julio Rodríguez, Sharay Meneses, Pilar Campos, Victor Buján, and Pedro Rodríguez who participated in the 1990s in a research project on the History of Mathematics in Costa Rica (under my lead), which provided important data used within this article.
- Patrick Scott (Emeritus Professor of the University of New México), for editing the English version of this paper.

- The project *Reform of Mathematics Education in Costa Rica* (2012–2015): a unique still living experience in the history of math education in Costa Rica (http://www.reformamatematica.net).

Bibliography

Adolio, N., González, C. & González, F. (1995). Una nueva Escuela en una nueva Universidad, in A. Ruiz (Ed.), *Historia de las Matemáticas en Costa Rica. Una introducción.* San José, Costa Rica: EUCR, EUNA. Retrieved from http://www.centroedumatematica.com/aruiz/libros/Historia%20de%20las%20matematicas%20en%20Costa%20Rica.pdf

Adolio, N., González, J., & Mora, F. (2003). El movimiento de las Olimpiadas de Matemática en secundaria: Un reto para Costa Rica. *Uniciencia (Revista de la Facultad de Ciencias Exactas y Naturales de la Universidad Nacional). 20*(2), 299–304, Heredia, Costa Rica: EUNA. Retrieved from http://www.centroedumatematica.com/aruiz/libros/Uniciencia/Articulos/Volumen2/Parte13/articulo23.html

Alfaro, A. L., Alpízar, M., Morales, Y., Ramírez, O., & Salas, O. (2013). La formación inicial y continua de docentes de Matemáticas en Costa Rica. *Cuadernos de Investigación y Formación en Educación Matemática.* Special Number, November. San José, Costa Rica.

Astorga, A., Meneses, S., & Rodríguez, J. (1995). Las Matemáticas en el Instituto Tecnológico de Costa Rica, in A. Ruiz (Ed.), *Historia de las Matemáticas en Costa Rica. Una introducción.* San José, Costa Rica: EUCR, EUNA. Retrieved from http://www.centroedumatematica.com/aruiz/libros/Historia%20de%20las%20matematicas%20en%20Costa%20Rica.pdf

Australian Council for Educational Research (2011). *PISA 2009 Plus Results.* Retrieved from http://research.acer.edu.au/cgi/viewcontent.cgi?article=1000&context=pisa

Azofeifa, I. F. (1973). *El viejo Liceo.* Costa Rica: Ministerio de Cultura Juventud y Deportes.

Barrantes, H., & Ruiz, A. (1995a). En la Escuela Normal y en los Colegios, in A. Ruiz (Ed.), *Historia de las Matemáticas en Costa Rica. Una introducción.* San José, Costa Rica: EUCR, EUNA. Retrieved from http://www.centroedumatematica.com/aruiz/libros/Historia%20de%20las%20matematicas%20en%20Costa%20Rica.pdf

Barrantes, H., & Ruiz, A. (1995b). La Carrera de Enseñanza en la Universidad de Costa Rica, in A. Ruiz (Ed.), *Historia de las Matemáticas en Costa Rica. Una introducción.* San José, Costa Rica: EUCR, EUNA. Retrieved from http://www.centroedumatematica.com/aruiz/libros/Historia%20de%20las%20matematicas%20en%20Costa%20Rica.pdf

Barrantes, H., & Ruiz, A. (1995c). La Reforma de Mauro Fernández y las Matemáticas, in A. Ruiz (Ed.), *Historia de las Matemáticas en Costa Rica. Una introducción.* San José, Costa Rica: EUCR, EUNA. Retrieved from http://www.centroedumatematica.com/aruiz/libros/Historia%20de%20las%20matematicas%20en%20Costa%20Rica.pdf

Barrantes, H., & Ruiz, A. (1995d). Los programas antes de la creación de la Universidad, in A. Ruiz (Ed.), *Historia de las Matemáticas en Costa Rica. Una introducción.* San José, Costa Rica: EUCR, EUNA. Retrieved from http://www.centroedumatematica.com/aruiz/libros/Historia%20de%20las%20matematicas%20en%20Costa%20Rica.pdf

Barrantes, H., & Ruiz, A. (1995e). 1964, in A. Ruiz (Ed.), *Historia de las Matemáticas en Costa Rica. Una introducción.* San José, Costa Rica: EUCR, EUNA. Retrieved from http://www.centroedumatematica.com/aruiz/libros/Historia%20de%20las%20matematicas%20en%20Costa%20Rica.pdf

Barrantes, H., & Ruiz, A. (1998). *The History of the Inter-American Committee on Mathematics Education.* Bogotá, Colombia: Academia Colombiana de Ciencias Exactas, Físicas y Naturales. Retrieved from http://ciaem-iacme.org

Brenes, R., & García, J. (1908). *Proyecto de Programas de Instrucción Primaria.* Tipografía Nacional, San José.

Buján, V. (1995). Olimpiadas matemáticas y colegios científicos, in A. Ruiz (Ed.), *Historia de las Matemáticas en Costa Rica. Una introducción.* San José, Costa Rica: EUCR, EUNA. Retrieved from http://www.centroedumatematica.com/aruiz/libros/Historia%20de%20las%20matematicas%20en%20Costa%20Rica.pdf

Buján, V., & Jiménez, M. (2003). Una Olimpiada Matemática para el segundo ciclo (OMCEP). *Uniciencia (Revista de la Facultad de Ciencias Exactas y Naturales de la Universidad Nacional), 20*(2), 313–319, Heredia, Costa Rica: EUNA. Retrieved from http://www.centroedumatematica.com/aruiz/libros/Uniciencia/Articulos/Volumen2/Parte13/articulo25.html

Chacón, E. (1984). *Influencia de las ideas del doctor Valeriano Fernández Ferraz en la vida cultural de Costa Rica.* San José, Costa Rica: EUNED.

Chaves, E. (2007). Inconstancia entre los programas de estudio y la realidad en el aula en la enseñanza de la estadística de secundaria, in *Actualidades Investigativas en Educación,* Instituto de Investigación en Educación, Universidad de Costa Rica, Costa Rica. Retrieved from http://revista.inie.ucr.ac.cr/uploads/tx_magazine/estadistica_01.pdf

Colegio Científico Costarricense de Pérez Zeledón (CCCPZ) (2013). *Brochure de información 2013–2.* Retrieved from http://www.circuito01pz.com/colegio-cientifico-de-perez-zeledon.html

Colegio Superior de Señoritas (1939). *Álbum del cincuentenario 1888–1938.* Imprenta y Librería Lehmann.

Conejo, A. (1975). *Henri Pittier.* San José Costa Rica: Ministerio de Cultura, Juventud y Deportes.

De Faria, E. (2003). La organización, promoción y proyección de las Matemáticas en Costa Rica. *Uniciencia.* Heredia, Costa Rica: Universidad Nacional. Retrieved from http://www.centroedumatematica.com/aruiz/libros/Uniciencia/Articulos/Volumen2/Parte14/articulo26.html

Dobles-Segreda, L. (1929). *Índice Bibliográfico de Costa Rica*. Tomo III. San José: Imprenta y Librería Lehmann.

Escuela Normal de Costa Rica (1940). *La Escuela Normal en sus Bodas de Plata*. Imprenta Nacional.

Fischel, A. (1987). *Consenso y represión, una interpretación sociopolítica de la Educación Costarricense*. Editorial Costa Rica, San José.

Fischel, A. (1991). Los estudios superiores en Costa Rica, 1888–1940. *Historia de la Educación Superior en Costa Rica*. Costa Rica: Centro de Investigaciones Históricas, UCR.

Gauss, C. (1995). *Disquisitiones Arithmeticae*. A Spanish commented translation from the Latin by A. Ruiz, H. Barrantes & M. Josephy. Bogotá, Colombia: Academia Colombiana de Ciencias Exactas, Físicas y Naturales. Retrieved from http://epsaleph.tripod.com/sitebuildercontent/sitebuilderfiles/disqui sitionesarithmeticae.pdf

Gómez, L. (1978). Contribuciones a la pteridología costarricense. XII. Carlos Wercklé. *BRENESIA* 14–15:361–393.

González-Flores, L. F. (1921). *Historia de la Influencia Extranjera en el Desenvolvimiento Educacional y Científico de Costa Rica*. San José, Costa Rica: Imprenta Nacional. (Reprinted by the University of Michigan Library). Retrieved from http://archive.org/stream/histori adelainfl00gonz/historiadelainfl00gonz_djvu.txt

González-Flores, L. F. (1945). *Historia del desarrollo de la instrucción pública en Costa Rica*. Tomo I. San José, Costa Rica: Imprenta Nacional.

González, L. P. (1989). *La Universidad de Santo Tomás*. Costa Rica: Editorial de la Universidad de Costa Rica.

Herrera, E. (1988). *Los alemanes y el Estado cafetalero*. Costa Rica: EUNED.

Herrera. R. (1993). Luis González: in memoriam. *Revista de Ingeniería*, Universidad de Costa Rica, 2(1), 20–24.

Herrera, R. (1995a). El Profesor Dirk J. Struik en Costa Rica, in A. Ruiz (Ed.), *Historia de las Matemáticas en Costa Rica. Una introducción*. San José, Costa Rica: Edit. UCR-UNA. Retrieved from http://www.centro edumatematica.com/aruiz/libros/Historia%20de%20las%20matematicas% 20en%20Costa%20Rica.pdf

Herrera, R. (1995b). La matemática en Costa Rica y la influencia del Dr. Biberstein, in A. Ruiz (Ed.), *Historia de las Matemáticas en Costa Rica. Una introducción*. San José, Costa Rica: Edit. UCR-UNA. Retrieved from http://www.centroedumatematica.com/aruiz/libros/Historia%20de%20las %20matematicas%20en%20Costa%20Rica.pdf

Herrera, R. (1995c). Luis González y la matemática en Costa Rica, in A. Ruiz, (Ed.). *Historia de las Matemáticas en Costa Rica. Una introducción*. San José, Costa Rica: Edit. UCR-UNA. Retrieved from http://www.centro edumatematica.com/aruiz/libros/Historia%20de%20las%20matematicas% 20en%20Costa%20Rica.pdf

Láscaris, C. (1982a). *Historia de las Ideas en Centro América*. San José, Costa Rica: EDUCA.

Láscaris, C. (1982b). *Historia de las Ideas Filosóficas en Costa Rica*. San José, Costa Rica: UACA.

Láscaris, T. (1976). *Luis González, una época en la matemática.* Thesis to obtain the degree of *Licenciado* in Mathematics, UCR.

Ledesma-Reyes, M. (1995). *Krausismo y educación en Costa Rica: la influencia de los educadores canarios Valeriano y Juan Fernández Ferraz.* Doctoral Thesis, Universidad La Laguna, Spain. Retrieved from ftp://tesis.bbtk.ull.es/ccssyhum/cs19.pdf

Liceo de Costa Rica (1930). *Memoria Anual del Liceo de Costa Rica.* San José, Costa Rica.

Meoño, R. (2008). *El derecho a la educación en Costa Rica.* UNESCO. (1ª ed). San José, Costa Rica.

Ministerio de Educación Pública de Costa Rica (MEP) (2012).*Programas de Estudio de Matemáticas. I, II y III Ciclos de la Educación General Básica y Ciclo Diversificado.* San José, Costa Rica: Autor. Retrieved from www.reformamatematica.net

Ministerio de Educación Pública de Costa Rica (MEP) (2013). Official web site http://www.mep.go.cr

Morales-Luna, G. (2013). *Olgierd Alf Biberstein.* Retrieved from http://cs.cinvestav.mx/~gmorales/Biberstein/indice.html

Organización de Estados Iberoamericanos (OEI) (2009). *24 Olimpiada Iberoamericana de Matemáticas.* Retrieved from: http://www.oei.es/oim/reporteOIM09.pdf

Organización de Estados Iberoamericanos (OEI) (2013). *Olimpiada Iberoamericana de Matemática.* Retrieved July 1, 2013 from http://www.oei.es/oim/oimhis.htm

Organization for Economical Co-operation and Development (OECD) (2010). *Pisa 2009 results: what students know and can do — student performance in reading, mathematics and science,* Vol. I. Retrieved from http://www.oecd.org/pisa/pisaproducts/48852548.pdf

Ossembach-Sauter, C. (2008). Adolphe Tonduz (1862–1921). *Biocenosis, 21*(2). Retrieved from http://web.uned.ac.cr/biocenosis/images/stories/articulos Vol21/Biocenosis21-05.pdf

Quesada, J. R. (1991). *Educación en Costa Rica 1821–1940.* Number 15 of the collection *Nuestra Historia.* Costa Rica: EUNA & EUNED.

Rodríguez-Vega, E. (1988). *Biografía de Costa Rica.* San José: Editorial Costa Rica.

Rodríguez, P., & Ruiz, A. (1995). Antes de la Reforma de Mauro Fernández, in A. Ruiz (Ed.), *Historia de las Matemáticas en Costa Rica. Una introducción.* San José, Costa Rica: Edit. UCR. Retrieved from http://www.centroedumatematica.com/aruiz/libros/Historia%20de%20las%20matematicas%20en%20Costa%20Rica.pdf

Rosero, L. (2011). *Presentación* to the article by H. Pittier: Impresiones y recuerdos: José Silverio Gómez 1801-1904. *Población y Salud en Mesoamérica, 8*(2 January–June). Retrieved from http://ccp.ucr.ac.cr/revista/volumenes/8/8-2/8-2-1a/8-2-1a.pdf

Ruiz, A. (1990). Matemáticas: una reconstrucción histórico-filosófica para una nueva enseñanza. In UNESCO, *Educación Matemática en las Américas*

VII *(Proceedings of the VII Interamerican Conference on Mathematics Education)*, Dominican Republic, 1987. Re-published: *Cuadernos de Investigación y Formación en Educación Matemática*, 7, 179–190, Costa Rica. Retrieved from http://revistas.ucr.ac.cr/index.php/cifem/article/view/6941/6627

Ruiz, A. (Ed.). (1995). *Historia de las Matemáticas en Costa Rica. Una introducción*. San José, Costa Rica: EUCR, EUNA. Retrieved from http://www.centroedumatematica.com/aruiz/libros/Historia%20de%20las%20matematicas%20en%20Costa%20Rica.pdf

Ruiz, A. (2000). Ideologías y extranjeros en la educación y las matemáticas de Costa Rica durante el siglo XIX. *LLULL, Revista española de Historia de las Ciencias y las Técnicas*, 23(48). Retrieved from http://dialnet.unirioja.es/servlet/listaarticulos?tipo_busqueda=EJEMPLAR&revista_busqueda=1511&clave_busqueda=221626

Ruiz, A. (2008). Los programas de formación de docentes de Matemáticas en Costa Rica: balance y perspectivas. Segundo *Informe del Estado de la Educación*, San José, Costa Rica: Programa Estado de la Nación. CONARE. Costa Rica. Retrieved http://www.estadonacion.or.cr/files/biblioteca_virtual/educacion/002/inf2educap4-edu02.pdf.

Ruiz, A. (2013). La reforma de la Educación Matemática en Costa Rica. *Perspectiva de la praxis. Cuadernos de Investigación y Formación en Educación Matemática*. Special Number, July. San José, Costa Rica. Retrieved from http://revistas.ucr.ac.cr/index.php/cifem/issue/view/1186

Ruiz, A., & Barrantes, H. (2000). La reforma liberal y las matemáticas en la Costa Rica del siglo XIX. *LLULL, Revista española de Historia de las Ciencias y las Técnicas*, 23(46). Retrieved from http://dialnet.unirioja.es/servlet/articulo?codigo=62244

Ruiz, A., Barrantes, H., & Gamboa R. (2009).*Encrucijada en la Enseñanza de la Matemática. La formación de educadores*. Cartago, Costa Rica: Editorial Tecnológica de Costa Rica. Retrieved from http://www.centroedumatematica.com/aruiz/libros/Encrucijada%20en%20Ensenanza%20de%20la%20Matematica%20La%20formacion%20de%20Educadores.pdf

Ruiz, A., Barrantes, H., & Campos, P. (2010). La Escuela de Matemática de la Universidad de Costa Rica: balance y perspectives.*Cuadernos de Investigación y Formación en Educación Matemática*. Number 10, April. San José, Costa Rica. Retrieved from http://revistas.ucr.ac.cr/index.php/cifem/article/view/6931

Ruiz, A., & Rodríguez, P. (2002). Educación y matemáticas en la Universidad de Santo Tomás, in G. Peraldo (Ed.), *Historia de las ciencias y las técnicas en Costa Rica durante el siglo XIX*. Cartago, Costa Rica: ITCR. Retrieved from http://www.centroedumatematica.com/aruiz/Articulos/Educacion%20y%20matematicas%20en%20la%20Universidad%20de%20Santo%20Tomas.pdf

Shulman, L. S. (1986). Those who understand: Knowledge growth in teaching. *Educational Researcher*, 15(2), 4–14. Retrieved from http://www.fisica.uniud.it/URDF/masterDidSciUD/materiali/pdf/Shulman_1986.pdf

Solano, D., & Ruiz, A. (1995a). Entre la creación de la Universidad y la Reforma de Facio, in A. Ruiz (Ed.), *Historia de las Matemáticas en Costa Rica. Una introducción*. San José, Costa Rica: EUCR, EUNA. Retrieved from http://www.centroedumatematica.com/aruiz/libros/Historia%20de%20las%20matematicas%20en%20Costa%20Rica.pdf

Solano, D., & Ruiz, A. (1995b). Matemáticos y Físicos juntos, in A. Ruiz (Ed.), *Historia de las Matemáticas en Costa Rica. Una introducción*. San José, Costa Rica: Edit. UCR. Retrieved from http://www.centroedumatematica.com/aruiz/libros/Historia%20de%20las%20matematicas%20en%20Costa%20Rica.pdf

Solano, D., & Ruiz, A. (1995c). El Dr. Bernardo Alfaro Sagot y las Matemáticas, in A. Ruiz (Ed.), *Historia de las Matemáticas en Costa Rica. Una introducción*. San José, Costa Rica: Edit. UCR. Retrieved from http://www.centroedumatematica.com/aruiz/libros/Historia%20de%20las%20matematicas%20en%20Costa%20Rica.pdf

Universidad Nacional (2013). Medallas para Costa Rica en Olimpiada Iberoamericana de Matemática, in *Boletín digital Campus*. Retrieved July 1, 2013 from http://www.hoyenelcampus.una.ac.cr//index.php?option=com_content&task=view&id=778&Itemid=41

Urbina-Vargas, S. (2005). Un científico costarricense de origen inglés: Charles Herbert Lankester (1879–1969). *Revista Biocenosis, 19*(2).

Valerio-Zamora, A. L. (2000). Henri Pittier Dormond (1857–1950). *Boletín del Museo Nacional de Costa Rica, 7*(1), 2–4. Costa Rica: Museo Nacional de Costa Rica, Departamento de Historia Natural.

Zelaya, C. (1973). *Rafael Francisco Osejo*, San José, Costa Rica: Ministerio de Cultura, Juventud y Deportes de Costa Rica.

About the Author

Ángel Ruiz is the President of the Inter-American Committee on Mathematics Education (2007–2015), Vice-President of the International Commission on Mathematical Instruction (2009–2016), Director of the Central American and Caribbean Network for Mathematics Education (founded in 2012), and founder and director of the Center for Research and Teacher Training in Mathematics Education (since 2006) in Costa Rica. He was the founder and director of the Center for Mathematics and Meta-Mathematics Research at the *Universidad de Costa Rica* for fourteen years. He is the author of more than 200 papers and 35 books (www.angelruizz.com).

Chapter 7

CUBA: Mathematics and Its Teaching

Otilio B. Mederos Anoceto, Miguel A. Jiménez Pozo, and José M. Sigarreta

Abstract: This chapter is divided into five sections corresponding to five periods: the Colonial (1510–1868), the Wars of Independence (1869–1898), the first military occupation (1899–1902), the Republican (1902–1958), and the Cuban Revolution (1959–2013). Mathematics has an extraordinary development in all directions, particularly in its teaching, in the last period.

Keywords: Education in Cuba; mathematics education in Cuba; primary education in Cuba; secondary education in Cuba; mathematics teacher training in Cuba; bachelor's degree in mathematics in Cuba; University of Havana; Academy of Sciences of Cuba.

The scientific and technological development of any country or region of the world will always be inextricably linked to the economic development and the social superstructure in the region in question. From this point of view, the study of the development of mathematics in Cuba can be established with exact bounds in five historical periods. They chronologically correspond to the Spanish colonial stage until 1868, the Wars of Independence in the Colonial Period (1869–1898), the American occupancy of Cuba at the beginning of twentieth century, the Republic of Cuba until 1959, and finally the so-called Cuban Revolution. Developments in each period are described below with an emphasis on the contemporary period.

The Colonial Period (1510–1868)

On October 27, 1492, the Spanish sailors captained by Christopher Columbus discovered the island called by its inhabitants Cubanacan, and hence the name Cuba. Three Cuban native cultures existed during the pre-Columbian period: Guanahatabey (Spanish: *Guanajatabey*), Ciboney (Siboney is more used in Cuba) and Taine (Spanish: *Taíno*). The Guanahatabey (Shell Age) and Ciboney (Stone Age) cultures had already disappeared by the arrival of Columbus while the Taínos were living in the Pottery Age. They reached a development equivalent to the Neolithic and Mesolithic, and in contrast to other cultures like Mayas, Incas and Aztecs, they did not leave visible traces in the cultural development of the country, except in some place names and words used by them, which persist to today. We could insist that their cultural, and consistently their mathematical status, was underdeveloped. This situation persisted until the sixteenth century, when the royal laws disposed that the Castilian language and the Christian doctrine were to be taught to the habitants of the island. The priests were mainly in charge of this instruction.

There is nothing worth mentioning about mathematical development in Cuba during the first two centuries of Spanish settlement. A few hundred colonizers commanded by Diego Velázquez started organizing permanent settlements in Cuba in 1511. A transformation began on the island. With not much gold in the Cuban lands or rivers, the colonizers turned to agriculture. To this aim they conceived an organization called "Encomienda", which consisted in taking care of the Christian education of one or more native families by so-called "encomenderos". As payment for their education, the Amerindians were obliged to work the land almost as slaves. Later on, as they decreased in number while more workers were needed, the scourge of black slavery covered the Caribbean island with shame. The development of science in Spain at this stage was not only limited but also viewed and studied through scholastic interpretations. Thus, without any Spanish scientific interest and with all mentioned facts at hand, what mathematical developments could be expected in the young Cuban colony?

The conquest and colonization of the island started in 1510. The colony was not interested in elementary education. In the period between 1510 until 1742 education was offered through the church and most of all by the family. According to De la Torre (1999), during the mid-sixteenth century to the mid-eighteenth century the Catholic Church was responsible for developing the little formal education provided to children of school age. In the second half of the eighteenth century some private schools arose. Those that were led by lay people offered instruction to the children from families that could afford the fee. In these schools women, men, whites and people of color were taught together.

In the year 1728 the Royal Pontifical University of Saint Jerome of Havana (RPUSJH) was founded in the San Juan de Letrán abbey, with four schools: Art and Philosophy, Theology, Doctrine, and Law and Medicine. The abbey, which does not exist anymore, was located in what is now known as the historical center of Havana.

One of the most relevant educational institutions from the colonial period was the Royal Conciliar San Carlos and San Ambrosio Seminary College. This institution was attended by young seminarians who aspired to the priesthood and young people who wished to obtain adequate college preparation. The year 1760 is considered the date of its inception because in that year its statutes were created. The first classes met in 1772.

The beginning of a scientific education in Cuba started with the establishment of the Economic Society of Friends of the Country that took place in Havana in 1793, five years after the founding of one in Santiago de Cuba. The peak period of this institution was between the end of the eighteenth century and the mid-nineteenth century. It had a large tradition of teachers that laid the basis for an autonomous theoretical-methodological conception of teaching. Among these educators we can highlight Father Félix Varela y Morales, José de la Luz y Caballero, Father José Agustín Caballero, and José Antonio Saco. These teachers were concerned about a comprehensive education for all society and about scientific learning, in particular.

Father Varela defended the scientific method as a tool for teaching, openly opposing the scholastic method that reigned in his time. The importance of scientific and mathematical development in the pedagogical theory formulated by Varela can be seen in the ideas that he put into practice in the cultivation of natural science, physics, chemistry, botany and moral education. To him, knowledge was based on nature and real science was the result of meditation and the good connection of ideas. Moreover, experimentation was essential for scientific work. Hence, knowledge of the arts and sciences became paramount.

Nevertheless, the development of mathematics was still slow as Cuba entered the nineteenth century. Jiménez and Sánchez (1993) highlight that although following the scholastic tradition there was a professorship of mathematics at the Faculty of Philosophy, the position remained vacant until 1816 due to lack of applicants.

With the support of the Economic Society of Friends of the Country during the first half of the nineteenth century, primary and secondary private schools were created in Havana, Matanzas, and Santiago de Cuba. "These schools were led by teachers with a Cuban identity and used the curriculum to help develop citizens who could think reflexively and who had a national identity" (García, 2005).

In 1824 the San Cristóbal de la Habana College was founded. It contributed for 40 years to the development of education in Cuba. It was run by Antonio Casas y Remón, José de la Luz y Caballero, and Rafael Navarro. The following mathematics courses were taught: elementary arithmetic; theoretical, practical and business arithmetic; and mathematical elements.

In 1829, Mariano Cubí founded Buenavista College, considered to be the first secondary teaching center on the island. The mathematical subjects that were taught in this college were: arithmetic, geometry, algebra, and trigonometry. Although at the beginning of 1840 some normal schools were established to educate teachers, these schools did not do much to promote advances in mathematics in the country, as mathematics was not given much importance in the curriculum. In 1841, the illustrious pedagogue Juan Bautista Sagarra contributed to the creation of Santiago College in the city of Santiago de Cuba in response to the lack of attention that the Spanish authorities were giving to education. In this college subjects like arithmetic, geometry, physics, mathematics, bookkeeping, and algebra were taught.

In 1842 the Spanish government, in an attempt to regularize and update primary and secondary teaching, presented a Plan for Public Instruction for Cuba and Puerto Rico. By means of this project they became responsible for public education, expressed the need for teacher certification, and proposed the creation of a normal school. The subjects to be included in the mathematical preparation for elementary teaching, according to this plan, were basic and advanced arithmetic, principles of geometry, and common applications.

In 1848, José de la Luz y Caballero founded the Colegio de Salvador in Havana. It had both elementary and secondary levels and was perhaps the most important institution of this type among the many that existed on the island. An equally important step of opening the university to scientific knowledge occurred in 1850, when the RPUSJH became a secular institution under the name of the Royal and Literary University of Havana.

In 1857 the Law of Public Instruction (LIP) was enacted. Through it, institutes of secondary education (ISE) were created in the cities of Havana, Matanzas, Puerto Príncipe and Santiago de Cuba, in which students could study to become technicians in surveying, accounting, and chemistry. This law established the division of education into two sectors: public education — directed by the colonial government — and private education, run by individuals but with public control of programs of instruction. The first normal school was created in the city of Guanabacoa in 1859. It functioned until 1868, when the Ten Years' War started.

The General Plan of Studies for the Island of Cuba was approved as a royal decree on August 25, 1863 based on the LIP. This plan precisely specified the mathematical subjects corresponding to elementary education: arithmetic, with the legal system of measures, weights and coins; geometry; and surveying. With regard to secondary education, the Plan noted that it should include general studies as well as applied studies in the industrial professions. It included the following mathematics courses: principles and exercises in arithmetic, principles and exercises in geometry, arithmetic, algebra (to second degree equations), elements of geometry, and linear trigonometry.

With the implementation of the 1863 Plan, the subjects of mathematics acquired greater weight and importance in the education of both students and teachers. The decree states

> ... that to aspire to the title of elementary teacher, in addition to at least two years at a normal school, the following courses were required: Fundamentals of Geometry, Technical Drawing and Surveying... high school teachers were required to pass a course in geometry and technical drawing (Barcia and Sardiñas, 2005).

With the creation of normal schools and support from economic societies, there was a favorable quantitative change in education. By 1863 there were 577 schools with more than 20,000 students. However, throughout this period the development of mathematics in Cuba was very slow, both at the pre-university and university levels.

According to Marchante and Merchán (2007), the Faculty of Science was created and authorized by the 1863 reform to grant bachelor's degrees, but not master's or doctoral degrees, for which students would have to attend the universities in Spain. This faculty began its activities during 1870–1871.

> The creation of the Faculty of Sciences in 1863 and the subsequent reform (and last) of the university during the colonial era, which occurred in 1880, would consolidate the design of studies in mathematics, in close relation with physics (Valiño, 2003).

The Wars of Independence in the Colonial Period (1869–1898)

This is a period marked by two wars that freed Cubans from Spanish rule. The Ten Years' War occurred between 1868 and 1878. It ended without achieving the independence of Cuba. During those years the number of schools decreased as did student attendance to the ones that were not closed. An important event of this period was the founding of the Preparatory School for Teachers, created by the Economic Society of Friends of the Country in the year 1872 with the objective of preparing teachers. After the war, this school was in very bad condition and contributed very little to the development of education in Cuba.

In 1880 through the Law of Public Instruction (LIP), reported in the *Gazette of Havana* on January 30 of the same year, a new curriculum was established on the island. Its purpose was to apply in Cuba the same curriculum that existed in Spain. They argued that teachers, to teach in public schools, should be appointed by the Spanish government and have graduated from normal schools. However, the creation of the first normal school that could provide compliance with the latter requirement was not established until 1892, and by the end of the War of Independence all the normal schools were closed. "Emma Perez in her book, *History of Education in Cuba*, raises the point that Article 19 of the Regulations for normal schools established Drawing as a subject for teachers and Article 20 required Applied Geometry and Drawing" (Barcia and Sardiñas, 2005).

With respect to university education, the LIP of 1880 indicated that the Royal and Literary University of Havana would have faculties of Arts; Exact, Physical and Natural Sciences; Pharmacy; Medicine; and Law. It also specified a set of changes, none of which referred to mathematical content. If we consider that Spain's purpose was to establish in Cuba the curriculum that existed in Spain — and to create the Faculty of Exact, Physical and Natural Sciences — then we must conclude that these changes represented progress for higher education and, in particular, for mathematics education. For more information, see Marchante and Merchán (2008).

After a long period of preparation, the War for Independence began in 1895. Its leader and instigator was the illustrious Cuban patriot José Martí: poet, journalist, writer and a man with one of the most enlightened minds of all the Americas. He died in combat during a military skirmish at the start of the war. The war ended suddenly in 1898 with American intervention in the island, justified by the military disputes between the

United States and Spain following the sinking of the USS Maine, without Cuba achieving independence.

The First Military Occupation (1899–1902)

The First US Military Occupation in Cuba was from 1899 to 1902. It restricted Cuba's sovereignty and prevented the War for Independence from ending with the creation of an independent republic. However, during this period some attention was given to education. In 1899 the Royal and Literary University of Havana received the name that persists until today, the University of Havana (UH). In May 1902, UH relocated to Aróstegui Hill — now known as University Hill — in the Vedado.

The position of Superintendent of Schools of Cuba was created in November 1899 and filled by the American educator Alexis Everett Frye. He organized elementary education, which improved both quantitatively and qualitatively. The number of schools (312) existing in 1898 increased to 3,316 in 1901. American books were translated for this level and 1,256 Cuban teachers received professional development at Harvard University. However, problems arose due to the contextualization of books and the teaching programs that were used, and a lack of teachers prepared for the new demands and the greater freedom they had in carrying out their work. The need for continuous and systematic professional development of teachers was evident.

The military governor of Cuba appointed the Cuban philosopher and educator Enrique José Varona as Secretary of Public Instruction and Fine Arts. Varona developed a program to reform secondary and university education, known as Plan Varona, by the Military Orders 266 and 267. The Plan Varona for high school began on June 30, 1900 and was supported by Military Order 267; it was in effect until 1939. It required three mathematics courses, two courses in physics, and one in chemistry for all high schools. Below are some of Varona's ideas that illustrate his pedagogical thinking:

> ... our teachers should only be teachers, and so in the modern sense: men dedicated to teaching how to learn, how to inquire, how to research; men that elicit and help students in their work; not men who will give recipes and formulas to learn as quickly as possible just a bit of science, with the objective that it be the most ostentatious (Varona, 1992).

José Antonio González Lanuza was in charge of the reforms in higher education that were part of Plan Varona. These reforms included the implementation of changes in the curriculum of several degree programs, and the creation of professorships and schools at the University of Havana. A degree program in pedagogy was created in the Faculty of Arts and Sciences. As part of these reforms, its teaching activities began during 1900–1901. This was the first university in the Spanish-speaking countries to train specialists in pedagogy.

The Republican Period (1902–1958)

The Republic of Cuba was created on May 20, 1902. At the beginning of the Republic there was only one high school in each of the capitals of six provinces. The purpose of these institutions was to prepare students for the university, so the level of teaching corresponded to a pre-university education. The curriculum of the high schools was designed for four years and it gave priority to science learning, which sought to respond urgently to the needs of the Republic — to the detriment of the humanities. This was despite the fact that Varona was probably the main humanist at the time in Cuba. In 1937, thirteen new high schools were created in various cities of the Republic.

In 1941 a new curriculum for secondary education was put into effect that gave more attention to the humanities as well as the sciences. Known as Plan Remos in honor of its author, John J. Remos, through this plan the duration of high school studies was extended to five years. Students interested in science programs were enrolled in the fifth year only in science subjects and received the title of Graduate in Science. Those interested in the humanities were enrolled during the fifth year only in non-science courses (arts) and received the title of Graduate in Arts and Letters.

In the first four years of high school, the mathematics courses were arithmetic, algebra, geometry and trigonometry. In the fifth year, students who received the title of Graduate in Science had to pass a course in advanced arithmetic and algebra, and another in differential and integral calculus. The teachers were university graduates. The math textbooks were written by renowned Cuban professors and contributed significantly to the development of mathematical culture in Cuba and other countries. Among the authors of these books are: Mario Octavio González Rodríguez (*Modern Elementary Algebra*, *Elements of Arithmetic* and *Algebra* and *Elements of Geometry and Differential Calculus Concepts*), Aurelio Angel Baldor Párraga, (*Arithmetic, Algebra, Geometry and Trigonometry*), Socrates Rosell Franco (*First Year Mathematics, Arithmetic, Third Year Mathematics,* and *Algebra*).

The mathematical level reached in the high schools was suited to the times and useful for success in university engineering and science. Secular and religious private schools also contributed to pre-university education across the country, particularly in mathematics. Among the denominational high schools the Marist College (for men) and Teresian College (for girls) were particularly noteworthy. Among the non-denominational high schools, Baldor College should be mentioned. The private high schools were required to have the same curriculum and final exams as the public high schools.

In the Republican era, high schools had good libraries as well as chemistry, physics and natural sciences laboratories, which were well-designed and equipped for the time. Some high schools even had museums. Attached to each of the high schools was a School of Surveying which gave diplomas in surveying, land appraisal, and drafting.

In addition to the traditional high school, Cuba also had secondary schools that gave specialized training. The Normal School for Teachers, whose main objective was to train elementary and lower secondary teachers (normalistas), were created in the capitals of provinces between 1915 and 1923, with university graduates as teachers. The existence of the normal schools made it possible for the teachers of all the elementary schools (first to sixth year) and lower secondary schools (seventh and eighth), to receive adequate training for teaching at those levels. Graduates of normal schools could enroll without examinations in the pedagogy degree programs in Cuban universities.

During the First US Military Occupation, American teachers created the first classrooms for pre-school education in Cuba. They were called Kindergartens. In 1902 the Kindergarten Normal School in Havana was created, specializing in pre-school education, and 32 years later began to spread to other provincial capitals. Among the entrance requirements, one had to be a woman and pass an examination in arithmetic.

In 1918 the Home School of Havana was created in order to train young women in proper home management. Among the subjects in these schools were practical arithmetic, accounting and commercial transactions. The Contreras Law changed Home Schools to Normal Home Schools whose aim was the training of teachers that would teach the subjects of crafts and home economics in elementary schools.

With the aim of contributing to the economic and commercial development of Cuba, the Trade Professional Schools were created in 1927 in several cities. Their curriculum included the following mathematics subjects: arithmetic, algebra, financial mathematics and statistics. These schools offered accountant, bookkeeper, and customs broker diplomas.

According to Valdés (2000),

> ... on February 23, 1942 the Cuban Society of Physical and Mathematical Sciences was founded and its Board of Directors elected. Pablo Miquel y Merino was elected President of the society. At the first meeting of this Board, among the agreements reached was the appointment of Dr. Pablo Miquel y Merino ... editor of the journal that would be published by the company according to the provisions of Article IX of the bylaws ... (Valdés, 2000).

In this journal, the *Journal of the Cuban Society of Physical and Mathematical Sciences*, the activities of the society were announced and research results were published, as well as work that contributed to the improvement of the teaching of mathematics and physics, especially at the high school level.

Luis J. Davidson San Juan (1921–2011) was a prominent Cuban educator and mathematician. For nearly 50 years he did much to promote and develop mathematics, national and international competitions in mathematics, and journals for teachers and mathematicians. He stated that "We believe that the *Journal of the Cuban Society of Physical and Mathematical Sciences* served its purpose during the 18 years of its existence. Upon the dissolution of the Society in 1960, the magazine went out of print" (Davidson, 2003).

> By the late 1940s, 156 doctors in physics and mathematics had graduated from the University of Havana. In most cases, they worked as high school teachers. As for the programs in physics and mathematics, the files confirm that while in the first 33 years of the Republic, 36 doctors of physical and mathematical sciences graduated from the University of Havana; in the next 15 years 120 graduated" (Sánchez, 2013).

The Central University "Marta Abreu" of Las Villas (UCLV) started its academic activities on November 30, 1952, in the northeast section of the city of Santa Clara, the capital of the former province of Las Villas. In 1955 a Doctorate in Physics and Mathematics was created (DFM) in the UCLV.

> The Central University was officially established by Law 16 of November 22, 1949 (Official Gazette No. 1, 1952), as cited in the Statutes of the Central University "Marta Abreu" of Las Villas, Article 1, Agreement 202 University Council, and the day August 26, 1957 (Official Gazette No. 46, 1957). The same document established the beginning of its activities in 1952, and formalized the Eastern University, which had functioned unofficially since 1947 (Morgado 2010).

During this period, there were qualitative and quantitative changes that helped Cuba escape the underdevelopment in education during the colonial period. Among the changes were several important books that were written for teaching mathematics in the high schools and universities. However, these changes did not guarantee access to education for large masses of the population.

The Cuban Revolution (1959–2013)

This period began with the triumph of the Cuban Revolution, which spent substantial resources on education at all levels. During this period, elementary education was divided into two cycles, the first cycle of first to fourth grade and the second from fifth to sixth grade. General secondary education consisted of the lower secondary school from seventh to ninth grade. Upper secondary education was then called Pre-University in tenth through twelfth grade.

The mathematics training of students, at all levels during the beginning of the 1960s, was very similar to the one developed in the early fifties, focusing on classical mathematics. By the mid-1960s, "New Mathematics" arrived in Cuba from Europe and was based essentially on set theory and mathematical logic.

Beginning in elementary school, students began to be introduced to intuitive notions of set theory and the first elements of Euclidean geometry. Textbooks for middle school and high school were developed with a focus on modern mathematics. Exercises, examples and proofs were presented with a formal structure. As the level of knowledge increased, the problems to be solved became much more complex. By secondary school students were expected to not only solve problems, but also to prove some elementary propositions and corollaries of classic theorems.

The introduction of the Bourbaki ideas in mathematics in Cuba did not have the expected results, mainly for the following reasons:

- The formality and purely deductive nature of the proofs caused them to be memorized without any meaning.
- The theoretical and methodological preparation of teachers was insufficient to carry out the programs.
- The rapid introduction in all levels did not allow for a gradual transition of students from previous levels of knowledge.

In the late 1970s, German and Cuban scientists began developing a new curriculum based on what was being done in East Germany. The East German influence on curriculum lasted until the late 1980s. This curriculum intended to give a deductive approach to all teaching, with mathematical proofs beginning in the elementary school. Textbooks in the beginning were literal translations of the German texts. They contained many definitions and theorems. The exercises were merely formal in terms of practical implementation. Actually, these texts did not help the students very much. However, it can be argued that the mathematical development at that time was at a very advanced level. To finish high school, students mastered the basics of differential and integral calculus, analytical geometry, theory of equations and combinatorics. At the beginning of the 1990s, new programs developed primarily by Cuban specialists were introduced, with textbooks and methodological orientations that were very well elaborated and contextualized. It is important to mention that although the goals and the mathematical level were the same as in the German program, the results were much higher. This was mainly because of the professional development that the teachers who were implementing the curriculum received on the mathematics and the related teaching methodology.

To discuss the development and evolution of mathematics competitions in Cuba it is necessary to mention the illustrious educator and mathematician Luis J. Davidson, winner of the Paul Erdös Award, who is recognized as the Founding Master of Mathematics Competitions and Olympiads. In 1963, while working as a national mathematics supervisor, Davidson proposed holding a national competition for pre-university students. This initial momentum favored the creation and development of a structured system for the comprehensive training of students with scientific talent, in order to create scientific leaders for the country. In addition, this system promoted the selection and development of students with an aptitude for mathematics for future participation in the mathematical Olympiads.

In Cuba, educational competitions are conducted at all levels. The selection and preparation of students for participating in mathematics competitions starts in elementary school and continues into higher education. The competitions are developed by the best teachers nationwide. These teachers receive specialized training in universities in their provinces. Once the talented students in the different provinces of the country are selected, they receive special training and compete to try to reach the national level. Each year a national team is formed with the best students. That team participates in various mathematical Olympiads and competitions at the international level.

To help in the development of scientific talent in Cuba, Vocational Pre-University Institutes of Exact Sciences were created in each province over 40 years ago. At their inception, students graduating from sixth grade entered these institutes and received lower and upper secondary education specializing mainly in physics, chemistry and mathematics. Beginning in the 1980s, students entered after completing lower secondary school.

Most Cuban representatives in international competitions in mathematics are students from the Vocational Pre-University Institutes of Exact Sciences across the country. This situation highlights the uniformity of the mathematics education they receive. The results of Cuba in the competitions held at the continental level can be classified as good. The most important event worldwide is the International Mathematical Olympiad. The Cuban team participated for the first time in the former Czechoslovakia in 1971, and since then has participated 39 times. The team has won one gold medal, seven silver medals, 36 bronze medals and 22 honorable mentions.

Mathematics Teacher Training

The first three pedagogical institutes were created in 1964 as faculties of the most important universities in the country at that time. They offered what were called higher level teaching degrees. They prepared students to become teachers in various subjects for lower and upper secondary schools. The preparation in some specialties was quite good. For example, in the case of mathematics, students from both the Pedagogical Institutes and the School of Mathematics attended the same classes for courses and modern mathematics seminars, such as modern algebra, topology, among others, taught by foreign teachers.

As universities were created in all provinces of the country, the network of Pedagogical Institutes was strengthened. By 1972 there was a significant increase in enrollment in lower secondary schools and a consequent need for more teachers. The "Manuel Ascunce Domenech" Pedagogical University was created. Its students were graduates of tenth grade. They studied for four years while at the same time working as teachers. Later they studied two more years to receive a Bachelor's Degree in Education in a specialty. That generation now composes the core of the most experienced teachers in the pedagogical universities of the country.

In 1976 pre-service and in-service education of teachers were integrated into a single subsystem of Preparation and Development of Pedagogical Personnel. At that time, institutions that had been engaged in pre-service and in-service education of teachers were part of a restructuring of higher education by Law No. 1307 of 1976. Rather than pedagogical faculties within existing universities, they became independent pedagogical universities known as Higher Pedagogical Institutes attached to the Ministry of Education.

During the academic year 1977–1978, Study Plan A for a Bachelor's in Education with a major in mathematics was implemented. It was designed for high school graduates and required four years of study. Upon considering the level of information to be processed and the little time devoted to it during 1982–1983, it was increased to five years. During 1983–1984, a new Study Plan, called B, was implemented. It is important to note that with the creation of the Ministry of Higher Education in July 1976 more emphasis was put on the pedagogical content knowledge in mathematics in the preparation of mathematics teachers.

The curriculum for the Bachelor's in Education with a major in mathematics, the original Plan A, focused on pure mathematics. In contrast, the pedagogical and methodological preparation was quite limited, basically, for a lack of subject specialists. Plan B retained the positive aspects of the previous plan in relation to the level of content to develop mathematical knowledge and increased the methodological preparation to meet the challenges of the so-called modern mathematics.

Until then, the theoretical and mathematical knowledge required for the Bachelor's in Education with a specialty in mathematics was very similar to that of the Bachelor's in Mathematics. Many notable mathematicians of the time, such as Marie Cottrell (French) and Diederich Hinrichsen (German), worked at the Higher Pedagogical Institute of Cuba. Most of the subjects and the depth of content were common to both groups of undergraduate students. However, their career paths were fundamentally different: those with a Bachelor's in Mathematics would be devoted to higher education and research in mathematics, and those with a Bachelor's in Education would be devoted to pre-university education and research in mathematics education. For the training of teachers, specialists from East Germany were incorporated. These German professors were commissioned to provide courses, seminars and professional development workshops. They worked diligently to prepare new instructional materials for an integral preparation of teachers. Within the field of mathematics methodology many of Cuba's best teachers received doctorates at universities in East Germany and the Soviet Union, ensuring the improvement of teacher preparation to meet the challenges of teaching modern mathematics.

Beginning with the academic year 1990–1991, Curriculum Plan C took effect. It required five years for a Bachelor's in Education and offered a mathematics computing specialty. By trying to form a more comprehensive professional, there was less emphasis on the mathematical knowledge of students, particularly as many hours were dedicated to the study of computing. Plan C also increased the time in student teaching practice beginning in the third year. In 2003–2004 there were several changes. A new program called "Integral General Teacher" appeared. Its graduates were prepared to teach the subjects of mathematics, physics and computer science at both the lower and upper secondary levels across the country.

In the 2010–2011 academic years Plan D went into effect. It aims to address the process of formation of mathematics and physics high school teachers. The level of mathematics that is expected to be achieved with this new plan can be termed basic, and is based on the level of knowledge that students at the upper secondary level are expected to reach in Cuba. It is important to note that the pedagogical and methodological emphases within this plan are very advanced and include the newest results in these areas.

The Bachelor's in Mathematics and its Contribution to Teaching

In 1962 there was a Higher Education Reform in Cuba. Its goal was to reorganize and update this level of education. That same year the Bachelor's degree in Mathematics at UH with specializations in pure mathematics, numerical analysis and statistics was implemented. In January of the following year, the degree was offered at the Central University of Las Villas (UCLV) with a specialty in pure mathematics, and in 1969 it was offered at the University of the East (UO).

The curriculum for the new Bachelor's in Mathematics began with a much higher level of mathematics than had been the case with the Doctorate in Physics and Mathematics. It was a five-year program rather than just four and the courses were more up-to-date. For example, in the course of mathematical analysis there was a formal construction of the real numbers. Modern algebra was taught using the textbook written by Roger Godement. Algebra, topology and set theory had books by Kazimierz Kuratowski. Functional analysis and measure theory used books written by Andrei Kolmogorov. The fundamental objective of the Bachelor's in Mathematics was to prepare research mathematicians.

During the early years of this period many educators emigrated from Cuba. The faculty left at UH and UCLV consisted of professors from the previous Doctorate in Physics and Mathematics, who had not migrated, and high school teachers. There were also professors from Latin America and European countries who came to Cuba to offer their assistance. Students played a significant role in the program as teaching assistants in what was called the Student Assistants' Movement. At this time the first groups of high school graduates with a socialist education began to arrive at the universities.

In the second half of the 1960s a group of professors, known as Liasson, arrived from France to contribute to Cuban higher education. They were led by Didier Dacunha-Castelle. Many of these French professors contributed to the development of mathematics in Cuban universities and pedagogical institutes. In the UCLV, besides the French Liasson professors, American professor Robert Sandling visited for one year. Moreover, West German professors created the support group KOWIZUKO, led by Klaus Kirkerberg. This committee had extraordinary impact on the development of numerical analysis and statistics, and the courses that were taught or restructured, contributed to what later became a major in operations research.

In the late 1960s and early 1970s, these two committees helped Cuban universities to recruit European professors to teach short courses. These summer courses contributed significantly to the professional development of mathematics professors and students, and put them in touch with advances in computer science. This helped with the subsequent creation of the specialization in computer science. At the same time, professors from Eastern Europe (Soviet Union, East Germany, Czechoslovakia, Hungary and Yugoslavia) arrived for both short and long stays at Cuban universities. They conducted research with professors and students, and in some cases, taught courses.

In the early 1970s the curricula of the three degree programs in mathematics were unified as a result of several meetings of professors from the three universities. This also facilitated the connection of students with other sectors, including production and services. These connections fostered the development of applied research and expanded job opportunities for students upon graduation.

The continuous improvement of curriculum and degree programs in mathematics also took place in the higher pedagogical institutes with the changes mentioned above called Plans A, B, C, C modified, and D. These plans were developed by Cuban professors, many of whom had earned doctorates in Cuba or abroad. Textbooks for these careers began to be written by Cuban professors in the late 1970s. In the 1980s, almost all subjects had a textbook written by Cuban professors.

Plan C had as one of its objectives to overcome a series of difficulties related to linking theory with practice and study with work. This effort was based on significant research conducted to determine what kind of mathematics was needed by the country. According to Valiño (2003), the results of this research provided a characterization of professional performance of the mathematicians with significant experience. This facilitated the definition of the content of the various areas within the program.

> In this process what was called "Professional Practice of the Mathematician" played an important integrative role. Its objective was to simulate mathematical professional activity by confronting real problems throughout the program. It culminated with a graduation thesis with a direct link to professional practice (Valiño, 2003).

Today at the University of Havana and the University of the East there is a Faculty of Mathematics and Computer with Bachelor's degree in Mathematics and Bachelor's degree in Computer Science. In the Central University of Las Villas the Bachelor's degree in Mathematics is one of the degrees in the Faculty of Mathematics, Physics and Computing. Others include a Bachelor's in Computer Science and another in Engineering Physics.

Currently there are 15 provinces in Cuba. In each there is at least one of the 16 existing universities, or a branch university that is run by a university in another province. Many of the faculty members in these universities have master's degrees or doctorates. For this reason, we believe the three Bachelor's degrees in Mathematics have contributed significantly to the development of mathematics education, and in particular, to the teaching of mathematics in the country.

Teaching of Mathematics in Other University Programs

The teaching of mathematics in programs other than the Bachelor's in Mathematics should have different objectives, not only in the syllabi, but also in teaching. For engineering, economics, chemistry and physics, among others, one of these objectives could be to serve as a tool for solving various problems of the discipline. To fulfill this objective, the following student skills and abilities should be developed:

- To use mathematical modeling to transform a problem of their discipline (raised in the language of the same) into a mathematical problem (posed in the symbolic language).
- To apply the mathematics needed to solve the model.
- To transform a mathematical (symbolic) solution into a solution relevant to the discipline.

However, these and other equally important objectives were often neither enunciated in the programs nor fulfilled. We will refer in more detail to the development of mathematics education in engineering programs since the situation in chemistry, physics, economics and others was not very different.

In the decade of the 1960s there was a great increase in the number of students enrolled in engineering programs at the same time that many university professors were emigrating. Consequently, the fundamental problem to be solved was to ensure that all mathematics courses had a teacher. Therefore, many engineers and students in the Bachelor's in Mathematics began to teach the mathematics courses needed in other programs. Obviously, the first thing that had to be done was to prepare these New Mathematics instructors both technically and pedagogically. This was done in each university in a more or less similar way.

The restructuring of plans and programs in engineering was also performed using Plans A, B, C, C modified, and D. In the Cuba of the early 1970s, the theoretical content was taught through lectures, and exercises and problem solving were taught through practical classes. The lectures were given by the better prepared professors from the departments, who were called master teachers, in auditoriums with up to 100 students. The practical classes had groups of no more than 25 students. Those who taught the practical classes received a preparation directed by a master teacher on how to teach the contents of those practical classes. The lectures were attended by students from different engineering disciplines, but practical classes were grouped by specific engineering disciplines.

In those years, mathematics courses were aimed at developing procedural techniques. It was very difficult to develop a learning-teaching process of mathematics through modeling (giving meaning to mathematical objects through problems from specific disciplines). Consequently, the skills developed in mathematics courses were mainly procedural techniques.

By the end of the 1970s there were mathematics professors who were well-prepared technically and pedagogically. They began to be concerned with the contextualization of the mathematics they were teaching and about the writing of textbooks. This situation continued to improve during the 1980s. In the 1990s the improvement of education in programs such as engineering, science, economics and others, took the contextualization of teaching more into account and the professional development of mathematics instructors to facilitate such contextualization. Master's thesis work on the contextualization of mathematics in various disciplines as a way to improve teaching was conducted. Among these theses are those of Ortega (1999), Ramírez (1996) and Taillacq (1996).

The contextualization of teaching is a problem that needs to be addressed in all courses, not just in mathematics courses. For example, if mathematics teaching is contextualized perfectly, little would be accomplished if professors of other subjects that use mathematics did not know or did not take into account what has been done in this sense, or if they did not follow up with the use of such mathematical thinking in specific disciplinary courses. Much in this regard is being done and much remains to be done.

The Academy of Sciences of Cuba and its Implications

After great efforts and negotiations were made by Cuban scientists from the Royal and Literary University of Havana (1728) and the Economic Society of Friends of the Country (1793) to Spanish authorities on May 19, 1861, the Royal Academy of Medical, Physical and Natural Sciences of Havana was established by decree of Queen Isabel II of Spain. One of the fundamental objectives of its founders — still a driving force — was to contribute to the development of Cuban science and publicize national and international scientific developments. In particular, their interests and initial studies were focused on medical and natural sciences. Mathematics was not considered to be within its domain. The creation of the Royal Academy of Havana marked a milestone not only for Cuba, but for all the Americas, since it was the first such institution established outside of Europe (the National Academy of Sciences in the United States was established two years later.) The founding President of the Royal Academy was the honorable Nicolás José Gutiérrez Hernández (1800–1890) who remained in office from 1861 to 1890. A sample of the natural relationship between the Royal and Literary University of Havana and the National Academy was that Gutiérrez was Rector of the University in the period 1879–1880. In conclusion it can be stated that the Academy from 1861 to 1898 accomplished its goals and was able to have as members the distinguished scientists of the time.

In 1902, the Royal Academy dropped the word Royal from its name, becoming the Academy of Medical, Physical and Natural Sciences. It maintained essentially the same goals, but experienced very little development until 1961. In 1962, a period of major structural and functional changes, the prominent Cuban scientist Antonio Núñez Jiménez (1923–1998), was elected President of the newly-named Academy of Sciences of Cuba and served until 1972. In order to organize and restructure the sciences and their implications for society, the National Commission for the Academy of Sciences of Cuba was created on February 20, 1962. For the first time, the Academy acquired a national character. One of the most important considerations of the Commission for the development of mathematics in Cuba was, first, to consider this science as an important element for the development of society and to encourage the creation of an institute dedicated to research in mathematics. At the end of 1963, a group dedicated to research in mathematics was created. The basic areas to be developed were algebra, geometry, functional analysis and statistical modeling. As a result of the diversification of the areas of research, and with the support of the Academy of Sciences of Cuba, the Institute of Mathematics and Computer Cybernetics was created in 1973.

Since 1980 the Academy of Sciences of Cuba has been the governing body for the development of science and technology in the country. According to Ruiz (1987) the objective conditions were given to develop a strong movement in mathematics research. The training of young mathematicians and support from the Soviet Union and East Germany in graduate training were pillars of this training. This favorable situation in mathematics research led to the unification in 1986 of the Institute of Mathematics and Computer Cybernetics and the Fundamental Technical Research Institute as the Institute of Cybernetics, Mathematics and Physics (ICIMAF). Currently, ICIMAF is attached to the Academy of Sciences of Cuba and senior specialists work there. The main areas of mathematics in which they are doing both pure and applied research are algebra, differential equations, analysis, statistics, geometry, numerical methods, optimization, and signal processing algorithms, among others.

Today the Academy of Sciences of Cuba is considered an official institution of the Cuban state. It plays an independent consultative role in science and technology. It is attached to the Ministry of Science, Technology and Environment. Composed of scientists with exceptional merit, it is organized to operate in the following sections: Agricultural Sciences and Fisheries, Biomedical Sciences, Humanities and Social Sciences, Technical Sciences and Natural Sciences. Among the main objectives of the Academy of Sciences of Cuba are the following: to contribute to the development of Cuban science and the dissemination of national and universal scientific advances; to promote excellent scientific research in the country; to raise the professional ethics and social value of science; and to provide closer links among scientists and their organizations, with society and with the rest of the world.

The Academy of Sciences of Cuba offers an award for the best results in scientific and technological research in the country. Two prominent Cuban mathematicians, Miguel Borges Quintana and Ricardo Abreu Blaya have won prizes from The World Academy of Sciences (TWAS — named Third World Academy of Sciences until 2004).

On February 20, 1962, Law 1011 created the National Commission for the Academy of Sciences of Cuba. This was the first effective step in realizing a strategic objective of the Cuban government to offer the scientific training needed to meet the challenges of society. In 1960 Fidel Castro stated: "The future of our country has to be necessarily a future of men of science ..." (Castro, 1960).

The creation of the National Commission for the Academy of Sciences of Cuba, university reform, and the fostering of research institutes in 1962 were the pillars for the current development of science in Cuba. In particular, the university reform constituted the legal framework which gave the university its function and role in graduate education. The graduate education in Cuba before 1965 was practically nil. The first defenses of master's and doctoral theses were performed at the National Center for Scientific Research in the late 1960s. Later, in the early 1970s, the first master's thesis in mathematics was defended at the University of Havana.

Graduate education in Cuba develops in two directions. The first, consisting of short courses, trainings and certificate programs, aims to continuously raise the professional competence of university graduates; the second is a graduate academic program leading to the Ph.D. or the Doctor of Science (Dr.Sc.). It should be noted that to obtain the Ph.D., two formal publications of the essential content of the dissertation, along with a presentation of the results at a minimum of two national or international scientific events must be achieved before the final defense. To obtain the degree of Doctor of Science, a professional career with national and international recognition is needed.

In 2000 Cuba had 5,378 individuals with doctorates and in recent years a very high level of graduate education has been achieved. As an example, consider the following data from the Ministry of Higher Education of Cuba. Since 2006, over 500 dissertations have been defended every year in the country. The record so far is 573 from 2009. The number of dissertations defended up to April of 2011 was 11,618. Of these, 313 are Doctors of Science. There are currently 58 institutions authorized to grant doctoral degrees and 58 permanent panels for dissertation defenses.

In Cuba, from the earliest years of the 1960s, courses were developed and seminars taught by foreign teachers from different countries. It has been mainly university professors who have attended those seminars. Moreover, as part of its science policy, Cuban specialists were sent abroad, mainly to the former socialist countries of Europe, in order to pursue doctoral studies. On December 2, 1974, Law 1281 came into force to regulate the National System of Scientific Degrees in the Republic of Cuba, and the National Commission for Scientific Degrees was established in 1997. By then, there were professors qualified to develop graduate education in the country, including more than 300 professionals with a doctoral degree, about 200 of them trained in foreign universities.

It is important to note that between 1991 and 1996, due to the fall of socialism and the adverse economic impact that this phenomenon produced in Cuba, scientific education in all disciplines decreased. An element for recovery beginning in 1997 has been the work done in Cuban universities to prepare gifted youth for scientific work.

The Faculty of Mathematics and Computer Science at the University of Havana is the leading center of mathematical sciences in Cuba and has more than 60 faculty members with master's or doctoral degrees. In addition, three master's degrees are offered, including the Master of Mathematics accredited for excellence in 2001 and 2009, and a Ph.D. in research mathematics which has a large group of domestic and foreign graduates. Currently, the faculty is successfully implementing a doctorate in mathematical sciences and is the official home of the prestigious National Panel for Scientific Degrees in the fields of mathematics and computer science.

There are several institutions in Cuba with authorization to grant doctorates in mathematics including UH, UO, UCLV and ICIMAF. The specific fields include mathematical analysis, differential and integral equations, algebra, mathematical statistics and probability theory, geometry, mathematical optimization, discrete mathematics, programming languages, information systems, artificial intelligence, computer science, numerical mathematics. Doctoral education in pedagogical sciences with a specialization in mathematics has been developed significantly, particularly with respect to the application of scientific results to school practice. Among the institutions that have developed both research and curricular doctorates are the University of Havana, University of Orient, Central University of Las Villas, the Higher Polytechnic Institute José Antonio Echeverría, the University of Camagüey, and all the pedagogical universities of the country.

As an example of graduate education in Cuba and the realization of the statements above, consider the year 1971 when Miguel A. Jiménez Pozo, anticipating the needs and regulations, presented a thesis for a Master's in Mathematical Sciences on a topic in algebras of functions to the authorities of the Faculty of Science at the University of Havana. It could not be defended until 1973 for lack of a competent panel. A panel was finally organized in Cuba taking advantage of the presence of Egbert Brieskorn, a German mathematician who had come to the University of Havana on a mission funded in cooperation with the International Mathematical Union, who agreed to serve as President. Brieskorn's presence in Cuba had a significant impact on Cuban mathematics. Jiménez also encountered difficulties with the lack of a panel in Cuba for the defense of his Ph.D. dissertation in 1979. In the spring of 1978, he had a successful pre-defense in the Department of Theory of Functions and Functional Analysis of the Lomonosov Moscow State University on approximation theory. Finally his thesis was accepted for a Doctor of Science in 1993, which by law must address several topics. But while Jiménez was always the first to defend all these degrees in Cuba, he did not always support other Cubans (Cuba residents) who were obtaining equivalent degrees in foreign universities. Indeed, Carlos Bouza obtained his degree of Master's of Science at the same time as Jiménez on topics related to the statistical sampling with a Yugoslavian advisor. In the mid-1970s, Mario Estrada, a member of the Institute of Mathematics of the Academy of Sciences, earned his Ph.D. on topics of algebra in East Germany. Shortly thereafter, the Candidature in Science, the Russian equivalent of the Ph.D., was obtained by Carlos Sánchez at Lomonosov Moscow State University, in algebra of functions. With the obvious exception of the Doctor in Science, which has a lot of academic requirements and is not essential for positions in the Americas, the defense of master's theses and first doctoral dissertations from many Cubans started to become quite common in Cuba and abroad.

The Cuban Society of Mathematics and Computer Science and its Implications

To speak of the Cuban Society of Mathematics and Computer Science, we must go back to 1942 when a group of students of physical and mathematical sciences, gathered in the city of Santiago de Cuba at the Third Congress, agreed to combine efforts for the development and dissemination of these sciences in Cuba. There were 121 founders of the society, on February 25, 1942, in the Poey Auditorium of the University of Havana when the Cuban Society of Physical and Mathematical Sciences was officially created. The first President of the fledgling society was Pablo Miquel y Merino. It can be suggested that his election as President was a recognition of his scientific and educational contributions to the development of mathematics in Cuba.

The Cuban Society of Physical and Mathematical Sciences remained active until 1960. The fourth and last President was Mario Orlando Gonzales, who worked hard to develop research in basic mathematics and to make the Cuban Society known in the international community. Keeping the essence of this society, the Cuban Society of Mathematics was established in 1978, having been made possible by the Associations Law 1320 in 1976. The first President of the Cuban Society of Mathematics was the outstanding algebraist and developer of new generations of mathematicians Mario Estrada Valdés (1930–2010). Ten years later, in recognition of developments in computer science, the name was changed to the Cuban Society of Mathematics and Computation. The current President of the society is Luis Ramiro Piñeiro. Extraordinary work has been accomplished despite the economic problems in the country.

Most members of the Society are professional mathematicians and computer scientists, some with specific interests in education. To meet the challenges of each of the areas of knowledge the Board is structured as follows: President, four Vice-Presidents and four at-large members per area (mathematics, mathematics education, computer science, and computer education). To achieve national objectives and take into account the different dimensions and scope of mathematics and computer science, the country has both institutional and provincial delegations.

Given the broad character and comprehensive nature of mathematics and computer science, there is a delegation in each of the provinces. There are also branches in the following institutions: University of Havana, UCLV and the Pedagogical University Enrique José Varona. Among the key achievements of the Society are the following: holding quality scientific events at both the national and international levels; organizing specialized scientific sessions; effective linking of all its members with the Society in general; collaborating internationally with universities, scientific publishing houses, and other companies; supporting both national and international exchange of scientific groups with scientific institutions; and promoting the identification and development of students with an aptitude for science.

The Cuban Society of Mathematics created a prize for research named after Pablo Miquel. It was first awarded to Miguel Antonio Jiménez Pozo in 1981, for his international contributions to the study of the convergence of sequences of linear operators in the context of the approximation theory. The second was received by Andrew Fraguela Collar in 1982, for his results in partial differential equations and differential operators applied to wave equations. Francisco Guerra Vázquez received the Miquel award in 1983 for his contributions to the subject of optimization, particularly in the treatment of multi-objective optimization problems. There has been difficulty in selecting recipients of the Miquel award given the large amount of work and varied topics treated by Cuban mathematicians, such as complex rational approach, computer science, pattern recognition, computational logic, inequalities in partial differential equations, linear programming with many variables, ordinary differential equations, pedagogy, mechanics, and applications.

In a General Assembly in July 1982 in Havana, it was agreed to recognize the work of the first Board of Directors and elect a new President of the Cuban Society of Mathematics: Miquel Awardee, Miguel A. Jiménez Pozo, who worked as a professor at the University of Havana. Jiménez was re-elected on two consecutive occasions until 1991, leaving a legacy of a Cuban Society of Mathematics and Computation that was vibrant and thriving. The Minister of Education, José R. Fernández, recognized the importance of the Society as a professional force organized for and interested in supporting the work of the Ministry in the teaching of mathematics at all levels. Consequently, there has always been fundamental support for the Society. The support received from the Ministry of Education has been reciprocated by the Society with serious work to support the plans and objectives of the Ministry.

The Second National Congress was held in 1986 in coordination with the University of Havana and its Rector, Fernando Rojas, with the strong support of the Cuban Society of Mathematics and Computer Science. The Minister of Education, José R. Fernández, was honored with the distinction of Honorary Member by the Congress. The Third Congress was held in 1988 in the city of Santiago de Cuba, with support from the University of Orient. The Fourth Congress was held in the city of Holguín, where the Cuban scientist and educator Manuel Mariño, a professor at the University of Holguín, was elected President. Thus, the Society passed for the first time to establish its headquarters outside of Havana. The Fourth Congress was held in conjunction with a scientific event called COMPUMAT, which is held every three years.

Another essential element developed by the Cuban Society of Mathematics and Computer Science was the organization of scientific activities throughout the country, taking advantage of the creation of its delegations and the powerful human wealth of some provinces, as was the case of Las Villas, with professors Eberto Morgado, Lorgio Batard, Otilio Mederos, and others. Holguín had counted on the enthusiasm of professors Manuel Mariño, Juan E. Nápoles and Mauro García, who eventually presided over the Society, as well as Héctor Pijeira in Matanzas.

The Society has also supported events of the Academy of Sciences of Cuba and the University of Havana. In coordination with the University of Havana and the Mathematical Cybernetics Institute of the Academy of Sciences, the Society participated in the organization in 1987 of an International Congress of Approximation and Optimization, which was attended by the leading scientists of the socialist countries in these specialties and other mathematicians of the Western world. For this event, support was received from the International Mathematical Union (IMU) and the Third World

Academy of Sciences (TWAS), among others. The proceedings were published by Springer Verlag in the collection *Lecture Notes in Mathematics*. The meeting was such a success that it was held again in 1993 and the proceedings were published by the Peter Lang Series. Since then every two or three years the International Series of Conferences on Approximation and Optimization in the Caribbean has been held in a Caribbean country.

Cuban Journals on Mathematics and its Teaching

In 1970, the *Epsilon* journal was published by the Department of Mathematics of the General Education Department of the Ministry of Education. In the 15 editions this magazine has published information related to the Mathematical Olympiads and other contests — and of general interest for teachers — as well as articles with topics on mathematics education.

The Cuban Society of Mathematics and Computer Science, along with the Ministry of Education, created the *Bulletin of the Cuban Society of Mathematics and Computer Science* in 1978 with the same purpose as *Epsilon*. Print-based publication stopped with Bulletin No. 16 in 1994, but it continues to be published electronically.

In 1966 the journal *Operations Research* was created in the Department of Applied Mathematics of the Central Planning Board, currently in the Faculty of Mathematics and Computer Science. This journal publishes papers with original results in optimization theory, mathematical statistics, numerical methods, game theory, and decision theory, as well as interesting applications. It is referenced, among others, by BINITI, *Mathematical Reviews*, and *Zentralblatt für Mathematik*.

The *Journal of Mathematical Sciences* was created at the University of Havana in 1980. It accepted articles in Spanish or English and was printed quarterly until 1999; it continues to be published electronically. It was registered in Latindex in 2001 and had a modification in February of 2013. Its main objective is to publish original research in mathematics and computer science. Currently, it also publishes works on mathematics education.

Concluding Remarks

The most significant leap in the development of mathematics in Cuba, and in particular of its teaching, began in 1962. The fundamental factor that allowed this to happen was the fact that the Cuban State established education as one of its core priorities. Many selfless specialists who came to Cuba to assist in this noble goal contributed to the monumental task of the extraordinary progress of mathematics education. These specialists created favorable conditions for the learning of and research in mathematics for those of us who were then — and those who are now, decades later — students of mathematics.

Bibliography

Barcia, R., & Sardiñas, H. (2005). Reseña histórica de la enseñanza de la Geometría en la formación de maestros primarios en Cuba antes del Triunfo de la Revolución. *Revista Electrónica Coronado*, *1*(1).

Castro F. (1960). Discurso pronunciado en el acto celebrado por la sociedad Espeleológica de Cuba, en la Academia de Ciencias, el 15 de enero de 1960.

Davidson, L. J. (2003). Primeras Palabras. *Boletín de la Sociedad Cubana de Matemática y Computación*, *1*(1), 1–3.

De la Torre, R. (1999). Historia de la Enseñanza en cuba. Proyecto Educativo de la Escuela de Hoy. Retrieved August 25, 2013 from https://sites.google.com/site/escueladehoy/historia-de-la-ensenanza-en-cuba

García, I. (2005). Contribución del movimiento educacional de la escuela privada de primera y segunda enseñanza en el período 1790 a 1868 al desarrollo de la teoría y la práctica pedagógica cubana. Doctoral dissertation. Instituto de Ciencias Pedagógicas "Enrique José Varona". La Habana. Cuba.

González, M. O. (1956). *Algebra Elemental Moderna*. La Habana: Editorial Selecta. Vols. I & II.

González, M. O. (1960). *Matemática Quinto Curso. Complementos de Aritmética y Álgebra*. La Habana: Editorial Selecta.

González, M. O. (1960). *Matemática Quinto curso. Complementos de Geometría y Nociones de Cálculo Diferencial e Integral*. La Habana: Editorial Selecta. Décima Edición.

Jiménez, M., & Sánchez, C. (1993). *Revista Ciencias Matemáticas*. *14*(2–3), 99–123.

Ley de Instrucción Pública. Gaceta de la Habana No. 233, 30 de septiembre de 1880, pp. 1–6.

Marchante, P., & Merchán, F. (2007). Enseñanza de la farmacia en la Real Universidad de La Habana según Plan de Estudio de 1863. Etapa 1871–1880. *Revista Cubana de Farmacia*, *41*(2).

Marchante, P., & Merchán, F. (2008). Reorganización de la enseñanza en la universidad de La Habana según el último plan de estudios de su etapa colonial (Plan de 1880). *Revista Cubana de Farmacia*, *42*(3).

Ministerio de Educación: Planes de Estudio "A" y "B" de la Licenciatura en Educación, carrera de Matemática, 1976, 1982.

Ministerio de Educación Superior (1987). Documento base para la elaboración de los Planes de Estudio "C".

Ministerio de Educación Superior (2009). Documento base para la elaboración de los Planes de Estudio "D".

Morgado, E. (2010). *Sucesos y anécdotas de la Universidad Central "Marta Abreu" de Las Villas.* Santa Clara: Editorial Feijoó.

Ortega, A. (1999). Perfeccionamiento de la Enseñanza de la Matemática en la Carrera de Agronomía. Master's degree thesis. Facultad de Matemática, Física y Computación. UCLV. CUBA.

Ramírez, E. (1996). Perfeccionamiento de la enseñanza de la Matemática en la Carrera de Licenciatura en Ciencias Farmacéuticas. Tesis de maestría Facultad de Matemática, Física y Computación. UCLV. CUBA.

Ruiz, J. (1987). Acerca de las perspectivas de las investigaciones matemáticas en Cuba (1ra parte). *CIENCIAS. Revista de Difusión,* 18–24.

Sócrates, R. (1941). *Matemática Tercer Curso. Álgebra.* La Habana: Editorial Selecta.

Sócrates, R. (1945). *Matemática Primer Curso. Aritmética.* La Habana: Editorial Selecta.

Sánchez, C., & Valdés, C. (2003). Bosquejo histórico de la actividad matemática en Cuba. La Habana. *Boletín de la Sociedad Cubana de Matemática y Computación, 1*(1), 17–29.

Sánchez, C. (2013). Discurso en Conmemoración del 70 Aniversario de la Fundación de la Sociedad Cubana de Ciencias Físicas y Matemáticas y de su Revista científica. *Revista Anales de la Academia de Ciencias de Cuba, 3*(1), 1–4.

Taillacq, A. (1996). Perfeccionamiento de la enseñanza en la Carrera de Ingeniería en Telecomunicaciones. Tesis de maestría. Facultad de Matemática, Física y Computación. UCLV. CUBA.

Valiño, B. (2003). La vinculación con la práctica en la formación del matemático. *Boletín de la Sociedad Cubana de Matemática y Computación, 1*(1), 17–29.

Valiño, B. (2012). Discurso pronunciado en el Aula Magna de la Universidad de La Habana en la conmemoración del 50° aniversario de la licenciatura en matemática. La Habana. 31 de octubre.

Valdés, C. (2000). La primera publicación periódica cubana de Ciencias físico-matemáticas (1942–1959): Noticias y consideraciones. *LLULL. 23,* 451–468.

Varona, E. J. (1992). *Trabajos de Educación y Enseñanza.* La Habana: Editorial Pueblo y Educación.

About the Authors

Otilio B. Mederos is a professor of mathematics at the Autonomous University of Coahuila, México, and holds a Ph.D. in physical and mathematical sciences from Odessa State University, Ukraine. He was a professor at the Central University of Las Villas, Cuba for over 40 years. During the prime of his career he conducted research on boundary value problems of analytical functions and differential equations. For the last 15 years, his research has been focused on mathematics education.

Miguel Antonio Jiménez Pozo is a mathematician with degrees from the University of Havana (B.S., M.S., Ph.D., and Dr. Scient.). He is a professor at the Meritorious Autonomous University of Puebla in México since 1995. He was a professor at the University of Havana until 1994, President of the Cuban Society of Mathematics and Computer Science from 1982 to 1991, President of international juries at international Olympiads, and the Cuban Representative at ICMI during the 1980s. He is also a member of the Mexican Academy of Sciences.

José M. Sigarreta is a professor of mathematics at the Autonomous University of Guerrero, México. He holds a Ph.D. in mathematics education from the Pedagogical University of Holguín, Cuba and a Ph.D. in mathematics from University Carlos III of Madrid, Spain. He has conducted research in geometry, graph theory, and the methodology of mathematics and its impact on mathematics education.

Chapter 8

GUYANA: The Mathematical Growth of an Emerging Nation

Mahendra Singh and Lenox Allicock

Abstract: A brief review of the mathematics educational development of a young nation still forging its identity is presented. Various periods are examined to determine how mathematics education was cultivated and changed. Mathematical reform projects that have been implemented are outlined in an effort to improve the quality of mathematics achievement in Guyana and its neighbors, Suriname and French Guiana.

Keywords: Mathematics education in Guyana; history of mathematics education in Guyana; Guyana Mathematics Project; Primary Education Improvement Project (PEIP); Escuela Nueva Project, Secondary School Reform Project (SSRP); University of Guyana; Interactive Radio Instruction (IRI); Guyana's Regional Education Departments (REDs); Education for All — Fast Track Initiative (EFA–FTI).

The northern coastline of South America was visited in the late fifteenth and early sixteenth centuries by European explorers, including Amerigo Vespucci. Little was done to develop this newly-found land until the late-sixteenth century, when explorers like Sir Walter Raleigh began to search for El Dorado, the mythical city of gold. It was not until the seventeenth century, however, that an interest in the area began to take place. While the Spaniards were busy colonizing the rest of South America, the Dutch, French and British all wanted part of the Caribbean basin and its products of sugar and tobacco.

The area was eventually divided up among the three nations. The British took the western part of the area and that became known as British Guiana; today it is an independent country called Guyana. The Dutch claimed the middle part of the area in exchange for giving up their share of the island of

Manhattan. That area became known as Dutch Guiana, which today is the independent country of Suriname. The French claimed the eastern portion of the area, called French Guiana. It remains under French authority to this day and maintains the same name. When slavery ended in the area during the early to mid-nineteenth century, the colonists used immigrants from some other colonies as indentured laborers, mostly from India. Indians and Asians, along with the indigenous people of the area, led to a unique cultural and genetic admixture. As a result of the European conquest and the ravages of new diseases and infections, little history of the indigenous people survived. Even their physical artifacts were either destroyed or lost.

Education under European Rule

Education during early colonization in all three countries was basically nonexistent. It was not until the early nineteenth century that even the most rudimentary educational programs appeared. The schools during this time mostly served the children of the European plantation owners, and were meant to initiate them into European culture. The slaves and native people were not allowed to go to school; schooling was forbidden by law (Bynoe, 1972). After emancipation, the British government established the Negro education grant, which allowed the freed slaves to attend school. They were still not permitted to go to the same schools as the British ruling class.

The education that they received was very basic mathematics and reading. It was not on the same level as the ruling British class (Gordon, 1963). With the basic education they were receiving, the freed slaves bought unused land from the plantation owners and formed small villages that survived by subsistence farming (Ishmael, 2012). The schools developed for the Africans under the Negro education grant were designed to educate them for life in an urban area, not a rural one (Callaway, 1968). As they became educated and trained, some of the Africans moved to the towns and took on various jobs within the government and, thus, became involved with the development of the country. To fulfill the need for workers since the freeing of the slaves, the British recruited "indentured laborers" from India. The Indians who were hired by the British, as laborers, were also allowed to attend school. Eventually they became educated, as were the Africans, to the point that they too began to move to the towns, and became involved in their own governance (Bynoe, 1972).

Most of the schools at this point were controlled by the Christian Church. They were mostly interested in making sure that the Africans and Indians remain loyal to the British and serve the British interests in Guiana (Ishmael, 2012). The educational curriculum was mirrored after the British. Anyone with higher educational aspirations would have to leave Guiana, usually to attend schools in the United States or England. This remained the norm until the 1950s as the ruling British expatriates began to be replaced by the Africans and Indians in significant numbers and in key positions. The Africans and Indians had more say in their governance. At the same time, a pro-independence political party, the People's Progressive Party (PPP), began to grow and gain support.

In the early 1950s, the PPP began calling for independence from Britain. The party began to grow extensively as more of the civil servants, teachers, and the middle class as a whole, became members. They pressed the legislature for better educational programs for teachers, expanding secondary and technical education. The PPP was elected as a self-governing body in 1953. They began to implement some of the programs that they lobbied for. However, just 133 days later the British forcibly removed the PPP from office under the pretext that they were communist (Robertson, 1954). The British imposed an interim government that was mostly made up of anti-PPP and Christian church leaders. They immediately stopped the PPP educational programs and re-established the pro-colonial ones.

In the late 1950s and early 1960s, the PPP was once again elected under a new constitution that allowed internal self-governance. The PPP once again began to openly implement their pro-socialist policies (Ishmael, 2012). This included state control of the educational system. They began a rapid expansion of the secondary and tertiary educational programs and developed a teacher-training program in several different locations throughout the country. They began the building of the University of Guyana. Over 15 percent of the national budget was appropriated to fund these educational programs (Ishmael, 2012). By 1964 the educational system and standards was one of the highest in the Caribbean (Reno, 1964). During this time the political landscape became unstable. The opposition party, the People's National Congress (PNC), claimed that because of the educational policies, the students were "brainwashed into becoming socialist and communist" ("Children May Be Sent to Factories: Vicious Plans Bared", Sun, April 14, 1963). There was so much political bickering and banter that the country was basically divided, most of it along racial lines. The incumbent PPP party was mostly supported by the Guianese of Indian descent, while the opposition PNC party was supported by those of African descent. As a result, the country was pushed to the brink of a full-fledged civil war (Despres, 1967). The British once again stepped in and when the dust was settled, the opposition PNC party gained control.

Education after Independence

It was under this self-government that Guiana gained independence from Britain and became Guyana (Despres, 1967). The newly-elected and independent government began to roll back most of the educational policies that were instituted under the PPP. The new PNC government did, however, continue the development of the technical educational programs as well as the University of Guyana. They did make some staffing changes at the University, most notably the dismissal of its first Principal, Lancelot Hogben, a world-renowned mathematician and educator (Ishmael, 2012). By the mid to late 1960s, the political instability and racial tension had taken its toll on the economy. Unemployment had reached 22%. All of this had a profound effect on the educational system (Jennings, 1999). By 1974 the government had difficulties in soliciting financial support from many capitalistic countries (United States, and England in particular). As a result, they turned to the so-called Socialist Bloc countries, Cuba, German Democratic Republic (GDR), the USSR, and even North Korea for aid. Part of this aid included academic scholarships for Guyanese students to study at the various higher educational institutions within the various countries. In return, the students would have to work for the government in the area or field of expertise. The leader of the government, L. F. S. Burnham even stated that he was now a socialist (Burnham, 1974). He stated in his Sophia Declaration in 1974 that education was going to be geared towards building a socialist society, and instituted full state control of the educational system.

One of the programs that came out of the State takeover was the Guyana Mathematics Project (GMP) (1974). The government had recognized that there was a need to improve the mathematics teaching and learning in the country, as they began to lag behind other Caribbean countries in scholastic achievement. Their goal was to improve mathematics teaching and learning, by providing better teacher preparation through teacher training and research, and developing better student materials. They designed a cooperative learning curriculum, while looking to develop critical thinking and problem-solving skills (Broomes, et al., 1975). Teachers used various techniques to enhance learning in the classroom. They used discovery learning, small groups, called cooperative groups, with guided discussions. Students were expected to work and communicate with each other in order to resolve the issues at hand and solve the problems they were tasked with (Broomes, et al., 1975). It was difficult to evaluate the effectiveness of the GMP. The project was bound to fail, because of the lack of teacher training to address the students' feelings of inhibition and inferiority in the classroom (Cumberbatch, 1972). A slowing economy and lack of trust in the government had finally caught up to the educational system (Jennings, 1999). Still, even though the socialist ideals and justifications may have been lost through the years, some of the materials developed for the GMP are still being used in the classroom today. The GMP remains one of the most significant attempts to reform mathematics education (Jennings, 1999).

A Decline in Educational and Academic Achievement

A few years after Guyana became an independent country in 1966, a number of things contributed to the decline of the education system. Some of the decisions taken by the government did not augur well for the nation. The espousal of socialism, the nationalization of the bauxite company — one of the mainstays of the economy — embittered countries who were chief sources of funding for Guyana (UNESCO, 2006/7). The economic viability of the country was shaken.

In the Executive Summary section of a World Bank document we find that, "In the early 1960s Guyana was among the education leaders in the Caribbean, with high rates of literacy, primary (elementary) school enrollment, and educational attainment of its population." However, since that time according to the same document, "Its education system deteriorated ... in all aspects" (The World Bank, 2005).

The 1970s witnessed a prolonged and serious economic decline in the country which was triggered by a complex of factors, including the socialist policies of the government, state ownership of key sectors of the economy, which in many instances resulted in poor management, an increase in oil prices worldwide, and a fall in the prices of key exports (UNESCO, 2006/7).

The takeover of the schools was also ill-timed. The decline in funds was felt by the education sector. Teachers' salaries were not commensurate with other workers at the same level. An exodus of teachers severely depleted the number of qualified and trained personnel in the teaching sector, and up to today this is still a problem. "High rates of teacher migration to other Caribbean territories where working conditions and salaries were better, had left the secondary system poorly staffed with many untrained and unqualified teachers" (UNESCO, 2006/7). This was a severe blow to education in the 1970s. The migration continued during the 1980s and beyond. The resulting scarcity of trained or qualified teachers led to the gradual decline in the quality of education. The school campuses also deteriorated; there was no running water or electricity and schools lacked adequate sanitary facilities.

Attendance levels dropped to a low of 50%. The Caribbean Examinations Council Secondary Education Certificate (CXCSEC) examination was taken by less than 20% of the age cohort and in 1992 only 18% of the students who took the Mathematics exam passed (UNESCO, 2006/7). This resulted in a drop in enrollment, and dropouts from secondary schools increased, particularly from Form 1 (seventh grade).

The economic decline also negatively affected the tertiary level of education. "Inadequate funding and poor salaries have made it impossible to attract sufficient numbers of highly qualified staff, given that their earnings would be one-fifteenth of what they could earn in similar jobs in other parts of the Caribbean" (Craig, 1993). The University had difficulty attracting students due to this problem.

By the mid-1980s, 15% of primary aged students did not attend school. Of those that did, 23% eventually dropped out (Progressive Youth Organization of Guyana, 1986).

Stemming the Tide of Decline

In the 1980s, the government, in an effort to stem the economic decline, began to diversify its economy, which led to a remarkable growth in revenues and a decline in inflation. In this new state of economic recovery, the government sought to resurrect the dying educational system (UNESCO, 2000).

The government, with the help of donor organizations, began to put a number of measures in place, such as:

(1) Rehabilitating or rebuilding school campuses.
(2) Upgrading the quality of education at primary and secondary levels.
(3) Training teachers so they may implement new strategies.

(4) Offering incentives to teachers to get them to stay and help education reach former levels of achievement.

Before this economic revival was realized, the Ministry of Education, in 1985, put in place a system they hoped would better serve the nation. They divided the country into 10 administrative regions to decentralize it from the capital Georgetown. They expected "to improve the efficiency and effectiveness with which education could be delivered to the rest of the nation" (UNESCO 2006/7). Some of their authority was transferred to the Regional Education Departments (REDs) of the Regional Democratic Councils. Georgetown was placed in a category of its own (UNESCO, 2006/7). Through this measure they hoped to:

(a) Promote the involvement of communities in the management of education.
(b) Enable the Ministry of Education through the REDs to respond more quickly to the needs of the community (Paul, *et al.*, 1986).

The work of improving the education system began with a training program for approximately 70 principals of nursery schools between the years 1987 to 1990 (UNESCO, 2006/7). This was followed by a one-year program (1990–1991) for training approximately 80 regional supervisors (UNESCO, 2000). This was meant to ensure that the nursery schools were well observed, and to enable the Ministry of Education, through the REDs, to make proper provisions. The feedback they gained was necessary to liaise with the Teachers Training College to prepare nursery school teachers with the skills needed to function at optimum. The information obtained by the field officers was given to the University of Guyana, which in 1991/1992 introduced a bachelor's degree program in Nursery Education (UNESCO, 2000; 2006/7).

The 1990s was a problem-solving decade for the education system in Guyana, with a number of projects experimented with to resuscitate the education system from the throes of academic death. In 1990, in an effort to ensure a smooth transfer for students from nursery to primary school, the Ministry of Education started the Primary Education Improvement Project (PEIP). The project was designed to last five years and was funded by a loan of $46.4 million USD from the Inter-American Development Bank. The aim of this project was to improve the quality of primary education. To this end, it focused on three areas that needed to be improved, which were impeding the efficiency and effectiveness of primary school education (UNESCO, 2000; 2006/7). These were:

(a) Improving human resources through their professional qualification and pedagogical skills development, as well as with practical skills training.
(b) Producing textbooks and other resources needed in the classroom and ensuring equitable distribution.
(c) Constructing and rehabilitating primary school campuses.

Although the three areas benefited from the project, the main thrust was aimed at the last two. However, the practical skills training helped to produce textbooks for all levels of primary school in the four core subject areas: English language, mathematics, science and reading. Twenty thousand books were produced at all six levels of primary (elementary) school (UNESCO, 2000; 2006/7).

The biggest, or most capital intensive, was the area of building, repairing and refurbishing schools. As mentioned earlier, the school campuses were dilapidated. However, because of shrewd financial management, the estimated construction of 27 new schools and rehabilitation of three existing structures, turned into 35 new schools built and 64 rehabilitated. As a result of this project, mathematics achievement began to reverse its downward trend (UNESCO, 2000; 2006/7).

The rural (jungle or hinterland) areas of Guyana did not see the same gains as the others. This was mostly due to the difficulties of transportation to and from these areas, which caused them to have a high percentage of unqualified and untrained teachers. The Escuela Nueva Project, a program that was started in Colombia, was implemented. In keeping with the culture of the region, this system was project-oriented. This meant that the students could, at times, be away from school helping their parents on their farms. When they returned to school, they would just continue their project. Promotions to the next level depended on the completion of the assigned projects and not just on continual presence in the classroom. The projects were based on issues that affected their own community and this motivated students to complete them (UNESCO, 2000; 2006/7).

Another project that came to the fore was the Secondary School Reform Project (SSRP). This project was by its very name aimed at doing the same thing for the secondary school system that had been done at the nursery and primary school levels (UNESCO, 2000; 2006/7).

In 1994, an Identification Mission conducted an in-depth fact-finding survey of the economic status and education system in Guyana. A report of this mission was made public in 1996. As a result, the Secondary School Reform Project (SSRP) was initiated. SSRP which commenced in December 1996, according to the Implementation Completion Report, had three broad objectives. These were to:

(1) Develop measures to improve the quality and efficiency of lower secondary education and test them at a limited number of schools in order to gain acceptance before extending the reform to the rest of the system.
(2) Improve the school environment by supporting rehabilitation and repair of schools.
(3) Enhance the ability of national and regional institutions to design, plan for, and implement sustainable education reforms.

The battle was against poor academic performance in Caribbean Examination Council exams (CXC) in Form Five (Grade 11) and the National Third Form Exam (NTFE) in Form Three (Grade 9). As a way to measure and gauge success, a non-elite high school was chosen from which they compared pre-intervention CXC exam scores to post implementation scores (World Bank, 2005).

In 1996 the chosen school had a CXC passing rate of 20.5%. By 2001 this had improved to an 85% passing rate, and in 2003 it was up to 92%. In 2001 a new NTFE exam, developed by the SSRP, was offered in the third form. In general, the pilot schools scores were lower than the national average in the NTFE. This was partially attributed to the displacement due to the school infrastructure improvements, as well as a teacher strike in 2003. Lower scores on the exam also suggested the need for better delivery and better quality assurance of both curriculum and the exam itself (World Bank, 2005). The core subjects tested were mathematics, English language, science and social studies. In the Executive Summary of the Implementation Completion Report, the committee stated that the quality of the education reform was in keeping with best practice in quality improvement. However, the most critical issue in quality reform was sustainability. Based on recent visits, test scores, and interviews, they observed that the pilot schools seemed to be "sliding back" to business as usual. The report on the improvement of the physical environment showed that:

> The 12 designated pilot schools were successfully upgraded. Repairs and renovations were made to 18 non-pilot schools as well. This represented 30% of the approximately 70 public secondary schools in

Guyana. Despite implementation difficulties, the renovated schools are of a reasonable quality, although because of poor scheduling, some facilities remain underutilized. Science laboratories are in this category and some underutilized equipment is already in disrepair. Libraries are staffed by fulltime librarians but are severely underutilized. The final objective of improving management at the national and regional institutions has had limited success, but they are moving in the right direction by analyzing data and learning how to evaluate schools by this means. The major obstacle, however, is not with the analysis of the data. It is rather a collection of problems, namely lack of incentives, inadequate leadership, stakeholders' accountability at all levels, and finally salaries. While they seem adequate in the Guyana context, salaries are far lower when compared to other countries to which many Guyanese can and do migrate to. These are far greater obstacles than training or underutilized equipment (World Bank, 2005).

Despite the PEIP, and Escuela Nueva projects at the primary level, students' achievement in mathematics and English, was still unsatisfactory. The number of students scoring above 50% was significantly lower in these two areas than the other core subjects. Mathematics was definitely the lowest of all the core subjects from 1992 to 1999, except in 1995. The Ministry of Education was still looking for ways to improve in these areas (UNESCO, 2000).

In 2005 the Ministry of Education implemented another project in an attempt to improve the quality of education. This one was aimed specifically at improving mathematics achievement at the foundational level, with the goal of strengthening the future of mathematics in the country. The name of this project was Interactive Radio Instruction (Wintz & Wintz, 2013).

Peter and Godryne Wintz found that "The paucity of trained and qualified teachers, and poor mathematical attainment at the primary school level in Guyana triggered the introduction of the cost effective Interactive Radio Instruction (IRI)." According to Wintz and Wintz,

> The IRI program requires a radio and a facilitator. Before each radio lesson commences, facilitators or classroom teachers are required to write exercises from the teachers' manual onto the chalkboard and pupils copy the exercises into their book. Pupils are guided through the exercises by radio teachers and characters

during the audio portion of the lesson. Approximately every 10 seconds, pupils answer questions or perform some other activity. Learners actively participate by responding orally, writing, drawing pictures, using counters and singing. Oral responses are used in mathematics lessons to encourage active mental participation. Children solved problems silently before giving the answer aloud (2013).

The IRI project was particularly aimed at improving mathematics in the primary schools. According to the two lecturers, it was geared at emphasizing active learning through meaningful interaction by distributive learning. Lessons were designed to help students to develop mathematical abilities and encourage active engagement in the learning process. In total, 106 lessons were designed for radio broadcast, each lasting 50 minutes. They were intended to engage the students sufficiently to ensure that both interest and development in mathematics were maintained (Wintz & Wintz, 2013).

Communications with some senior teachers in Guyana at the primary school level support the findings of Wintz and Wintz. They concur that the IRI program has been working well and has the potential to improve mathematics learning in the classroom. Additionally, Wintz and Wintz have made some recommendations that will, according to their study, help IRI benefit students more and have a greater impact on the successful teaching of mathematics in the classroom. The following are excerpts of some of their recommendations:

(1) Children were not exposed to words on the chalkboard during the IRI lessons. Teachers should write appropriate vocabulary on the chalkboard where possible to help pupils develop their reading skills.
(2) During any mathematics lesson there could be misconceptions. It is the teacher's responsibility to address misconceptions appropriately at the earliest opportunity.
(3) Children requested more engagement from the teachers during the program. During the IRI program, teachers should be guiding children or monitoring their work.

Finally we look at the tertiary level of education and see how it is developing especially in mathematics. Tertiary (university) education in Guyana had its beginnings a little before gaining independence in 1966. The people who had access to university education in the nineteenth and early twentieth centuries were the children of expatriate British officials and the planter class. This type of education could only be obtained by traveling to Europe or the United Kingdom. For those who did not fit into these classes, only one scholarship was offered per year. Alternatively, a student could have done correspondence courses to acquire external degrees from the University of London, but at an extremely high cost (Bernard, et al., 2002).

This all changed in 1955 when the University of the West Indies (UWI) was established. More Guyanese who had completed secondary school could now afford a university education. An added advantage was that it was in a geographic region they were familiar with, without the cultural and ethnic alienation of the UK or Europe. Additionally, the Government of Guyana offered more scholarships. However, the Government of Guyana soon realized that UWI was not as relevant to Guyana's needs in terms of course content and orientation of research. In addition, the entry requirements were based on British universities' standards, making it difficult for many students in Guyana who were prepared by life experiences or secondary achievements to be enrolled (Bernard, et al., 2002).

In 1963, despite all the regional debate against Guyana creating its own university, the University of Guyana was instituted by an Act of Parliament. The University, as was expected, developed a strong Faculty in Agriculture, since the economy was essentially agricultural based. This Faculty has many programs leading to a bachelor's degree in science in agriculture, microbiology, biochemistry, and agricultural economics. There were many other faculties, but the one that is of interest here is the Faculty of Arts, which has a degree in mathematics. The Mathematics Department of this faculty is important to the whole system, since many of the departments of other faculties are dependent on mathematics. That is, they are mathematically based, such as engineering, economics, chemistry, architecture and building technology, and applied soil science, among others. With the diversity of degrees available that require mathematics, the future of mathematics was well positioned for future advancement (Bernard, et al., 2002).

In 2011 the Ministry of Education mandated all primary and secondary schools implement a remediation program (Performance Enhancement Project) for students who scored below 50% in their year-end exams in mathematics. The remediation included after school activities as well as summer classes (Ministry of Education, 2011). Additionally, the Minister of Education, Shaik Baksh, in an article in a local Guyanese newspaper, said that the Ministry was looking into recruiting teachers from foreign countries, particularly for mathematics and science (Baksh, 2011). However, without proper compensation, this attempt to attract teachers would be futile. Today the current Minister of Education, Priya Manickchand, also recognized this. Soon after assuming her position as Minister, she started a pilot teacher-training program specifically in the mathematics program with 36 schools. She brought in subject matter specialists to train the teachers, and instructors at the teacher training facilities as well. The program also supplied students at pilot schools with required texts, study guides, and other relevant resources, including past exams and study guides. The results of the CSEC exams after the implementation of these programs showed that even though national mathematics scores had decreased from the year before, the 36 pilot schools scores were significantly higher (39.85) than the national average of 29.69%, (Stabroek News, 2012). Encouraged by the results at the pilot schools, Minister Manickchand is looking to expand and grow the program, to include more schools.

Conclusion

All levels of mathematics education are being helped by these projects either directly or indirectly. Whatever is done at the nursery level affects the quality of students at the primary level. Further, the primary level projects, which are aimed at re-establishing the standard of education in the four core areas — among which is mathematics — will redound to the benefit at the secondary level.

These projects are directed at restoring the pride that Guyana once felt in being amongst the leaders in education in the region. When the quality of education is improved up to the secondary level, students will be equipped to handle mathematics, or mathematics-related careers, at the tertiary level. The one problem that still has to be dealt with realistically is the remuneration of teachers. Teachers are still migrating to other countries where salaries and benefits are greater. One idea that can be looked at carefully is to ensure that benefits can be greatly increased and salaries can be made more attractive, even if not at the level of competing regional territories. This is an important consideration, since no country can really succeed if a high percentage of its professional citizens leave the homeland. This brain drain will always cause the progress of the nation, and particularly that of mathematics, to be stunted.

Much can still be done to help in sustainability of the measures achieved by projects at all levels. Guyana needs to look at the education picture from a more universal standpoint rather than just a narrow local or regional outlook. We need not be kept back by poor political decisions of the past. The main goal of mathematics development should be to apply the principles of this subject area to solve some of the problems in agriculture, engineering to deal with flooding, deforestation, and air, land and river transportation, to name a few. They can use the present to develop attitudes where each member of the country sees himself as a vital link in helping his country to rise out of the "sliding down" seventies (The World Bank, 2005) in education, to higher mathematical achievements in the third millennium.

Bibliography

Baird, C. (1972). *Education for development*, Paper presented at the 15th Annual Delegates Congress of the People's National Congress, Guyana.
Baird, C. (1975). Opening address to the annual Delegates Conference of the Guyana Teachers' Association, Guyana Lithographic.
Baksh, S. (2011). Importing math and science teachers, *Stabroek News*, February 24, 2011.
Bernard, D., et al. (2002). *National report on higher education*, University of Guyana.
Broomes, D., Rahman, J., & Richards, C. (1975). *Mathematics teaching laboratories*, Ministry of Education and Social Development.
Burnham, L. F. S. (1974). *The Sophia declaration*, Ministry of Information.
Bynoe, J. (1972). *Education for political socialization*, University of Guyana.
Bynoe, J. (1973). *Education and opportunity at the post primary level in Guyana — some problems, practices and proposals*, New tasks for Guyanese education, op. cit.
Callaway, A. (1968). Unemployment among African school-leavers, in A. M. Kazaimas (Ed.), *Schools in Transition*. Boston: Allen and Deacon, Inc.
Children may be sent to factories: Vicious plans bared (1963). Sun, Guyana, April 14, 1963.
Craig, D. R. (1993). *Society, economy and the university, the case of Guyana*, London Associates of Commonwealth University.
Cumberbatch, H. (1972). *The changing role of the classroom teacher, Crossroads in Teacher Education: A Search for Relevance*. Report of Conference on Teacher Education in Guyana.
Despres, L. (1967). *Cultural Pluralism and Nationalist Politics in British Guiana*. Chicago & New York: Rand McNally.
Gordon, S. C. (1963). *A Century of West Indian Education*. London: Longmans.
Guyana Ministry of Education (2011). *Remediation programmes in mathematics, English now compulsory*.
Illich, I. (1968). The futility of schooling in Latin America, *Saturday Review*. Reprinted (1973) in *New Tasks for Guyanese Education, op. cit.*

Ishmael, O. (2012). *The transition of Guyanese education (in the twentieth century)*, GNI Publication.
Jagan, C. (1980). *The West on Trial*. Berlin: Seven Seas Books.
Jennings, Z. (1999). Educational reform in Guyana in the post-war period, *Education Reform in the Commonwealth Caribbean. Interamer 54 Educational Series.*
Paul, U. M., Hamilton, E. M., & Williams, R. A. (1986). Participation of local communities in the financing of education in Guyana. *Prospects, XV*(3).
Progressive Youth Organisation (PYO) of Guyana (1986). *Education Crisis*, New Guyana Company, Georgetown.
Reno, P. (1964). *The Ordeal of British Guiana*. New York: Monthly Review Press.
Robertson, J., et al. (1954). *Report of the British Guiana constitutional commission 1954*. London: Colonial Office.
Stabroek News (2012). The pilot project in Math and English A, August 13, 2012.
UNESCO (2000). *EFA 2000 assessment: country reports Guyana*, UNESCO.
UNESCO (2006/7). *World data on education 6th edition*, UNESCO.
Wintz, P., & Wintz, G. (2013). *Old technology — new experience: Interactive Radio Instruction (IRI) in grade two mathematics classrooms in Guyana*, Paper presented at the Biennial Conference of The University of the West Indies Schools of Education, St. Augustine, Trinidad and Tobago.
World Bank (1996). *Staff appraisal report, Guyana, secondary school reform project*, The World Bank.
World Bank (2005). *Implementation completion report (IDA-28790-PPFI-P8580)*, The World Bank.
World Bank (2013). *Implementation completion and result report (TF-053679)*, The World Bank.

About the Authors

Mahendra Singh is doctoral candidate at Teachers College, Columbia University. As a former adjunct professor at St. John's University's School of Education, he was involved with numerous projects and grants for incorporating technology in the classroom, and was a member of the President's Committee on Multicultural Affairs. With over 20 years in the Information Technology field, he is currently the Lead Engineer for application and desktop virtualization for the Metropolitan Life Insurance Co. (MetLife).

Lenox Allicock is a doctoral student at Teachers College, Columbia University. He holds a master's degree in pure mathematics from Morgan State University in Baltimore, Maryland. He has been an assistant professor and the Chair of the Mathematics Department at the University of the Southern Caribbean. Prior to that, he was a lecturer in mathematics at the Teacher's Training Department at the Caribbean Union College in Trinidad.

Chapter 9

HAITI: History of Mathematics Education

Jean W. Richard

Abstract: This article sets out to trace the history of mathematics education in Haiti from the post-slavery period to the present. Haiti's history of mathematics education involves not only the advancement of mathematical knowledge, but the impact of mathematics education on the development of Haitian society. Numerous political factors elucidate the current state of mathematics education in Haiti. Those factors explain the inadequacy of educational facilities, the inferior training of instructors, their low wages, and the absence of strong governments to create a socially transformative mathematics education. At the end of the chapter, perspectives for mathematics education in Haiti are explored.

Keywords: Mathematics education in Haiti; neo-colonialism and scientific knowledge; mathematics education and underdevelopment.

Many studies (Vincent & Lherisson, 1895; Logan, 1930; Clement, 1979; Cook, 1948; Salmi, 1998) have explored different aspects and periods of Haiti's educational history. Such studies emphasize the roles of poverty, political instability, and language (since the majority of Haitians only speak Creole, whereas French is the official language for education[1]). None of these studies, however, has focused specifically on mathematics education. This task is significant not only because of its interdisciplinary aspects but also because of the importance of mathematics' role in the development of a society.

---

[1] French was the official language in education until the Bernard Reform. Currently it is partly french, partly Creole.

One of the poorest countries in the world — and the poorest in Latin America and the Caribbean — Haiti remains the least industrialized region in the hemisphere and the least stable politically. Many authors (Clement, 1979; Cook, 1948) have emphasized Haiti's economic deficiencies as the reason for the stagnation of the Haitian education system. A recent study (Salmi, 1998) is more critical of the role of the state or its absence in the disintegration of the system of education. Salmi (1998), to a certain extent, distinguished the important role that private schools are playing in the absence of the state.

In accordance with its objectives, the chapter is organized into five sections. The first section covers the period which can be called the slavery period (1697–1803). The second section covers the period 1804–1860, which can be classified as the period of consolidation of independence. The third section covers the period 1861–1914, when there was a struggle between two political parties: the Liberals and the Nationals. This struggle had a crucial effect on Haitian education. The fourth section covers the period 1915–1934, when education in Haiti shifted and followed the economic and political agenda of the United States, which occupied Haiti at that time. The fifth section covers the period of 1935 to the present, during which there were two major educational reforms: the Dartigue Reform (1940) and the Bernard Reform (1982).

The Slavery Period (1697–1803)

On December 6, 1492, Christopher Columbus landed on the island that the Indians called Quisqueia or Haiti. Though there are several claims about the size of the Indian population, by 1507 it had been reduced to about 60,000. Later the colonialists would import blacks turned into slaves from Africa. Dupuy (1976) posited the Spanish invasion of Haiti as the origin of underdevelopment. He used a Marxist approach in establishing the relationship between the level of cultural development of a society and the level of productive forces. According to Dupuy, at the time of the Spanish conquest, the Haitian Arawaks[2] were producing for consumption and not for exchange. Furthermore, with the Spanish invasion, Haiti was pulled into the mercantile system.

---

[2]The name of the Indians living in Haiti before the Spanish arrived.

In the Treaty of Ryswick in 1697,[3] Spain gave up the western part of the island to France. By 1791, "there were some 32,000 whites, nearly all Frenchmen, 24,000 freedmen, most of whom were mulattoes, and half a million slaves, almost all of whom were blacks" (Logan, 1930, p. 402).

Education in Haiti during this period must be understood in the context of economic forces, in particular, that of the mercantile system. Through colonialism and exchange relations and multiple modes of production,[4] the leading colonialist powers created in their colonies economic dislocations. The goal was accumulation of capital, via exploitation of human resources, in particular through the "plantation system" in the Caribbean nations (Frank, 2009a). This kind of economic culture in Haiti did not contemplate schools for the slaves.

Nevertheless, there were some scientific activities during this time, as shown by the existence of *Le Cercle des Philadelphes* that would become later *La Société Royale des Sciences et des Arts du Cap Français*. Many authors (Tardieu-Dehoux, 1990; Moreau de Saint-Méry, 2004) signaled that there was the embryo of education during that period. Moreau de Saint-Méry who lived in Haiti during that period and wrote about that period mentioned that there existed boarding schools. The teaching included "mathematics, history, geography and sometimes Latin" (Moreau de Saint-Méry, 2004, p. 530). Also, in the constitution of 1801 of Saint Domingue also called the constitution of Toussaint Louverture, Article 68 stated that: "All persons with the knowledge can create specific places capable of providing education and instruction for the youth under the authority and the watch of municipal administrations" (Madiou, 1989, p. 53).

The Consolidation of Independence (1804–1860)

The abolition of slavery created several handicaps for the creation and maintenance of an educational system. The first handicap was the destruction

---

[3]The Treaty of Ryswick of 1697 was an accord made between France, Spain, Holland, Germany, and England.
[4]The colonialist powers (France, Spain, England, Portugal) would implement several variants of modes of production in Latin America and the Caribbean: (a) Mining economies in Perú, México; (b) Latifundia — minifundia in Chile; (c) Mining economy and plantation system in Brazil; plantation system in Haiti, Jamaica and the rest of the Caribbean. For further readings, see Frank (2009).

of the existing structure for producing goods. The sugar plantations were destroyed and coffee grown in the mountains became one of the major crops (Hubert, 1947). Furthermore, after independence, Haiti's economy was drawn into the expanding markets of Britain, France, and the United States. The second handicap was the nature of the state that emerged after independence: militarist and authoritarian. One of the first acts of the leaders of the newly free country was to build fortifications on hilltops for protection against an invasion by France (Bellegarde-Smith, 1974).

The third handicap was the isolation of Haiti by the major European countries and the United States. The American government also banned trade with Haiti, though the American merchants did not completely adhere to this prohibition. Moreover the United States refused to recognize Haitian independence after 1804. Even though they were trading with Haiti they refused to recognize its independence. "By the 1820s, US trade with Haiti revived, but nonrecognition persisted. In 1813, a commercial agent was appointed to Haiti, but northern politicians did not challenge the South for recognition. Defenders of slavery went on the attack whenever Haiti was raised in national forums" (Matthewson, 1996, p. 37). At the time, there was internal conflict among the Haitian leaders that divided the country into two: Henry Christophe in the North and Alexandre Pétion in the South.

1804-1818

Henry Christophe would approach England for technical help to develop education by establishing close relationships with British philosophers and abolitionists William Wilberforce and Thomas Clarkson (Bellegarde-Smith, 1974). Christophe's model was a project based after England's system of education. He established the Lancastrian system of education with the help of William Wilberforce, Thomas Clarkson, and an educated African-American, Prince Saunders from Boston. Six teachers were sent to Haiti who taught arithmetic, writing, reading, and English (Clement, 1979). Christophe also created schools of higher learning. According to Tardieu-Dehoux (1990, p. 131), there were "four English Lancastrian professors sent by the British and Foreign School with the goal to form Haitian teachers. The Royal College, created in 1816 had two English university professors. [...] Besides mathematics, they also taught physical sciences as well as arts."[5]

Pétion's vision was to structure education after the French model, since he studied ballistics in France. Early after independence, Pétion established the principle of free elementary education and believed that intellectual development was a human right. According to Logan (1930, p. 412), "Article 36 of the Constitution of 1816 proclaimed the great principle of free primary education, but President Pétion, because of lack of funds and a scarcity of trained teachers, deemed it advisable to concentrate his energies on secondary education. Thus it is that the first real step toward education in Haiti was a high school, the famous Lycée Pétion". Two professors from France, Durrive and Lapree, were to teach all the subjects: arithmetic, geometry, trigonometry, algebra, navigation as well as Latin, French, English (Logan, 1930). When both leaders died, the country was again reunited but education did not advance.

1818–1859

Following unification, the government of Jean-Pierre Boyer (1820–1843) did not do much to advance education. In Haiti the President did not invest in education and during his invasion of the Dominican Republic, he closed the University of Santo Domingo. He also diverted funds to support his military rule: "With a population of under 800,000 in the mid-1820s, the regular army numbered 32,000 men... in the highest ranks were mulattoes in the majority" (Nicholls, 1996, p. 68). Furthermore Boyer agreed to pay France for the recognition of Haiti's independence by taking loans from France banks.

The opposition and the majority of the population did not approve of Boyer's decision to pay an indemnity to France. Furthermore, his authoritarian rule excluded completely the opposition which campaigned for his resignation. Boyer would leave for exile after an earthquake destroyed most of the northern part of Haiti.

It was under the government of Philippe Guerrier (1844–1845) that Honoré Fery was named Secretary of State of Justice, Culture and Public Instruction. In a note dated February 8, 1845, he described the general

---

[5] Author's translation.

mathematics program for the first three years. Arithmetic would be taught for the first and second year. For the third year the program was more intense. Mathematics teaching would include arithmetic, algebra, geometry, and analytic geometry (Vincent & Lherisson, 1895).

The Struggle between the Liberals and the Nationals (1860–1914)

After independence, color as an ideology would play a greater role in the control of the state apparatus. Members of the better educated mulatto elite formed the Liberal Party, while black politicians would form the National Party. Both parties emphasized public education, but it was the Liberal Party under the guidance of the intellectuals Boyer Bazelais (1833–1883) and Edmond Paul (1837–1893) who advocated science education. Paul's ideas can be summarized the following way: (a) the government should be active in promoting local industries; (b) the government should form alliances with the private sector to invest in new economic enterprises; (c) the government should prioritize scientific and technical education. Paul's ideas can be understood as an expression of political thought and debates raging in the Haitian society at the time (Nicholls, 1996).

1860–1893

According to the legislation of 1860, the Haitian school system was divided into different categories: (1) rural primary schools; (2) urban primary schools; (3) special secondary schools for boys; (4) secondary schools for girls. There was also a medical school and a law school. While there was no normal school at this time, around 1875 there developed a program of study that aspiring school teachers would be tested for. In 1877, a proposition for two normal colleges (colleges for training secondary school teachers) was made. The project had to wait until October of 1913 for the Normal School for girls to be opened (Cook, 1948).

Under the Géffrard government, Elie Dubois became Minister of Education. According to Clement (1979), Dubois's reforms were the most significant in the nineteenth century. At this time, the Haitian school system was based on the French system. Classes in primary schools were organized in descending order, unlike the American system (where classes are organized in ascending order, so that 8th grade in the French system is 1st grade in the American system). Minister Dubois ordered mathematics books from France. Among the materials ordered there were mathematics books written by Dumouchel, such as *Dumouchel, Problèmes et exercises*, and *Dumouchel, Arithmétique*. The government also brought professors from France to teach in the high schools (Dubois, 1867).

During the government of Fabre-Nicholas Géffrard in March 1860, special agreements were made with the Vatican and this would impact mathematics education (Bellegarde-Smith, 1974). With the arrival of priests from the Vatican, mathematics education improved and Haiti had its first coherent mathematical program. Around May 1864, *Les Frères de l'Instruction Chrétienne* came to Haiti and by 1894 there would be 78 members of the congregation in Haiti (Vincent & Lherisson, 1895). Other religious congregations also came to Haiti: the *Sisters of St Joseph of Cluny* and *Lés Filles de la Sagesse*. French priests were responsible for *Le Petit Séminaire Collège Saint Martial* and the *Institution of Saint Louis of Gonzague*, which followed the mathematics program designed by the Ministry of Education. The mathematics program was centered on algebra, trigonometry, descriptive geometry, and mechanics.

The contents for elementary classes were fundamentally arithmetic, starting with basic operations of whole and decimals numbers. The curriculum also included operations on fractions, percentage, and interest rule, including word problems. The subjects for secondary schools were algebra and geometry as well applications. From 1860 to 1895, the number of schools and the number of students increased as can be seen in Figure 1.

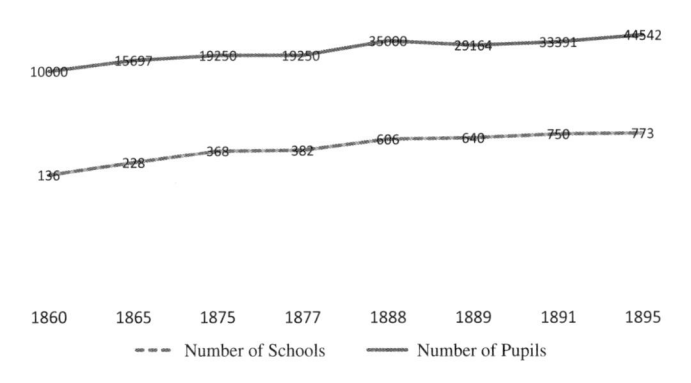

Figure 1. Number of schools and number of students from 1860 to 1895. *Source*: Data from Clement (1979) and Cook (1948, pp. 53 and 55).

1894–1914

The highlight of this period regarding mathematical education was the creation of the first private engineering and science university. In 1902, a group of six Haitian professionals, mostly engineers, created *l'Ecole des Sciences Appliquées* (ESA). This university would later become a public university with the objective of training civil engineers. The curriculum of the university would explain the different changes made in the new reform, called Reform Bellegarde, regarding mathematics education. The science and mathematics curriculum of the university included algebra and mathematical analysis, organic chemistry, industrial chemistry, geometry (analytic, descriptive), hydraulics, bridges, strength of materials, and mining. Among the exams administered in the schools, mathematics had the highest coefficient (Bouchereau & Héraux, 1933). Admission to this college was very selective. Data regarding the number of students admitted to this school are not available.

Of equal importance was the law of August 24, 1913 that the Minister of Public Instruction Tertulien Guilbaud introduced regarding the creation of two normal colleges to train pre-service teachers. At the beginning of the post-slavery period, the law regarding high school teachers was very simple: it required that all instructors take an exam administered by the Commission of Public Instruction. This exam was to assure that the instructors had the required knowledge to teach. But in 1913 all persons who desired to become instructors received, besides the fundamental areas (mathematics, chemistry, physics), training in psychology and pedagogy.

The duration of the teacher training program was three years. The mathematics curriculum for pre-service teachers in primary schools was as follows: for the first year, the subjects were pre-algebra, algebra, and geometry. For pre-algebra, the topics were operations of natural numbers, fractions, and operations as well as the metric system. The geometry curriculum focused mostly on geometrical figures and constructions with ruler and compass.[6] For the second year, only algebra and geometry were taught. Regarding algebra, the program emphasized logarithms, interest, and a review of the first year program. The third year of study of algebra was spent on an intensive study of logarithms and its applications (Bouchereau & Héraux, 1933).

---

[6]The professor can only demonstrate the most important theorems.

For the normal school, training for female pre-service teachers, the number of hours of classes per week was 30 hours for the first, second, and third year. Of the 30 hours, for the first year 5 hours were spent on mathematical concepts, 4 hours the second year, and 3 hours the third year.[7] For the first time, teachers were exposed to concepts of psychology and pedagogy. According to Bouchereau and Héraux (1933, pp. 96–97), "The practical instruction and the professional education of the pre-service teachers in the third year include the following knowledge: Pedagogical applications of psychology and moral courses; practical pedagogy, school's legislation and administration; introduction to the principal modern pedagogical concepts."

The History of Mathematics Education During the American Occupation (1915–1934)

It is from the nature and objectives of the American occupation that one can best understand the changes to the mathematical programs in Haiti during this period. On July 28, 1915, the American marines invaded Haiti, and the Americans would stay there for nineteen years. The American policy during this period was to achieve hegemony in certain regions of the globe, and the Caribbean was one of them (Schmidt, 1971). According to Schmidt (1971, pp. 5–6), "the United States sought to destroy trade barriers, colonial enclaves, and economic spheres of influence established by rival powers on the premise that the United States could best realize its full political and economic potential in an open world."

The first change was made in the 1918 Haitian constitution, permitting foreigners to own land in Haiti. Several agricultural and industrial projects implanted in Haiti were geared towards exportation: the Haitian-American Sugar Company (HASCO), which displaced peasants from their land and destroyed the local distilleries (Dubois, 2012). According to Dubois (2012, p. 269), "they paid their workers no more than thirty cents a day (the equivalent of about $4 in modern currency)."

These low wages soon attracted other companies that acquired peasants' land in the north and the northwest of Haiti: The North Haytian Sugar Company, the Haytian Pineapple Company, and the Haytian Agricultural Corporation. Later the Standard Fruit Company, a US company, would

---

[7] For further study, see Bouchereau & Héraux (1933), and Cook (1948).

export bananas from Haiti. During World War II, Haiti became one the countries experimenting with the production of rubber. *Societé Haitiano-Americaine de Développement Agricole* (SHADA)[8] was established in Haiti, a move which displaced more rural farmers. "Nearly fifty thousand acres of land throughout Haiti were cleared to make way for imported *hevea* seedlings. In the process, decades of agricultural work by rural farmers was destroyed" (Dubois, 2012, p. 314).

From the beginning of the occupation, the Americans proposed to change the structure of the Haitian educational system so as to help suppress a revolution (Pamphile, 1985). Because American expansionism was prioritized during this period — as well its needs for agricultural products — the emphasis was placed on training a cheap labor force for agricultural development by implementing a curriculum of vocational and industrial schools. These objectives would be felt in the mathematics program formulated for the Haitian school system.

During this period there were two systems of education, one controlled by the Americans and another by the Haitians under the Minister of Education, Dantès Bellegarde. At the time of the American occupation, the duration of the primary school system was six years. The subjects that formed the fundamentals of primary education in Haiti were religious instruction, moral and civic instruction, reading, writing, French, notions of general history and geography, drawing, elements of natural sciences and physics, notions of agriculture, hygiene, music, gymnastic, handicrafts for boys and needle works for girls. The amount of time allocated to study mathematics in primary schools was: 2 hours and 15 minutes for elementary classes, 2.5 hours for middle classes and 3 hours for superior classes.

The official curriculum for mathematics in the first three years of secondary school (beginning in 1920) (Bouchereau & Héraux, 1933, pp. 147–162) was pre-algebra, algebra, and geometry. The concept of the derivative and other calculus concepts first appeared in the curriculum during this period, in the last three years of secondary school. Probably this was due to the creation of the School of Applied Sciences in 1920. This school had a written examination in general mathematics as well an oral examination in theoretical arithmetic, algebra, geometry, trigonometry, analysis and descriptive geometry.

---

[8](In English), Haitian-American Society of Agricultural Development.

The Normal College for teacher training was also reorganized. To be admitted to the Normal School, three exams were required: (a) a French language exam; (b) a grammar and writing exam; (c) a mathematics exam. The duration of the program was three years. According to a regulation of October 26, 1923 signed by Dr. Paul Salomon, Secretary of State for Education, during the duration of the program, all students at the end of the school year were to take general exams in French grammar and vocabulary, composition and analysis, mathematics, physical sciences, history, geology, and geography. The mathematics program for the first year included: (a) pre-algebra with one theoretical question and two problems; (b) a geometry and algebra exam: two algebra problems and a geometry problem that involved ruler and compass. The mathematics exam for the second year included: (a) a theoretical question in pre-algebra; (b) an algebra problem and a geometry problem that involved a proof (Bouchereau & Héraux, 1933). By 1929, though the primary education covered a period of six years, the classes were divided into primary classes, middle classes, and superior classes. Each one of them covered a two-year period. Around the same year, there was a new curriculum for mathematics education. For primary schools, the mathematics curriculum emphasized the teaching of arithmetic at all levels using visual objects such as matches, marbles, and beans.

Regarding professional training, the Americans offered middle class Haitians scholarships to American universities in different areas such as medical sciences, arts and education. Several Haitians received scholarships for short periods of time to Teachers College, Columbia University. One person, Maurice Dartigue, would have a profound impact on educational reforms in Haiti. Some of his precocious ideas are present in current educational reforms as we will see later. A graduate in the Masters program in the Department of Rural Education at Teachers College, Columbia University, Dartigue was committed to the American approach to education. This approach was based on the fact that 90% of the Haitian population lived in rural areas, and because of their objectives in Haiti, the Americans implemented the philosophy of Booker T. Washington in Haiti. "In 1922, a Select Committee of Inquiry created by the US Senate .... concluded that the Haitians needed elementary, agricultural, and industrial schools patterned after the system developed by Booker T. Washington at Tuskegee Institute" (Pamphile, 1985, p. 102).

The History of Mathematics Education from 1935 to the Present

1935–1946

When the Americans were forced to withdraw from Haiti, the Haitian economy and the political structure had been completely under their control. During the occupation, Americans created a new military structure, controlled the treasury department, the customs apparatus, and introduced the *corvée*[9] labor. There was an emergence of Haitian nationalism during the American occupation. Nicholls (1974) assessing the Haitian political scene distinguished three different protest movements: (a) the *noirisme* movement, also known internationally as the *negritude* movement; (b) Marxism; and (c) the technocratic socialists, who believed in specialists capable of taking control of different structures of the government.

After the Americans left, Haiti was more indebted. Education in the country would suffer severely. A glance at the graph in Figure 2 will show how from 1916 to 1947, there was a negative balance between imports and exports.

From 1916 to 1944, spending exceeded revenue. "From 1916 to 1944, the total yearly revenue ranged between 16,048,390.75 *gourdes* (his emphasis) ($3,209,678.15) and 50,421,016.49 *gourdes* (his emphasis) ($10,084,203.29) [...] With so little money available down through the years, it is not surprising that educational facilities in Haiti have never been adequate. On less than the average income of a first-class American university, the Haitians must provide and maintain elementary, secondary, professional, rural, and vocational schools for a population of approximately 3 million" (Cook, 1948, p. 5).

Of course there are other missing links that can explain this economic anomaly, but given the importance of education and economic development, it is worth mentioning. From 1941 to 1945, Maurice Dartigue became the

---

[9]The *corvée* was an Article in the 1864 Code Rural which allowed "the government to conscript men as laborers on public projects" (Dubois, 2012, p. 239). According to Dubois (2012, pp. 239–240), "within a year of the arrival of the marines, Haitian men found themselves taken from their homes, sometimes tied together in coffles, and put to forced labor." Dubois continued to explain that "the *corvée* as it was actually practiced involved tremendous abuse, which echoed the historical horrors of colonial slavery."

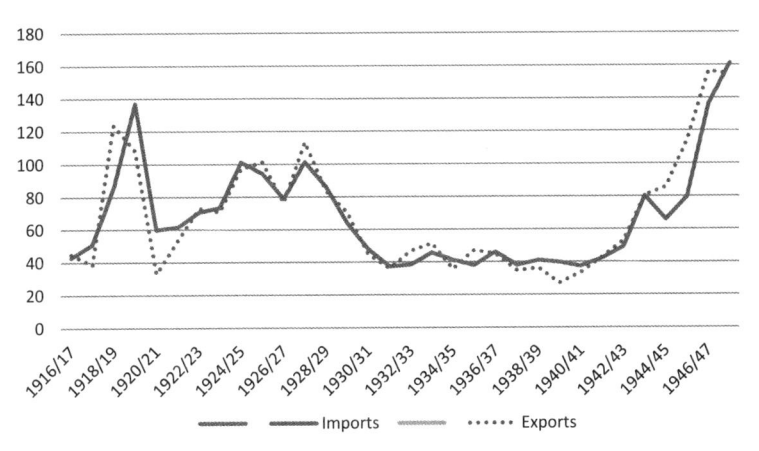

External Trade
(Fiscal years beginning 1 October)
(million gourdes)

Figure 2. Imports vs exports in Haiti from 1916 to 1947.
Source: Data from United Nations (1949). *Mission to Haiti*, p. 210.

Minister of Public Instruction, Agriculture and Employment. Two of his professors and friends at Teachers College were Leon Kandel and Mabel Carney. The first professed experimentalism in education and the second the role of socio-economic factors in curriculum development. From Dartigue's practices in education we can posit that he was also implementing some aspects of experimentalism in his approach of education. "For Dartigue, [...] only the constructive transformation of the education system can contribute to create favorable conditions to transform the socio-economic situation of the country" (Pierre-Jacques, 2002, p. 22).

During that period the reform in education also affected mathematics education. Under his administration, emphasis was placed on teacher training, the reorganization of the normal schools, and scholarships for Haitian professors to study abroad. He also organized summer schools focusing on continuing education for teachers. One professor who participated and taught in this program was Lucien Hibbert, a Haitian mathematician who later became the Dean of the Faculté des Sciences (FDS), and engineering school. During the Dartigue period the teaching of mathematics was not only traditional but also more applied to agriculture and industrial arts.

Two years after the disappearance of the two normal schools in 1942, two new normal schools were established during the Lescot government: the

Girl's Normal School at Martissant and the Normal School for secondary school teachers. In the Normal School for secondary school teachers established in 1944, the following mathematics courses were taught: mathematical analysis (2 hours of teaching per week), analytical geometry (2 hours), mechanics (2 hours), complements of geometry (1 hour), descriptive geometry (2 hours), review of the last three years of mathematics taught in secondary schools (2 hours), drawing (2 hours) (Cook, 1948, p. 68).

Again economic factors were at the core of the problems of Haitian education. In a report prepared by the United Nations Mission of Technical Assistance (1949), it was already concluded that as an agricultural and subsistent economy, Haiti's capacity to provide different types of services to the population depended on the export of its agricultural products. "Customs duties, chiefly levied on imports, account for the major part of the Government revenues. Hence, the amount of government receipts available for public investment and for foreign debt service is in large measure correlated with the volume of trade" (p. 209). In terms of growth rate, the Haitian economy did poorly during the period of 1916 to 1946. The situation did not change with the next governments.

1947–1986

Notwithstanding their political ideologies, many of the intellectuals who formed the core of the resistance against the American occupation would be government officials in the following three governments: the governments of Stenio Vincent (1930–1941), Elie Lescot (1941–1945), and Dumarsais Estimé (1946–1950). The School of Applied Sciences created in 1902 became in 1947 the Polytechnic School of Haiti and the College of Sciences. The *Ecole Normale Superieure* was also created in 1947. From 1950 to 1980, Haiti accumulated more debt. By analyzing the graph in Figure 3 we can conclude that the Haitian economy was already heavily dependent on foreign assistance.

There are two characteristics of the Haitian economy that create this imbalance in foreign assistance. The economy was based on commerce controlled by the Haitian bourgeoisie and exports was based on agricultural products. A non-diversified and open economy, Haiti had no regulation that guaranteed protection for farmers. Furthermore, most of its essentials and energy products were imported. This explains the government incapacity to invest in education. Figure 4 shows the rise in number of students in private schools over the years. Compare the data for the number of students in the

Foreign Assistance Disbursements 1970-1981

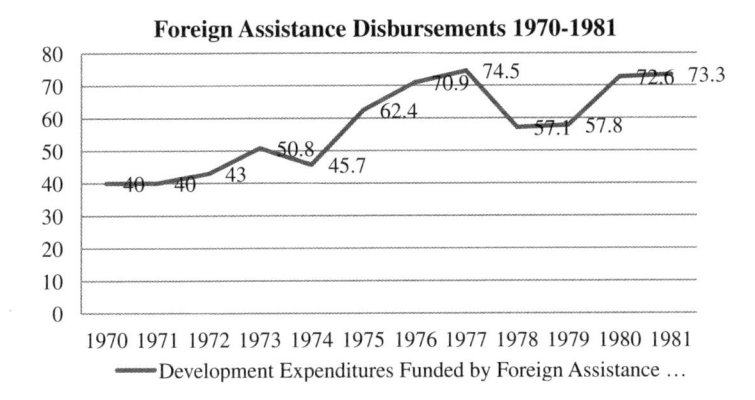

[Chart showing values: 1970: 40, 1971: 40, 1972: 43, 1973: 50.8, 1974: 45.7, 1975: 62.4, 1976: 70.9, 1977: 74.5, 1978: 57.1, 1979: 57.8, 1980: 72.6, 1981: 73.3 — Development Expenditures Funded by Foreign Assistance ...]

Figure 3. Development expenditures funded by foreign assistance.
Source: Data from Walker (1984, p. 208).

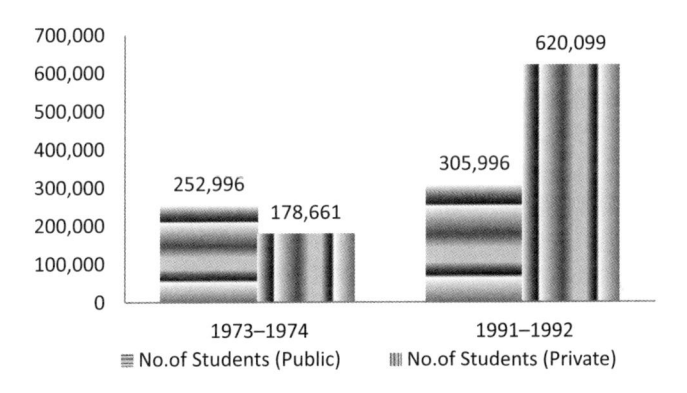

[Bar chart:
1973–1974: No. of Students (Public) 252,996; No. of Students (Private) 178,661
1991–1992: No. of Students (Public) 305,996; No. of Students (Private) 620,099]

Figure 4. Number of students in the private and public sectors.
Source: Data from Saint-Germain (1997, p. 624).

public sector and the private. From 1973/1974 to 1991/1992 the increase in the public sector was simply 21% whereas in the private sector for the same period the increase is 247% (Saint-Germain, 1997, p. 624).

Around 1980 there were several events that had significant consequences both in education as well as in the social, political and economic spheres. Though they appeared to be separate events, their interconnection should be seen in a more global context. The applications of neoliberalist policies through the Caribbean Basin Initiative (CBI) should not be separated from the Bernard Reform in education which is still being implemented today.[10]

Bernard Reform and Mathematics Education

According to Saint-Germain (1997), the justification for the Bernard Reform in Haitian education stemmed from the facts that there was limited access to education as well as a problem of functionality in the system. Though there is truth to this analysis, there is a more important factor in the reform. If Haiti was to be fully integrated into the new approach to world economy, its system of education had to be restructured. The objective of the Bernard Reform was to change the structures of the educational system preparing people capable of entering the job market with basic education.

After the reform, the educational system was divided into fundamental education having three cycles: first cycle (4 years), second cycle (2 years), third cycle (3 years). Even before finishing the third cycle, students can already opt for technical formation. After fundamental education, students can go to secondary schools which have three tracks: one leading to university, one to technical schools facilitating entering the job market, and another to normal schools preparing pre-service teachers for schools specializing in fundamental education. The mathematics curriculum for the first year of the third cycle included the following areas:

(1) Algebra: set theory, natural numbers, relations, fractions.
(2) Geometry: geometrical concepts such as planes, lines, construction, computations on geometrical figures.
(3) Mathematical applications: elementary statistics and basic financial mathematics.

According to this reform, an adolescent who finishes seven years of schooling can go to professional schools where only basic mathematics is required. According to a Research Triangle Institute document (2004): "The structure of the reformed educational system that is the articulation of educational tracks and programs as defined by the Reform of

---

[10]The Caribbean Basin Initiative, a result of the neoliberal policies, is based on trade openness advantageous to major industrial countries, decline of agricultural and traditional exports for the small countries, a service economy and reduction of the role of government. For further readings, see Palmer (2009).

1982[11] which aimed to eliminate the rigidity of the traditional structure by opening options *toward employment at different levels of the system* (my emphasis), was not fully implemented as anticipated" (Research Triangle Institute, 2004).

Mathematics Education after 1986

In spite of the wretched nature of Haiti's economy, before 1986 the government received neoliberal directives from international agencies.[12] Some of these directives included curbing wage increases, eliminating price controls, devaluing the Haitian currency in order to attract foreign capital, imposing deregulation on the Haitian economy, transferring state enterprises to private hands (Dupuy, 1997). Since then, all the state-owned enterprises have been transferred to the private sector (international or national). These policies have completely reduced the government's capacity to fulfill its

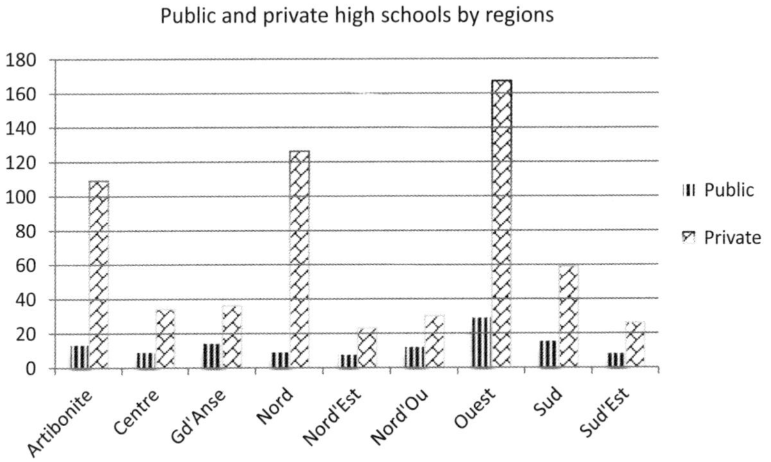

Figure 5. Enrollment for private and public schools by region.

Source: Data from *Ministère de l'Education Nationale et de la Jeunesse et des Sports* (1995). Diagnostic technique du système éducatif haïtien, p. 118.

---

[11] Also known as the Bernard Reform.
[12] USAID, The World Bank, and the International Monetary Fund (IMF).

obligations. Due to the government's weakness in investing in education, there was a rise in private schools.

There are several factors for the rise in private education. One is the rise of the non-governmental organizations (NGOs) supported by the United States and Europe. This new approach is a result of the neoliberal agenda to privatize education. Education would become the affair of NGOs, the Catholic Church, and the private sector. Secondly, governments are incapable of investing in different educational areas.

Currently, the State University of Haiti has eleven colleges grouped according to subject areas. The *Ecole Normale Superieure* is one of the colleges. The college created in 1947 has changed its objectives three times. In 1947, it had objectives to train secondary school teachers as well as university professors in arts and sciences. During the Duvalier regime, using the law of January 23, 1969, the college became the College of Arts and Pedagogy. At the time its objectives were "to dispense theoretical and practical teachings" (my translation). In 1973, the objectives and the name were changed again, back to *L'Ecole Normale Superieure*.

Mathematics is one of the majors offered at the college. The mathematics program requires three years of study, and is very theoretical. The first-year students are exposed to abstract algebra, and analysis using Walter Rudin, *Principles of Mathematical Analysis*. Those are courses that students in the United States are exposed to in their fourth year in college. The college admits 50–60 students in the first year and only 10–15 students the third year. Because the students are ill-prepared, the rigor imposed knocked out most of the students. The reason for the low retention rate can be partly attributed to a low budget and a lack of qualified professors. The prospective teachers do not receive any courses either in pedagogy or techniques of teaching.

To minimize the lack of preparation of the graduating students, a recent program was prepared with the full collaboration of several organizations representing public and private sectors of education.[13] According to this

---

[13] The list of organizations preparing this document were: The Multisectorial Commission for the Implementation of New High Schools (COMINS); The National Institute of Professional Formation (INFP); Help and Action, Haiti; Julmiste Joseph High School; Leo Defay College; Valparaiso University; Haitian Association of Professors of the French Language (ASHAPROF); Ministry of Youth, Sports and Civic Action; National School of Arts; Center for Secondary Studies; Catts Pressoir High School; Blaise Pascal High School; Bird New High School;

document (Ministère, 2008–2009, p. 4), the objectives of this new program are to "consolidate the philosophical, sociological, pedagogical and psychological foundations of education of the students during their high school years." But the program had a much larger objective: to be an instrument to promote democracy through mathematics for all.

This document includes the following set of objectives: (1) developing competency in methods of reasoning, abstraction and rigor, (2) forming students' character in perseverance as well as intellectual honesty; (3) linking the other scientific disciplines by adapting mathematics to them; (4) continuing increasing mathematical knowledge using the approach of constructivism; and (5) helping students approach the cultural aspects of mathematics as well as preparing them for university studies. To attain these objectives: (a) classes will be student-centered; (b) emphasis will be on problem-solving to reinforce theoretical knowledge; and (c) there will be use of technology in solving problems.

The curriculum for secondary schools has several routes: literature and artistic section (LAS), social and economics sciences section (SESS), experimental sciences section (ESS), mathematical and physics section (MPS) and a section for teacher training (TT). Let us explore the contents of the mathematics programs for the literature and arts section.[14]

The mathematics SESS and the ESS programs differ from the LAS program only in the third and fourth year. In the third year for algebra, the SES program contains the following topics: second degree equations, graphical solutions of system of equations or inequalities, numerical solutions of a system of equations. The curriculum for analysis includes polynomial functions of second degree, composition of functions, limits, asymptotes, derivatives and approximation. The curriculum for trigonometry includes the study of sine and cosine functions. The probability and statistics program includes more advanced topics such as probability of an event, conditional probability, the notion of independence, random variables, expectation, variance as well as binomial formulas. In the ESS program, there is an extensive study of limits, continuity derivatives of polynomial functions. Also included is a numerical approach to solving problems. The program for mathematics

---

Sainte Rose of Lima Institute; Ketnel Vernet Studies center; Joakim Etienne High School; Quisqueya University; and the Society of Engineering and Technology (INGIETEK).
[14]For further readings, see *Ministère de l'Education Nationale et de la Formation Professionnelle* (2008–2009).

Table 1. Mathematics curriculum for the literature and arts section.

Programs	1st Year	2nd Year	3rd Year	4th Year
Algebra Numbers	natural numbers exponents, abs value	absolute value, distance, lower limits, upper limits		logarithms and exponents, numerical solutions
Equations, Inequalities and applications	equations and inequalities with one unknown, applications and graphical representations	2nd degree equations, inequalities with two unknowns, graphical representations	word problems involving second degree equations	
Analysis (Functions, Sequences)		notions of logic, functions, polynomials functions, rational functions	notions of limits, continuity, derivatives, arithmetic sequences, geometric sequences	derivatives of functions, of composite functions, logarithmic and exponential functions
Trigonometry	right angle trigonometry	trigonometric functions: the unit circle		
Geometry (Vectors)	equality of two vectors, sum, orthogonality, isometrics, projection	scalar product of two vectors, norm, parametric representation of a line and a plane	review of program of previous year, applications	
Statistics and Probability	graphs, histogram, frequency, cumulative frequencies	cumulative frequencies, measures of center, measures of variations	applications of a finite set, arrangements, permutations, combinations	measures of positions and applications, binomial distributions and binomial law

and physical sciences (MPS) is different from all the other programs due to the level of abstraction and complexity in the mathematics taught. The document emphasized that this program should have the following characteristics: (a) the students should be engaged in intensive mathematical work; (b) they should feel comfortable in abstract reasoning; (c) they should be at ease in understanding the rigor necessary for written and oral expression, and problem solving and experimentation will be present at all levels.

The Earthquake of January 12, 2010 and Conclusive Perspectives

On January 12, 2010 a 7.0 Mw earthquake destroyed most of Port-au-Prince, the capital of Haiti. International support poured into Haiti assisting different areas. Among the issues considered one can list access to education for handicapped children, retention, teacher training, curriculum and teaching materials (WISE, 2011; MENFP, 2010). The State University of Haiti (UEH) also received much assistance.

Among the recommendations one can list: to reorganize the State University of Haiti, to build other public universities in several departments of Haiti, and to create an environment that facilitates research in different areas of study (MENFP, 2010). All the major documents (MENFP, 2010; INURED, 2010; WISE, 2011) addressed the problems of mathematics education more specifically. Besides a proposal to create a ministry of higher education to supervise reform in this area, there is also one to create a Haitian Institute of Advanced Mathematics. This institute will be a member of the Caribbean Institute of Mathematical Sciences (CIMS) incorporated in Dominica.

There is also a proposal to create an International Inter-University Institute of Advanced Studies to serve as a link with other universities of Latin America and the Caribbean. Before that, there was also a proposal coming from France to create a doctoral program in mathematics in partnership with the State University of Haiti. This partnership is not new. From year 2000, there has been collaboration between the *University des Antilles et de la Guyane* (UAG) and *l'Ecole Normale Superieure*. The mathematics program at the ENS requires three years of study. Because of lack of preparation, the students do not possess the required knowledge to enter a master's program at the UAG. To remedy this lack of preparation, UAG and ENS created a program of continuing education to train the Haitian students. The Haitian students in Haiti spent a year studying mathematical analysis and measure theory, complex analysis, abstract algebra and numerical methods.

A more recent program[15] (for the period of 2010–2015) developed after the earthquake of January 12, 2010 reemphasized the objectives of providing to Haitian education "the curricular and programmable framework that will help to form modern citizens rooted in their culture. This will also make them aware of human values, social, cultural as well as the politics of justice and progress necessary for a harmonious development of the society."[16] In this program, higher education is supposed to play a major role, especially the "promotion of scientific research and innovation" (MENFP, 2010, p. 95).

There is a recent effort developed by the Massachusetts Institute of Technology (MIT) to create open-based educational resources in creole[17] (Dizikes, 2013). Funded by the US National Science Foundation and MIT, the objective is to promote active learning in science, technology, engineering and math (STEM) disciplines.

Conclusion

This chapter has focused on exploring the history of mathematics education in Haiti from the post-slavery period to the present. Prior to independence, there was some form of education in the colony, but it was minimal. Independence of course destroyed the means of production, which means the sugar plantation and the refineries. Moreover, it destroyed for the emerging nation the capacity to create a new type of model of development as

---

[15]For further readings see (a) *Ministère de l'Education Nationale et de la Formation Professionnelle* (MENFP) (2010). (b) World Innovation Summit for Education (WISE) (2011).

[16]My translation of: *"Fournir à l'Ecole haïtienne le cadre curriculaire et programmmatique qui l'habilitera à for mer des citoyennes et citoyens modernes et ancrés dans leur culture, et imprégnés des valeurs humaines, sociales, culturelles et politiques de justice et de progrès nécessaires au développement harmonieux de la société."* (MENFP, 2010, p. 2).

[17]This project is carried out in collaboration with the State University of Haiti, University Caraibe, Ecole Superieure d'Infotronique d'Haiti, Université Quisqueya, NATCOM and the Foundation for Knowledge and Liberty.

well as the capacity to invest in education. After independence there was a consensus among European nations and the United States that Haiti should remain isolated despite the numerous demands to be recognized as an independent nation. The initial appearance of mathematics education, under both Christophe in the north and Pétion in the South, was limited to arithmetic and geometry.

In the succeeding governments, mathematics and science education did not advance much. Edmond Paul, one of the leaders of the Liberal Party, advocated the importance of science education, the relationship between scientific knowledge, and industrial development. But a power struggle between the two parties (the Nationals and the Liberals) sidetracked governments from formulating a coherent educational and science program to foster economic and industrial development. In that same period, in an accord between the Vatican and the Géffrard government, religious congregations like *Les Frères de l'Instruction Chrétienne* and *Les Filles de la Sagesse* came to Haiti. Their arrival had a positive effect on mathematics education in Haiti.

However, around 1915 — using political events in Haiti — the United States occupied Haiti and implemented a model of education corresponding with their political and economic agenda. The Haitians fought through various channels for their second independence but the political era was different. The capitalist countries reacting to the independence of many colonies created a neo-colonialist approach to ensure continuous exploitation. Through the years, due to unequal exchange, Haiti's economy sank deeper into deficit and never fully developed for several reasons. Some of the reasons are the context of the world economy, the weakness of the dominant classes, and the lack of a coherent development program for linking education — specifically science education — with the advancement of Haitian society. Then the earthquake of January 12, 2010 occurred. The international community reacted positively by addressing several issues including mathematics education, teacher training, curriculum development, as well as infrastructure. Perhaps this is the time for a new beginning.

Bibliography

Bellegarde-Smith, P. (1974). Haiti: Perspectives of foreign policy: An essay on the international relations of a small state. *Caribbean Quarterly, 20*(3/4), 21–38.

Bouchereau, C., & Héraux, H. (1933). *La Législation scolaire d'Haïti.* Rue Roux: Imprimerie Aug A. Héraux.

Clement, J. B. (1979). History of education in Haiti: 1804–1915 (First part). *Revista de Historia de America, 87,* 141–181.

Cook, M. (1948). Education in Haiti. *US Office of Education, Bulletin,* No. 1.

Dartigue, E. (1994). *An Outstanding Haitian Maurice Dartigue: The Contribution of Maurice Dartigue in the Field of Education in Haiti, in the United Nations, and UNESCO.* New York: Vantage Press.

De Regt, J. P. (1984). Basic education in Haiti, in C. R. Foster & A. Valdman (Eds.), *Haiti — Today and Tomorrow: An Interdisciplinary Study.* New York: University Press of America, pp. 119–139.

Dehasse, J. G., Lumarque, J., & Spratt, J. (1995). *Diagnostic technique du système éducatif Haïtien: Rapport de synthèse (Technical diagnosis of the Haitian education system: Synthesis report): Policy Support to Haiti's National Educational Policy Reform and Planning Process project.* Port-au-Prince, Haiti. Research Triangle Institute; The Academy for Educational Development; Educat S.A. US Agency for International Development.

Dizikes, P. (2013). *Program to teach the sciences in Haiti in kréyòl language.* Canada Haiti Action Network. Retrieved September 25, 2013 from http://www.canadahaitiaction.ca/content/program-teach-sciences-haiti-kreyol-language

Dubois, F.-E. (1867). *Deux ans et demi de ministère.* Paris: Imprimerie de P-A, Bourdier et Cie.

Dubois, L. (2012). *Haiti: The Aftershocks of History.* New York: Henry Holt and Company, Metropolitan Books.

Dupuy, A. (1976). Spanish colonialism and the origin of underdevelopment in Haiti. *Latin American Perspectives, 3*(2) 5–29.

Dupuy, A. (1997). *Haiti in the World Order: The Limits of the Democratic Revolution.* Boulder, CO: Westview Press.

Moreau de Saint-Méry, M. L. (2004). *Description topographique, physique, civile, politique et historique de la partie française de l'isle Saint Domingue.* Paris: Publication de la Société Française d'Histoire d'Outre-mer.

Fass, S. M. (2004). *Political Economy in Haiti: The Drama of Survival.* New Brunswick, NJ: Transaction Publishers.

Frank, A. G. (2009). *Capitalism and Underdevelopment in Latin America.* New York: Monthly Review Press.

Frank, A. G. (2009a). *Dependent Accumulation and Underdevelopment.* New York: Monthly Review Press.

Hubert, G. A. (1947). War and the trade orientation of Haiti. *Southern Economic Journal, 13*(3), 276–284.

INURED (Interuniversity Institute for Research and Development) (2010). *The challenge for Haitian higher education: A post-earthquake assessment of higher education institutions in the Port-au-Prince metropolitan area.* Port-au-Prince, Haiti: Retrieved September 25, 2013 from http://webarchive.ssrc.org/challenge-haiti-report.pdf

Logan, R. W. (1930). Education in Haiti. *The Journal of Negro History*, 15(4), 401–460.
Lundahl, M. (1991). Underdevelopment in Haiti: Some recent contributions. *Journal of Latin America*, 23(2), 411–429.
McClellan III, J. E. (2000). L'historiographie d'une académie coloniale: Le Cercle des Philadelphes (1784–1793). *Annales historiques de la Révolution française*, 320, 77–88.
Madiou, T. (1989). *Histoire d'Haïti: 1492–1799*. Tome 1. Port-au-Prince, Haiti: Editions Henri Deschamps.
Matthewson, T. (1996). Jefferson and nonrecognition of Haiti. *Proceedings of the American Philosophical Society*, 140(1), 22–48.
Ministère de l'Education Nationale et de la Formation Professionnelle — Direction de l'Enseignement Secondaire (DES). (2007–2008). *Mathématiques Module I*. République d'Haïti.
Ministère de l'Education Nationale et de la Formation Professionnelle (2008–2009). *Curriculum de l'école secondaire — Programme Pédagogique Opérationnel*. République d'Haïti.
Ministère de l'Education Nationale et de la Formation Professionnelle (MENFP) (2010). *Vers la refondation du système éducatif Haïtien: Plan opérationnel 2010–2015. Des recommandations de la Commission Présidentielle Education et Formation*. Port-au-Prince, Haiti: Retrieved September 25, 2013 from http://datatopics.worldbank.org/hnp/files/edstats/HTIpla10.pdf
Nicholls, D. (1974). Ideology and political protest in Haiti, 1930–46. *Journal of Contemporary History*, 9(4), 3–26.
Nicholls, D. (1996). *From Dessalines to Duvalier: Race, Colour and National Independence in Haiti*. New Brunswick, NJ: Rutgers University Press.
Orrill, R. (2001). Mathematics, numeracy, and democracy, in L. A. Steen (Ed.), *Mathematics and Democracy: The Case for Quantitative Literacy*. Princeton, NJ: The Woodrow Wilson National Fellowship Foundation, pp. xiiv–xx.
Pierre-Jacques, C. (2002). *D'Haiti a l'Afrique, initinéraire de Maurice Dartigue, un éducateur visionnaire*. Montreal, Quebec: Les Editions Images.
Palmer, W. R. (2009). *The Caribbean Economy in the Age of Globalization*. New York: Palgrave MacMillan.
Pamphile, L. D. (1985). America's policy-making in Haitian Education, 1915–1934. *The Journal of Negro Education*, 54(1), 99–108.
Research Triangle Institute (2004). *Technical diagnosis of the Haitian educational system: Summary of conclusions and recommendations*. USAID-funded programme of Technical Assistance to the National Education Planning and Reform Process. Retrieved September 25, 2013 from http://www.rti.org/pubs/haiti_education_sector_diagnosis_-_english-language_summary_1995_.pdf
Saint-Germain, M. (1997). Problématique linguistisque en Haïti et réforme éducative: Quelques constats. *Revue des sciences de l'éducation*, 23(3), 611–642.

Salmi, J. (1998). Equity and quality in private education: The Haitian paradox. *LCSHD Paper series, no. 18.* Washington, DC: The World Bank, Latin America and the Caribbean Regional Office.

Schmidt, H. (1971). *The United States Occupation of Haiti, 1915–1934.* New Brunswick, NJ: Rutgers University Press.

Steen, L. A. (Ed.) (2001). *Mathematics and democracy: The case for quantitative literacy*, prepared by the National Council on Education and the Disciplines. Princeton, NJ: The Woodrow Wilson National Fellowship Foundation.

Tardieu-Dehoux, C. (1990). *L'éducation en Haïti de la période coloniale à nos jours. (1980).* Port-au-Prince: Imprimerie Henri Deschamps.

United Nations (1949). *Mission to Haiti: Report of the United Nations Mission of technical assistance to the Republic of Haiti.* United Nations, Lake Success, NY.

UEH (Université d'Etat d'Haïti) (2009). *Guide du postulant à l'université d'état d'Haïti (UEH).* Port-au-Prince, Haiti: 21, Rue Rivière. Retrieved September 25, 2013 from http://ueh.edu.ht/etudes/Inscriptions/Guide2.pdf

Vincent S., & Lherisson, L. C. (Eds.) (1895). *La Législation de l'instruction publique: 1804–1895.* Paris: Vve Ch. Dunod & P. Vicq.

Walker, J. L. (1984). Foreign assistance and Haiti's economic development, in C. R. Foster & A. Valdman (Eds.), *Haiti — Today and Tomorrow: An Interdisciplinary Study.* New York: University Press of America, pp. 205–229.

World Innovation Summit for Education (WISE) (2011). *Education in Haiti: An overview of trends, issues and plans.* Port-au-Prince: WISE Haiti Workshop. Retrieved September 25, 2013 from http://www.wise-qatar.org/content/education-haiti-overview-trends-issues-and-sept-2011

About the Author

Jean W. Richard holds a Ph.D. in mathematics education from Teachers College, Columbia University. He is currently an associate professor in the Mathematics Department at the Borough of Manhattan Community College (BMCC), of the City University of New York (CUNY). Before coming to BMCC, Jean W. Richard taught mathematics at the Normal College of the State University of Haiti. His main research interests are statistics, international and comparative mathematics education, statistical learning theory, and the history of mathematics. Besides teaching different courses at BMCC, Jean W. Richard is the acting chair of the department.

Chapter 10

HONDURAS: Origins, Development, and Challenges in the Teaching of Mathematics

Marvin Roberto Mendoza Valencia

Abstract: This article presents an overview of the teaching of mathematics in Honduras throughout history. It tries to capture the events, personalities and institutions that have set standards in the conceptualization of mathematics and its teaching, recognizing turning points, demands and challenges of mathematics education in Honduras.

Keywords: Teaching mathematics in Honduras; mathematics education; historical perspective; aspirations and challenges of education in Honduras.

Events of a historical, political, social and cultural nature, and of the educational system itself, have affected the teaching of mathematics in Honduras over time. It began with informally transmitted instruction (mainly oral) and was later institutionalized.

Contributions of the Maya and other ethnic groups, the creation of educational centers by religious orders during the Colonial period, the Liberal Reform, the founding of the National Autonomous University of Honduras (UNAH), and the creation of the Mathematics Department at the UNAH, Centers for Mathematics Teacher Preparation, and Professional Development Programs in Mathematics Teaching are relevant reference points in the conceptualization of the teaching of mathematics in Honduras.

Pre-Colonial Period

The indigenous groups that inhabited what was to become Honduras before the arrival of the Spanish in 1502 — and that still exist — can be characterized as non-nomadic. Many were dedicated mainly to agriculture: the cultivation of corn, beans and tubers. Others were involved in hunting, fishing or navigation.

Their social organization was based on the family, then on small groups. These small communities were governed by a chief or *cacique*, responsible for watching over the community. The *caciques* enjoyed significant respect and received, generally in kind, tributes from their groups.

In these native communities there was a division of roles in the activities of the family. The mother was responsible for transmitting language, values, and the development of abilities for making clothing using either plant or animal fibers. The father was in charge of teaching farming, hunting and fishing.

Research on the native groups in Honduras (Durón, 1956; Martínez & Edén, 2002; Rivas, 1993) has reported that before the arrival of the Spanish only the Maya used writing. There are no records of the mathematical practices of the non-Mayan groups, but nevertheless there are conjectures on their mathematical practices in the selection of the amount of tribute for the *cacique*, the ideal combinations for weaving garments, the quantity of a crop to be saved for the next cultivation, among others. There is also evidence that they utilized various counting strategies and other arithmetical knowledge that was transmitted orally.

Mayan civilization

The Mayan civilization developed in Mesoamerica, in a region comprising the current countries of México, Guatemala, Belize, Honduras and El Salvador. In Honduras, the civilization came to prominence in the western part of the country, in the city of Copán.

Researchers such as Bahena (2012), and Docter, Reents Budet and Agurcia (2005) claim that this city was an important hub during the classical period (400 to 800 CE) and the main site of the economic, administrative, religious and cultural activity of the time.

The city had a number of public squares as well as temples, city halls and palaces built atop platforms. These structures were colorful, well-formed and contained a considerable amount of symmetry in terms of their dimensions, size and placement. Not only were these structures temples, stelae[1]

---

[1]Monuments carved from stone, with hieroglyphic engravings and generally several meters tall.

and altars, part of the record of the cycles of celestial bodies such as the sun, the moon, the planets, and the stars, but they were used as observatories and/or calendars (Agurcia, et al., 2010).

The Maya had a writing system based on hieroglyphics, many carved in stone. (In Copán the longest hieroglyphics in the Mayan world can be found on one of its stairways; it consists of approximately 1500 glyphs).

To leave a written record, they also wrote on cloth — a parchment-like material, made of fibers, nowadays known as the Codices[2] — and stone stairways. In this record, the Maya tell of their advances in mathematics, astronomy, the arts and in other areas, as well.

As for their numbering system, the Maya used a vigesimal system with 20 symbols; this is different from our decimal system that uses 10. They discovered the concept of zero, wrote their numbers from 1 to 19 using dots and bars — a dot was worth 1 and a bar 5 (Bahena, 2012a; Docter, et al., 2005). Their vigesimal system allowed them to carry out arithmetical calculations such as additions and multiplications as are seen in the Dresden Codex (Tonda & Noreña, 1991, cited in Bahena, 2012a).

Mayan arithmetic was used a great deal in daily life. For example, Bahena (2012b) and Docter, et al. (2005) assert that the Mayans tallied days with their various calendars known as *Tzolkin*, *Haab* and *Calendario Redondo*.

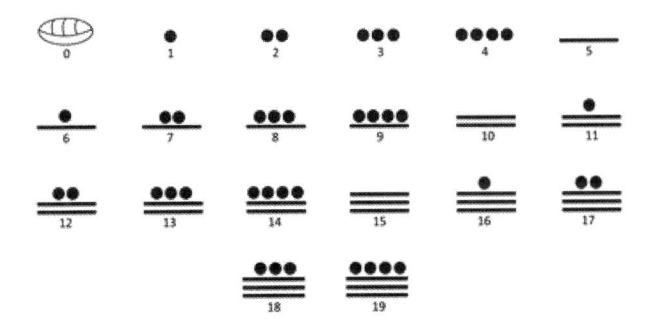

Figure 1. Mayan numeration system.

---

[2]The Mayan Codices are books written before the Conquest and used hieroglyphic characters that show some of the features of Mayan civilization and have preserved the Mayan legacy. Three that are known are the Dresdensis (or Dresden), Peresian (or Paris) and the Tro-Cortesianus (or Madrid) Codices.

The first of these was 260 days long, divided in 13 months, each with 20 days. The second calendar consisted of 365 days, distributed among 18 months of 20 days and one additional month of 5 days. The third integrates the first two, with a total of 18,980 days, or 52 years. This way, each day of this last calendar had its own designation: 4 *Ajaw* 8 *K'umk'u* (these represent an interpretation of sounds in Mayan) would be as if some said "Thursday, January 8" as "Thursday the 4th day of the week and 8 days in January, the first month of the year" (Docter, et al., 2005, p. 34).

The Maya recorded celestial events over extended periods of time, which is the reason that they invented another calendar system called *la Cuenta Larga* (it is a tally of days similar to how the Gregorian calendar counts.

The *Cuenta Larga* has, as its base, the vigesimal system and is a sequence of as many as 13 vigesimal digits, in which a unit of time is represented by one of the 13 vigesimal digits and each unit of time is made up of twenty of the following unit of time. Typically only 5 vigesimal digits were used to express the time and they were called, starting with the smallest, the *k'in*, then came the *uinal* (or *winal*), next the *tun* and the *k'atun* and finally the *bak'tun*. In Copán, on Stela 2, dates such as Sunday, January 1, 2002 of our calendar have been found written as 12.19.6.15.0 or 9 *Ajaw* 8 *Kank'in* (Docter, et al., 2005).

Another Mayan legacy involving their numbering system is its application to astronomy, a discipline that enjoyed a significant scientific development. Bahena (2012b) claims that in Copán a number of important discoveries were made, such as: the length of a tropical year being 365.2420 days (a result that differs from the actual result by one tenmillionth), the use of lunar formulas they developed starting with the 15th *K'Atun* and the equivalence between the days and the moon, namely, 149 moons corresponded to 4400 days, 235 moons to 19 years or 6939.597315 days, one moon to 29.53020134 days, two moons are close to 59 days, 6 moons are close to 177 days, 17 moons are close to 502 days and 21 moons to 620 days.

Bahena (2012b, p. 20) contends that another Mayan contribution was the calculation of the synodic revolution of Venus (the amount of time that elapses between two passes of Venus either in front of or behind the sun, as seen from earth). This has an oscillation that varies between 580 and 588 days (583.92 days). The Maya calculated it at approximately 584 days. That is, the alignment of the sun, earth and Venus is every 584 days.

Another of Copán's mathematical contributions is its buildings and the precision, accuracy and geometric form with which they were built. This indicates advances in geometry. Typically these buildings were built in order to observe celestial bodies. Docter, *et al.* (2005) suggest that the buildings in this Mayan city were placed to indicate specific astronomical events, such as: the dawn, the spring equinox or the winter solstice. They also served as observatories to identify specific dates of the year. Phenomena like the lengthening or shortening of a shadow can be observed at specific dates and times in some of the stelae of Copán. This provides evidence of the extent of their advances in their scientific understanding of celestial bodies. Mayan mathematics has been a part of the elementary school mathematics curriculum since the 1960s. However, since 2000 there has been increased attention due to the work of various new centers such as *Casa K'inich* and *Mundo Maya*. They have promoted a rebirth in the study of Mayan culture and its transcendence in the educational, cultural and social life of Honduras.

The Colonial Period (1502–1821)

The first efforts to establish some type of formal education were not attempted until years after the arrival of the Spanish in Honduras.

Cruz-Reyes (2005) contends that both laypersons and members of various religious orders were in charge of the first schools in Honduras. Although they directed their efforts to transmitting Spanish culture and in evangelizing, none of them had formal preparation for teaching. Those from the religious orders did have religious training, but it was basic literacy, rudimentary numeracy, and some notions of culture that were the only qualifications for teaching for both groups.

The first school in Honduras was established in 1539 when a parochial primary school was established in the city of Gracias a Dios by Don Cristóbal de Pedraza to teach the Christian doctrine and literacy. Presbítero Álvarez was the school's first teacher.

In 1564 Obispo Fray Jerónimo de Corella began teaching grammar and in 1602 Presbítero Fray Esteban Verdelete began teaching Latin grammar. The *Colegio Seminario San Agustín* was founded by Obispo Fray Alonso de Vargas y Abarca in 1662. In this school they taught Spanish and Latin grammar, history and theology, as well as Christian doctrine for the preparation of clergymen (Cruz-Reyes, 2005; Pérez, 1996, 1997; Zuniga, 1987).

In 1731, bishop Fray Antonio López de Guadalupe, founded the *Colegio Tridentino de San Agustín de Comayagua*. It opened two years later to teach Latin language and culture, religion and mathematics. The *Colegio Tridentino* closed in 1827, reopened in 1843, and closed definitively in 1856. Pérez (1996, 1997) suggests that this school can be considered the origin of higher education in Honduras.

Formal education reached a new level in Tegucigalpa in 1804 when Fray José Antonio Murga began to teach classes in grammar and Latin to more than 20 students. However, in 1779 the first primary school was founded in Tegucigalpa's Villa de San Miguel de Heredia.

During the colonial period Honduras was part of the Captaincy General of Guatemala. As such, it depended in many ways on Guatemala — including for education — given that Guatemala was the political, administrative, and economic center of Central America.

In 1676 the first university in Central America, San Carlos University, was founded in Guatemala. Many who were born in Central America and México attended San Carlos. Among them was José Cecilio del Valle,[3] who studied the French Enlightenment and then played a leading role in the movement for independence from Spain (Amaya, 2009; Castro, 2008; Cruz-Reyes, 2005).

The founding in 1814 of the University of Leon in Nicaragua was another important antecedent for higher education in Honduras. In accordance with the thinking at the time it emphasized Latin grammar, religion and some notions of arithmetic (Pérez, 1996, 1997). This educational institution had great importance for higher education in Honduras. It was there that Padre José Trinidad Reyes (the father of higher education in Honduras), Máximo Soto, Alejandro Flores, Miguel A. Rovelo, Yanuario Girón and Pedro Chirinos, founders of the institution that was to lead to higher education in Honduras, received their education.

Beginning of the Republic (1821–1875)

Throughout Latin America and specifically in Central America, independence led to an awakening of an educational consciousness and was the starting point for what became in various ways and times the first

---

[3]Honduran philosopher, politician, lawyer and journalist (1777–1834). Del Valle was the author of works on mathematics, philosophy, geography, history, botany, mineralogy, religion and law.

organized educational system. Important figures such as Francisco Morazán, José Cecilio del Valle, José Trinidad Cabañas, José Trinidad Reyes, among others, were influential in these important efforts (Amaya, 2009; Cruz-Reyes, 2005).

Three particularly relevant events for Honduras occurred after independence from Spain: the annexation of the Captaincy General of Guatemala to the short-lived First Mexican Empire, the creation of the Federal Republic of Central America, and then in 1838, the formation of the independent republics of Guatemala, El Salvador, Honduras and Costa Rica. Nevertheless, during the period of the Federal Republic of Central America legislation was initiated in the region that led to the first national laws to guide education.

During the colonial period there had not been any formal public and secular educational structure (Pérez, 1996) established by the government. Education had been in the purview of the Church. Private and denominational schools that offered limited access to a few had existed during those years. The creation of the Federal Republic of Central America marked the beginning of the responsibility of the State to address the educational needs of a new era, anchored in a liberating and progressive ideology throughout the region.

A landmark in education for Honduras was 1830, when Francisco Morazán became the first head of state and enacted the First Educational Policies, from which the First Educational Statutes were derived. They established an organization for public education that was free, non-denominational and universal. Constant civil wars during that time prevented the realization of those aspirations (Zuniga, 1987). Nevertheless, at that time educational institutions were created that contributed to the progress of the nation. Among them were the first normal school based on the ideas of Joseph Lancaster (which functioned from 1836 to 1840), the *Liceo of Honduras* (1834), the Literary Societies of Tegucigalpa and Santa Rosa de Copán. These two Literary Societies became post-primary educational institutions (Pérez, 1996, 1997; Salgado, 2004a).

Amaya (2009) reviews the production and distribution of textbooks and other written materials during this period. It was limited mainly to works of a religious nature and to a lesser extent philosophical. Personal libraries such as those of Dionisio de Herrera and José Cecilio del Valle existed. They contained a variety of works including arithmetic, algebra, geometry, physics, philosophy, history and the arts.

First book published in the country

Mathematics teaching in Honduras during this time was limited to fundamental arithmetic, with an emphasis on counting and basic operations. The classes were taught by priests. The textbooks they used usually came from Spain, but they did write a few themselves. General Francisco Morazán brought the first printing press to Honduras in 1829 and with it began a national production and distribution of printed material. In 1848 the first national Honduran textbook was published. It was entitled *Rudimentos de Aritmética* (Basics of Arithmetic) and was written by Domingo Dárdano, a teacher at the *Colegio Tridentino de San Agustín de Comayagua*.

The Liberal Reform

The Liberal Reform marked a new direction for education and the teaching of mathematics in Honduras. Dr. Marco Aurelio Soto was elected President of Honduras in 1876. He and Ramón Rosa, his Minister of Education, began a significant transformation of the government of Honduras in many areas: economic, social, political and educational, influenced by positivism, liberalism and the ideas of Guatemala's President Justo Rufino Barrios (Amaya, 2009).

During this period (1876–1883) the First Statutes of Public Instruction were created and declared that education would be non-denominational and obligatory, thus transitioning the schools from the scholastic thinking of the church toward a more scientific and technical point of view (Amaya, 2009). Part of this reform was the organization of schools into the elementary, secondary and university levels.

Also, at that time many important schools were created. In 1878 the National Secondary School in Tegucigalpa (today known as the Vicente Cáceres Central Institute) was created. Salgado (2004a, 2004b) has reported on the development of normal schools throughout the country: Santa Bárbara in 1877, Choluteca in 1878, Juticalpa in 1880, Gracias in 1880, Ocotepeque in 1882, La Esperanza in 1883, and Danlí in 1884. The Western National University was also created (Pérez, 1996, 1997).

The Public Instruction Act produced changes in the secondary curriculum. These changes included an increase in religious aspects and the arts. With respect to mathematics beyond the rudiments of arithmetic, there was an introduction of more advanced arithmetic, plane geometry, algebra and trigonometry. There were applications to surveying, and magnitude conversions in the metric system.

At the university level the Liberal Reform brought about a transformation in the teaching of science (Pérez, 1996, 1997) which will be described below. Another contribution during the government of President Soto was the creation of the National Library in Tegucigalpa in 1880. It is now called the Juan Ramón Molina National Library (Amaya, 2009; Pérez, 1996, 1997).

Founding of Higher Education Centers

As was mentioned above, during the colonial period there was an increasing interest and need to develop institutions of higher education. On the one hand, this was because Guatemala and Nicaragua already had them, and, on the other, Honduras had individuals who had been well-trained in those two countries.

During the colonial period, Comayagua, the capital of Honduras, and Tegucigalpa, a city with important mining activity, both had the conditions needed to develop institutions of higher education. Comayagua had its *Colegio Tridentino de San Agustín* (Pérez, 1996). Latin grammar had been taught in Tegucigalpa and in 1842 Mariano Castejón petitioned the authorities to create an educational institution that could offer university studies (Cruz-Reyes, 2005; Pérez, 1996, 1997; Zuniga, 1987), but the petition was denied (Fernández, 1992).

The efforts to create an institution of higher education continued and on December 14, 1845, they were finally fruitful with the inauguration of the Society of the Enterprising Genius of Good Taste (*Sociedad del Genio Emprendedor y del Buen Gusto*). It was charged with the teaching of Latin grammar and basic arithmetic with Máximo Soto, Alejandro Flores, Miguel Rovelo, Yanuario Girón and Pedro Chirinos among its teachers (Pérez, 1996, 1997).

In 1847, during the government of Juan Lindo, the Literary Academy of Tegucigalpa was founded. It later became the National Autonomous University of Honduras (UNAH, https://www.unah.edu.hn/) thus marking the starting point for higher education and scientific development in the country. The Faculty of Jurisprudence and Political Science was the first created. It offered bachelors and doctorates with three specializations, one of which incorporated mathematics, geometry, in addition to physics and subjects from the humanities in its program of studies. Arithmetic and some notions of algebra, taught by a Belgian engineer whose name is unknown, were included in the curriculum for political science. This first effort to institutionalize higher education can be considered the first stage in the development of UNAH.

The second stage, according to Pérez (1996), rose during the Liberal Reform of Marco Aurelio Soto with the Statutes of Public Instruction establishing new educational liberal and scientific policies that guided higher education. Education was declared to be non-denominational and science was added to what was considered integral to education.

The university grew and two more faculties were added: Medicine and Surgery, and Sciences. Also, programs were created for jurisprudence and political science; medicine and surgery; pharmacy; medical practitioner; midwifery; surveying; agronomy; and civil, mining, construction, chemical and mechanical engineering. These programs took into account the physico-mathematical objectives that were considered fundamental and useful in industry, agriculture and commerce. It was with the University President at the time, Dr. Adolfo Zúñiga, that mathematics acquired an important role in engineering programs.

Thus the teaching of mathematics was expanded during the period of the Liberal Reform. The earlier vision was limited to arithmetic and a few rules of algebra. That vision expanded with the creation of technical and engineering programs. The first programs that were created for engineers included more advanced algebra, geometry and trigonometry, and differential and integral calculus.

The program of studies for surveying included arithmetic, algebra I and algebra II, geometry and trigonometry I, trigonometry II, elements of descriptive geometry and their application to shadows, surveying and practical problems. Those studying mining engineering took courses in algebra, geometry and trigonometry, statistics, and differential and integral calculus. Construction engineers had courses in algebra, geometry and trigonometry, elements of descriptive geometry, as well as differential and integral calculus. Future agronomists studied arithmetic, algebra, geometry and trigonometry.

The mathematics textbooks used at that time were classic books mainly from Europe. While Soto was president there was considerable immigration to the country, intellectual interchange was very common, and bookstores were appearing in many cities. There was also the growth of educational institutions in other cities to meet educational demand far from where power was concentrated. In 1874 the Institute of San Carlos del Occidente was created as a secondary school. In 1879 it became the Western National University, but was closed in 1884 (Pérez, 1996, 1997). While it lasted it taught courses on algebra, geometry and trigonometry, three-dimensional geometry, plane geometry, logic, arithmetic and surveying.

Department of Mathematics at UNAH

The creation of Departments of Mathematics was a fundamental aspect of the development of the teaching of mathematics in Honduras. The first was created at UNAH and the second at the Higher Normal School. Those departments led to the development of academic majors in mathematics at the university and secondary levels, respectively. The origins of the Department of Mathematics at UNAH (Portillo, 2003) can be traced to April 28, 1960 and the approval by the University Council led by President Raúl Corrales Padilla of the Department of Basic Sciences. It offered students general courses in mathematics, biology, chemistry, physics, languages including Spanish and English, humanities, national history, principles of philosophy, and others on general culture. These courses were required for admission to programs in medicine and surgery, chemistry, pharmacy and dentistry. In 1962 the regulations for the University Center for General Studies were approved. They proposed an educational reform that included the creation of programs for pedagogy and the science of education so that normal school trained teachers could enter the university. That same year the departments of Biology, Physics, Chemistry, Mathematics, Philosophy, Pedagogy, Social Sciences, and Languages and Arts were created. Later in 1967 teaching degrees (*profesorados*) and bachelor's degrees (*licenciaturas*) were created in mathematics, physics and biology.

Contributions of the Department of Mathematics at UNAH

Portillo (2003) suggests that during the 1980s and 1990s the Department of Mathematics (DM) at UNAH had a leading role in educational reforms. It reached out to Honduran society with activities related to diffusion and research. These were all done in coordination with the Higher Normal School, the Ministry of Education and with the Honduran Committee on Mathematics Teaching (CHEM). Many professional development opportunities were provided to elementary and secondary teachers. They also wrote books for the first three primary years and for the first three years of secondary school.

For the university level the DM gave seminars related to mathematics and its teaching focusing on such topics as linear algebra, analytical geometry, linear programming and optimization theory, elliptical partial differential equations, and stability in the solution of the Cauchy problems of parabolic equations. They also presented workshops on the teaching of calculus. They organized national and international scientific conferences, including the 4th Central American Congress on Mathematics, the IXth Central American Seminar on Mathematics, and the 1st National Meeting on Mathematics. This last meeting served as the basis for recognizing all aspects of the situation concerning mathematics and its teaching, problems related to teaching mathematics, the needs and interests of students, and the integration of different branches of mathematics in a coherent way by posing scientific solutions to said problems (Portillo, 2003).

All this involvement led the DM to realize the need for mathematics curricular reform in the entire national educational system. A reform was proposed that took into account the interrelationship that exists between mathematics and the other disciplines in the school curriculum. For elementary schools it was proposed that the teaching of the arithmetic of positive numbers should be strengthened and that geometry should be introduced from an intuitive perspective (Portillo, 2003).

One research project realized by the DM in the last two decades of the twentieth century was Project Nine. It consisted of a curricular proposal to strengthen the mathematical knowledge that had been developed in elementary and secondary schools. This initiative had its origins in an analysis of the results of an exploratory test that was given to university students. The purpose of that test was to discover the mathematical understanding with respect to the use of elementary algorithms, operations with integers and fractions, solution of simple equations and problems, and operations with algebraic expressions (Portillo, 2003). Another research project from the DM was on the "Relationship between certain operators of harmonic analysis, probability theory, and the geometry of Banach spaces". Similarly Richard Bourret and Gustavo A. Pérez investigated "Integral calculus by matrix substitution" in which they presented a method of approximation for calculating integrals that involved square roots that were linearized using Clifford algebras.

Another important development in the DM at UNAH was in 1996 with the organization of the department into areas that dealt with specific academic areas in the university: physics and mathematics, economic and administration, and social sciences and health. Each one of those areas had a coordinator assigned to oversee the planning processes for each course.

The Mathematics Degree Program at UNAH

UNAH (2008) asserts that the degree program in mathematics at UNAH has its origins in the Department of Basic Sciences (DCB). The first coordinator was José Adán Cueva who was in charge of the areas of Spanish, English, mathematics, physics, chemistry, and biology. In the beginning, Edgardo Sevilla taught the classes. He had three collaborators who handled the practicums and test corrections. In 1961 Salvador Llopis, of French origin, joined the DCB. Collaborating with Sevilla he taught a double schedule of classes from Monday through Saturday for numerous groups of students from various faculties.

By the end of the 1960s a group of Honduran teachers joined Llopis and other professors from France (working in Honduras in a program sponsored by the French government) in proposing the creation of a bachelor's degree program in mathematics, which was approved in 1968. From its beginning the mathematics degree at UNAH has had professors from many different countries, including France, Cuba, México and Honduras. A partial list of those who might be noted includes Hipólito Martínez, Salvador Llopis, Raquel Angulo, Norma Funes, Luz Estela Sarmiento, Ernest Meneano, Róger y Cristina Tropiano, Guillermo Casco, Elisa I. Villareal de Hasbum, Silvia Mercedes Alcerro, Felipe Vinicio Espinosa, Alejandro Pineda, and Trinidad Hernández. Oscar Montes, Ibrahím Pineda Guzmán, Claudia Martínez, Adalid Gutiérrez, Ramón Cue, Nicolás Pineda Guzmán, Carlos Benjamín Ustariz, Jean P. Henry Bonnet, Evert Cristoff Miranda, Concepción Ferrufino, Guillermo Cristoff Miranda, Alejandra Banegas, Roberto Guevara, Alma Rodas, Rosibel Pacheco, Gilberto Gálvez, Raúl Martínez and Fidel Ordóñez. Most of them had undergraduate or graduate degrees in mathematics with different specializations from universities in the United States, Europe or Latin America. The year 1967 was not only important for the proposal for the creation of the degree in mathematics in the DM at UNAH, it was also the year that a mathematics teaching degree (*profesorado*) was created at Northern Regional University Center (CURN) in San Pedro Sula.

The program of studies for the degree in mathematics at the UNAH has changed over the years. The current program (2013) is designed to be a five-year program with required basic and advanced courses. It offers specializations in engineering mathematics, informatics, statistics, systems, and operations. This program is oriented to develop higher level thinking in analysis, synthesis, and critical abstract thinking. The objective is to enable students to face situations in different contexts and be capable of problem solving with mathematic tools that become more sophisticated as they advance through the program. Also, the mathematics program seeks to prepare students with a solid background so that they can pursue graduate studies or succeed in various academic and job-related environments.

The Education System in Honduras

The precedents for the Honduran educational system were set during the Liberal Reform (Amaya, 2009). A Ministry of Public Instruction was created during the government of Marco Aurelio Soto and the First Statutes of Public Instruction decreed that education was a right of all the population, and should be non-denominational and obligatory (Amaya, 2009; Cruz-Reyes, 2005; Pérez, 1996, 1997; Zuniga, 1987).

In 1889 the name of the Ministry of Public Instruction was changed to the Ministry of Public Education (MEP). It is now called the Secretariat of Public Education (SEP http://www.se.gob.hn/). It regulates all public education in Honduras, except higher education, which is regulated by UNAH. Since its creation it has faced changes and challenges in trying to strengthen Honduran education at all levels (FAO, *et al.*, 2004; FEDERAMA, 2010).

In 1957 a new educational system was created. Later, on November 14, 1966, the Constitutional Act of Education was passed. It was modified on February 22, 2012, with the Fundamental Act of Education. There are other statutes that govern certain aspects of education: the Teachers' Salary Scale Act, the Teachers Retirement and Pensions Act, the Constitutional Act for the Association of Honduran Secondary Teachers (COPEMH), and the Teachers and their Regulations Act (UNESCO, 2011; Secretaría de Educación Pública-OEI, 2001).

Pre-Service Preparation of Teachers

By the middle of the 19th century, normal schools were established throughout Latin America to prepare elementary school teachers. These normal schools were at the secondary level and had many philosophical influences. The main one in Honduras was the Lancastrian[4] view.

Besides the arts, the normal schools provided basic knowledge of mathematics, social studies, teaching of trades and artistic education, so that graduates could teach the students from the new democratic societies that were being created.

By 1972 there were 42 normal schools in Honduras. At one point the number was reduced to four, but currently there are 13 (Salgado, 2004b). While in some countries educational changes led to the closing of normal schools, in Honduras they became institutions of higher education. In the case of Honduras the Francisco Morazán National Pedagogical University (UPNFM) supervises them in the preparation of elementary school teachers. This transformation has not been easily accomplished in the country. In 2005 the last elementary school teachers were graduated from the traditional normal schools. Since then, the UPNFM has overseen the preparation of teachers for the national education system (Salgado, 2006).

Founding of the Higher Normal School of Honduras

Internal and external forces contributed to a movement to create an institution that would specialize in teacher preparation for all levels in the country's educational system (Portillo, 2011), but especially elementary and secondary. On December 5, 1956 the Higher Normal School (ESP) was founded with the support of UNESCO. Part of that support was a commission that conducted a needs-assessment of the educational system focused on teacher preparation. That commission included such eminent Latin American educators as Luis Beltrán Prieto (Venezuela, head of the mission), Daniel Navea Acevedo (Chile), Jorge Arancibia (Chile), Aída

---

[4]Pedagogical method developed initially in Europe by the English educational innovator Joseph Lancaster (1779–1838) who influenced public education in various ways.

Migone (Chile), Doris Ruth Lerner de Almea (Venezuela), and Luz Vieira Méndez (Argentina). With the priorities determined by the commission, the ESP was created with an emphasis on the preparation of secondary school teachers, principals, technicians, and other administrators.

In 1989 the ESP became the Francisco Morazán National Pedagogical University (UPNFM): a semi-autonomous entity charged with directing the preparation of teachers. Since its founding it has expanded its coverage of aspects of teacher preparation and its presence in various cities in the country. It offers both face-to-face and hybrid programs. As of 2013, UPNFM is the only pedagogical university in Central America. It prepares teachers for all school levels and offers a number of degree programs in various fields at the following levels: technical (5), bachelor's (35), master's (14) and doctoral (1) (UPNFM, 2011).

Preparation of Mathematics Teachers at the UPNFM

The preparation of mathematics teachers at the bachelor's and master's levels at the UPNFM has been very important for secondary education in the country. According to the Program of Studies for Mathematics Teaching at the Bachelor's Level (*Plan de estudio de la carrera de profesorado en matemáticas en el grado de licenciatura*) (Departamento Ciencias Matemáticas UPNFM, 2008) the program has passed through various stages. It began in the period from 1956–1960 as the Program in Mathematical Sciences at a pre-university level. At that time the program was part of the area of exact and natural sciences and was offered as what is called a *profesorado* (teaching degree). It took three years to complete and had three main areas: cultural, pedagogical and "preferential" (i.e. mathematics content) with 92, 67 and 92 contact hours respectively.

The cultural training came from the humanities. In the pedagogical area there were courses on teaching methodology, and the mathematics major had courses in pedagogical studies of arithmetic, plane and three-dimensional geometry, elementary algebra, trigonometry, advanced algebra and analytical geometry. In 1963 the "Modern Mathematics" movement had its repercussions on the program as courses in set theory, relations and functions, algebraic structures and fundamentals of Euclidean geometry were introduced.

During the 1970s there were new changes to the curriculum. They included adjustments in the number of credits necessary in each area and other regulations. In 1972, the preparation of secondary mathematics teachers was divided into two strands: one of common general culture and another of mathematics content courses. In 1973 specific programs were designed for future mathematics teachers in normal schools and another for high schools with a business emphasis. The mathematics courses in that decade included mathematical logic, inferential statistics, computing and geometry, vectors, algebra I and II, calculus I and II, and numerical systems. This last course replaced advanced geometry and philosophy of mathematics from the previous programs.

During the following two decades the program experienced other curricular adjustments. A plan was approved for core courses, as well as three specializations: pure mathematics, physics, and computer science. In 1990 the degree for secondary mathematics was changed from the *profesorado* to a bachelor's granted by the Higher Education Council. There were also some changes in courses: vector geometry was substituted with differential equations, and logic and set theory were integrated in a set theory course. In 1993 the following courses were incorporated into the bachelor's degree: history of mathematics, topics in mathematics, physics or computing, technology applied to the teaching of the sciences, and research in mathematics education.

The current bachelor's level mathematics teacher preparation program responds to various international agreements as well as national initiatives such as the Fund for National Convergence (FONAC) and the National Basic Curriculum (CNB). It is also influenced by social, economic and political factors, as well as external variables. The program is based on competencies and has two parts: foundational and specific. This curricular design took into account the importance of interdisciplinary integration of mathematics, physics and computing, as well as the integration of content and pedagogical knowledge.

The curricular design proposes general, instrumental, interpersonal, systemic and specific professional competencies. There are specific objectives for each competency. Problem solving, mathematical thinking and the use of technology are considered to be cross-cutting competencies. There are four basic areas in the program of studies: basic mathematics, foundations of mathematics, research and outreach, and pedagogy/instruction.

The basic mathematics courses are two in algebra, three in calculus, two in geometry, trigonometry and analytic geometry, vectors and matrices, two in physics, mathematical statistics, two in programming, differential equations, and numerical analysis. The area of foundations of mathematics focuses on deepening abilities in mathematical thinking with elements of mathematical language, theory of numbers, algebraic structures, real analysis, linear algebra, topics in mathematics, and history and nature of mathematics. The third area develops curricular, pedagogical, instructional and evaluative competencies. The courses are design and development of mathematics curriculum, mathematics education seminar, evaluation of the learning of mathematics, mathematics teaching methods, and technology applied to the teaching of mathematics. This last area also deals with connections to the community and the development of educational projects. It develops competencies in curriculum, pedagogy, instruction and evaluation.

For this new curriculum for teacher preparation, UPNFM has faculty with teaching (*profesorado*), bachelor's, master's and doctoral degrees. Among their academic specialties are teacher education, mathematics teaching, mathematics education, computational mathematics, curriculum, educational technology, administration, research, statistics and online education.

Opportunities for mathematics teacher preparation and professional development

Today, besides the bachelor's in mathematics at UNAH and the bachelor's level teaching degree in mathematics at UPNFM, there is a master's degree program in mathematics education that was created in 2001 to improve the academic and scientific level of mathematics in Honduras. The objective of this program was to create reflection and research on the teaching of mathematics by both pre-service and in-service teachers of mathematics and related areas from all levels of the educational system. As of 2013, there have been 18 graduates from the program. The graduation rate has been very low and there have not been enough faculty advisors for student research projects.

Also at UPNFM there is a master's degree in "Training of Trainers" for elementary teachers. Some graduates of that program have written their thesis in research topics related to mathematics. Among them are "Attitudes and Perceptions of Students who have Failed Mathematics: A Case Study in a High School in Danlí" (Godoy, 2012), "The Use of Mathematics Textbooks in Teaching Mathematics: A Comparative Analysis of Cases" (Cárcamo, 2012), "Systemization of the Mathematics Professional Development Experience for Twelve Teachers from Normal Schools: PROMETAM Phase II" (Meléndez, 2012).

Professional development for mathematics teachers in Honduras has been offered for many decades. In 1983 a group of mathematics professors at UNAH and UPNFM (Oscar Montes Rosales, Francisco Figeac, Gustavo Cerrato, Armando Vásquez, Ibrahím Pineda and Rigoberto Gómez Madrid) created the Honduran Committee on Mathematics Teaching (CHEM). CHEM functioned until the end of the decade. Its purpose was to create a curricular reform across the elementary and secondary levels. CHEM's plan of action included professional development for teachers, the writing of mathematics textbooks for the Honduran context, and collaboration in the preparation of other teaching resources. During the years that CHEM functioned, approximately 2000 Honduran elementary and secondary school teachers received professional development (Portillo, 2003; Salgado, 2004a).

Based on an evaluation carried out by UPNFM's External Evaluation Unit (UMCE) and on results from tests given to elementary and secondary students, concrete actions were taken with the implementation of face-to-face seminars/workshops. Targeted at elementary school teachers, their principal products were guides with strategies for teaching mathematics (Meléndez, 2012).

Parallel to the efforts already mentioned, international aid groups made significant contributions to strengthening the teaching methods of Honduran elementary and secondary teachers, thus improving the quality of mathematics teaching. In 1984 an agreement was signed with the Japan International Cooperation Agency (JICA). In the first phase, called PROMETAM JICA-Japan, mathematics teachers from three states received professional development (Godoy, 2012; Meléndez, 2012; Salgado, 2004a). A second phase lasted from 2006 to 2011. Among the products from both phases of that effort are teacher guides and student workbooks for grades one to six. Teachers were also introduced to the Japanese Lesson Study strategies. The books that were written were distributed to all the public schools in the country and have been the official textbooks since 2005.

During the second phase of this project there have been extensive professional development sessions: four focused on the mathematical content and two focused on associated teaching methods. Meléndez (2012, p. 22) reports that the topics presented have been teaching natural numbers, decimals, fractions, geometry, area, volume, measurement, and statistics; educational evaluation; lesson planning; demonstration classes; lesson study; and blackboard use. Teaching materials have been developed through these sessions and teacher educators at the institutions where the sessions were held have participated.

The current National Basic Curriculum

During the 1990s national and international events converged to indicate the need to transform education, particularly for the most disenfranchised. Three antecedents influenced the conceptualization of the New Curricular Proposal. The first was the creation of FONAC to promote civil society dialogue. The second was the establishment of a strategy to reduce poverty as part of a plan for National Transformation. The third was the "Education for All Project", an international effort that provided financial support to Honduras. Among the goals of this project were improving the quality of education and increasing achievement levels in Spanish and mathematics, two areas that had shown poor results in national studies (Cárcamo, 2012; Meléndez, 2012; SEP, 2003a, 2003b). This project also served as the basis for a new curricular proposal called the National Basic Curriculum (CNB). The CNB was based on a National Curricular Design for Basic Education (CNB-EB) which was embedded in the national reality with a desire to prepare Hondurans for that reality and for the challenges of globalization.

The Honduran Educational System (SEP, 2003a) is composed of the following levels: pre-school from 4 to 6 year-olds, elementary from 6 to 14 years old, lower secondary from 15 to 17 years old, and upper secondary from 17 years old on up. The CNB indicates all the aptitudes, competencies, concepts, skills, abilities and attitudes that all students in the National Education System should have achieved by the end of secondary school (Enamorado, 2009).

Objectives for mathematics in the CNB include the development of abilities, competencies, and critical thinking. To achieve those objectives the CNB encourages reflection on situations in various contexts, development of structures for logical thinking, increasing capacity for abstraction using deductive and inductive processes, and the use and interpretation of symbols to express and communicate both quantitative and qualitative information in other disciplines.

The strands that the SEP (2003b) has identified for the three pre-university levels are as described below:

(1) **Numbers and operations** form the fundamental concept of mathematics. They are used to formally represent regularities, and to order, classify and describe quantitative relationships. This strand combines set theory, relations and functions, and the base-ten number system.
(2) **Geometry** is the theory of plane and solid shapes and figures. Because of the nature of its concepts, it lends itself to graphical representations,

and is therefore one of the most accessible for students. When combined with numbers, operations and measurement, it has many applications to technical careers such as architecture, carpentry, bricklaying, among others.

(3) **Measurement** is used to model concrete situations. It is also where there are strong connections to other disciplines such as physics, chemistry, economics, etc., thus facilitating a connection to daily life and careers.

(4) **Descriptive statistics and discrete probability** serve as tools to interpret, evaluate and judge particular events. This strand connects with mathematical statistics and was selected because it is useful in technical and financial careers.

(5) **Algebra** in lower and upper secondary offers methods for solving equations and inequalities of one and more variables.

Mathematical Olympiads

The Mathematical Olympiads are events for secondary school students under 18 years old (the ages vary according to the rules of each competition). The goal of the Olympiads is to encourage the development of higher-order competencies such as analysis, reflection, and problem-solving in various non-routine contexts. The Olympiads began in Romania at the end of the 1950s, when students from mostly Eastern European countries were brought together to demonstrate their mathematical knowledge. Since then Olympiads have spread to many other parts of the world (Ramos, 2006).

Some Olympiads have a regional focus: the Southern Cone, Iberoamerica (OIM), Central America and the Caribbean (OMCC). There are other national and regional Olympiads culminating in the International Mathematical Olympiad (IMO). Every year Honduras is invited to participate in the OMCC, the OIM, and the IMO.

In Honduras there are two stages in the history of the development of Mathematical Olympiads: 1985 to 2002 and 2003 to 2013. Various institutions provided leadership during the first stage: the Institute of Application (IDA), the Honduran Mathematics Education Committee (CHEM), the Ministry of Education, and others. There was very little participation in the first Olympiads. There was participation in only six of the 19 Iberoamerican competitions, two in those of Central America and the Caribbean, and none in the international competitions. During the first stage, no medals or honorable mentions were won (Ramos, 2006).

The second stage has been an important landmark in the history of mathematics in the country. The Mathematical Olympiad became a special project of a new permanent organization: the Honduran Mathematical Olympiad (OHM, www.honmatsps.com). It has a National Committee coordinated by faculty members from UPNFM: Luis Ramos, Mariano Solórzano, Juan Iglesias, and Mario Canales. The OHM is considered to be one of the research projects of the UPNFM (2007) with the following objectives that should impact secondary education: (a) encourage teachers, students and parents from throughout the country to participate in a more dynamic study of mathematics, (b) provide an early identification of students with special abilities and encourage them in the further study of mathematics, and (c) select the students and teachers who will be on the Honduran national team at the IMO.

To facilitate the OHM at the national level, local mathematics committees have been developed in each state of the country. The National Committee has delegated specific responsibilities to the state committees. They help organize Olympiads in all the secondary schools that want to participate. Then they work with the National Committee to organize yearly State Olympiads to select teams to participate in the national competition. Another rule of the OHM is that the local competitions should follow the international norms. Thus Honduras has competitions for students from three age groups: the first age group is for students from 11 to 13 years old, the second from 13 to 15, and the third from 16 to 18.

The Honduran Mathematical Olympiads project has three phases. In the first phase the teachers at each school choose students to participate who take tests with non-routine items that were either written by the local teachers or selected from the OHM national database. In the second phase, school teams compete at the state level. At the state level competitions, students from each of the three age groups take nationally-developed written tests. The most successful are invited to participate in a second round from which participants in the National Olympiad are selected. At this level, each of the 18 states in the country has the right to send two from the first age group, four from the second age group and three from the third age group to the OHM. It is usually held sometime between October and November, each year in a different city.

The current process of state participation in the Mathematics Olympiads in Honduras began in 2003. At the state level, the international norm of awarding gold, silver and bronze medals, as well as honorable mentions, is followed.

A third stage of the OHM is to form the groups who will participate in the international competitions. The students in these groups are called "preolympians". Their selection is based on medals and other awards they received during the most recent OHM. They are selected to make sure that all the age groups are adequately represented.

Beginning in 2005 Honduras has had some success in international competitions. In the Iberoamerican Mathematical Olympiad, Honduran students received honorable mentions in 2004, 2006, 2008 and 2009; bronze medals in 2005, 2009 and 2010; and a silver medal in 2007. Also, in 2007 Honduras received a cup for showing the most improvement over the previous three years. In the Central American and Caribbean Olympiad, Honduran students received honorable mentions in 2006 and 2007; bronze medals in 2008 and 2012; and silver in 2008, 2009 and 2010. In 2007 Honduras received a cup as the most improved team, similar to what it received that year in the Iberoamerican competition. At the Xth OMCC in 2008, Honduras came in fourth place overall, behind Colombia, México and Cuba.

Honduras participated in the International Mathematical Olympiad for the first time in 2008. The two Honduran students (up to six are permitted from each country) who participated that year won two honorable mentions. Another honorable mention was won in 2012. Bronze medals were won in 2009, 2011 and 2012. A high point was reached for the Central American region in 2010 when José Ramón Madrid won the silver medal for Honduras. Because of that triumph he was given a scholarship by Brazil's National Institute of Pure and Applied Mathematics (IMPA). Other olympians from Honduras have received scholarships to prestigious centers of mathematics at universities in México and Brazil. The Honduran Mathematical Olympiad Project is promoting student interest in the discipline. Many students who have participated in the mathematics competitions at various levels are now studying mathematics at national or international universities.

Tegucigalpa and San Pedro have been important centers for the Mathematical Olympiads. In those cities there are weekend programs to improve the problem-solving ability of students from throughout the country. They also provide workshops on specific mathematical topics: algebra, number theory, geometry, and combinatorics. Many past participants of Olympiads help with these sessions. Also, mathematics students from UPNFM work at the sessions as part of a practicum. Thus, they not only help the younger students, but reinforce their own mathematical knowledge.

The major challenges for the project have been to integrate distant regions of the country and increase the participation of ethnic minorities. Other challenges have included providing professional development on specific topics to teachers, integrating more individuals into the process, and opening new centers in order to improve the achievement levels throughout the country. Initially, most of the medals were won by students from Cortés and Francisco Morazán, the states with the most population and major cities. Lately, other states have had important successes. The silver medal won in the international competition was by a student from a small village in the municipality of Trinidad in the state of Santa Bárbara. It is also worth mentioning that from 2003 to 2013 most of the states have participated. The Xth Olympiad was held in Puerto Cortés, the major port of Honduras, with 16 teams from the 18 states. Each team had nine students and three teachers, as well as parents.

Research in Mathematics Education

The scientific production in Honduras of articles, conference proceedings, books, and research reports on mathematics education in Honduras has not been substantial. However, it has increased during the last decade mainly with graduation projects for bachelor's and master's degrees in mathematics, as well as articles in the journal, *Paradigma*, of UPNFM, and journals from other institutions (INIEES, 2011). Professors who have graduated from mathematics education programs in various institutions have developed, published and directed research projects for theses that are part of the master's in mathematics education at UPNFM. Among the topics have been works on mathematical thinking, designs for instructional engineering, visualization, curricular improvement, student beliefs, and mathematical representations.

The hosting of national and international conferences on mathematics teaching has contributed to the production and dissemination of scientific knowledge about mathematics education in Honduras (Portillo, 2003). One of those was the Fifth Meeting on Teacher Preparation and Research in Educational Mathematics in Tegucigalpa in 1991. At that conference many aspects of elementary, secondary and university curriculum, and in particular new teaching strategies and methodologies, were considered. Similarly, other contributions to the discipline have been the publication of textbooks for various grade levels. Examples are the series *Mi Honduras* for the elementary level, and the publications from the Project PROMETAM JICA-Japan (Cárcamo, 2012; Meléndez, 2012; Sekiya, 2011).

There has also been an increase in the publication of works related to pre-Columbian civilizations: the teaching of Mayan arithmetic and valuing "what is ours" (Díaz, 2006), *Casa K'inich* (Docter, et al., 2005), and *Manual de los monumentos de Copán* (Agurcia, et al., 2010). These and other similar works have contributed to the Honduran educational community's knowledge of Mayan mathematics and its contribution to the western world.

Concluding Remarks

In the historical becoming of education in Honduras, we have mainly observed the predominance of foreign curricula poorly adapted to the Honduran reality, with the exception of the Mayan splendor, when Copán stood out as an important research center in mathematics, astronomy, and the arts. We need a reformulation of the current curriculum that vindicates the multiethnic richness of Honduras, along with incorporating the historical, cultural, social, and political aspects of these diverse groups. A new curriculum should also incorporate subject areas like information science, peace education, and the protection of the environment. Given that current trends in mathematics education problematize teaching methodologies, we should foster a view of mathematics as a social construct in continuous change which includes practices, realities, and aspirations.

Other challenges that mathematics education faces in Honduras are the training of human resources and the need to increase the scientific output. Both aspects can be addressed with the creation of new well-designed teacher training programs at the undergraduate and graduate levels across the country. Moreover, we can increase scientific research through the creation and support of research institutes and refereed journals. These challenges must be addressed in order to allow mathematics education to be fully consolidated as a scientific discipline at the service of Honduran society.

Bibliography

Agurcia, R., Veliz, V., Biro, P., Gastélum, A., Gastélum, C., Mathews, P., & Reent, D. (2010). Manual de los monumentos de Copán, Honduras R. A. y. V. Veliz (Ed.), pp. 1–387. Retrieved from http://www.famsi.org/research/copan/monuments/CopanMonumentManual.pdf

Amaya, J. (2009). Historia de la lectura en Honduras: libros, lectores, bibliotecas, librerías, clase letrada y la nación imaginada, 1876–1930, pp. 1–192. Retrieved from http://www.er-saguier.org/crisisyestado-nacion.org/archivo/lecturas/Historia_de_la_lectura(final).pdf

Bahena, D. (2012a). Pláticas Populares sobre Astronomía, Astrofísica y Cosmología. 2012, fin e inicio de un ciclo en la Cuenta Larga de los Mayas, *Frente de trabajadores de la Energía*, pp. 19–23. Retrieved from http://fte-energia.org/pdf/e236-19-23.pdf

Bahena, D. (2012b). Pláticas Populares sobre Astronomía, Astrofísica y Cosmología. 2012, fin e inicio de un ciclo en la Cuenta Larga de los Mayas, *frente de trabajadores de la energía de México*, pp. 12–18. Retrieved from http://www.fte-energia.org/pdf/e236-12-18.pdf

Cárcamo, D. (2012). *Uso de los Libros de Texto de matemática en el proceso de enseñanza: Un análisis de casos comparado.*(Máster), UPNFM. Retrieved from http://www.upnfm.edu.hn/bibliod/images/stories/tesnov/Formaciondeformadores/Donaldo%20Carcamo.pdf

Castro, H. S. (2008). El legado de Valle. Retrieved from http://www.cna.hn/archivos/educacion_valores/Jose_Cecilio_del_Valle.pdf

Cruz-Reyes, V. (2005). Reseña Histórica de la Universidad Nacional Autónoma de Honduras. *Revista Historia de la Educación Latinoamericana*, 39–50.

Departamento Ciencias Matemáticas UPNFM (2008). Plan de estudio de la carrera de profesorado en matemáticas en el grado de licenciatura. Retrieved from http://www.google.cl/url?sa=t&rct=j&q=plan%20de%20estudio%20de%20profesorado%20en%20matem%C3%A1ticas%20en%20el%20grado%20de%20licenciatura%20upnfm&source=web&cd=1&ved=0CCsQFjAA&url=http%3A%2F%2Fcompetencias2010.files.wordpress.com%2F2010%2F04%2Fplan-de-estudios-de-matematicas-21-de-nov-del-2008-educacion-super ior1.doc&ei=zOspUvfKLM_0igKmroCoCw&usg=AFQjCNFyUwudc7tfnr2tEYsz6CWElpaFUw&bvm=bv.51773540,d.cGE

Díaz, R. (2006). Apuntes sobre la aritmética Maya. *Educere*, *10*(35). Retrieved from http://www.scielo.org.ve/scielo.php?script=sci_arttext&pid=S1316-49102006000400007&lng=en&nrm=iso&tlng=es

Docter, C., Reents Budet, D., & Agurcia, R. (2005). Casa K'inich C. M. Foundation (Ed.) *Guía de Estudio*, pp. 1–64. Retrieved from http://www.famsi.org/reports/03075es/CKguidebook_spanish.pdf

Durón, R. E. (1956). Bosquejo histórico de Honduras (No. 1). Ministerio de Educación Pública.

Enamorado, N. (2009). *Evaluación de impacto del currículo basado en competencias en el logro acadèmico de los estudiantes de refrigeración y aire acondicionado de bachillerato técnico profesional en Honduras.* (Magíster), Universidad de Chile. Retrieved from http://www.tesis.uchile. cl/tesis/uchile/2009/cs-enamorado_n/pdfAmont/cs-enamorado_n.pdf

FAO, UNESCO, ITALIA, D., CIDE, & REDUC (2004). Educación para la población rural en Brasil, Chile, Colombia, Honduras, México, Paraguay y Perú, pp. 1–442.

FEDEREMA (2010). Los Informes de Progreso Educativo del Programa de Promoción de la Reforma Educativa en América Latina y el Caribe (PREAL), pp. 1–34. Retrieved from http://www.thedialogue.org/PublicationFiles/HondurasRC2010-Final.pdf

Fernández, B. (1992). Aspiraciones administrativas de Tegucigalpa en el tránsito del siglo XVIII al XIX. *Temas Americanistas*, pp. 75–86.

Godoy, F. (2012). *Actitudes y percepciones de los estudiantes reprobados hacia las Matemáticas: Un estudio de caso en el Tercer Ciclo del Centro de Educación Básica Francisco Morazán, Municipio de Danlí, Departamento de El Paraíso.*(Máster), UPNFM. Retrieved from http://www.upnfm.edu.hn/bibliod/images/stories/tesnov/Formaciondeformadores/Felipe%20Marthell%20Proyecto%20de%20Tesis.pdf

Gómez, G. (2008). Profesores del departamento de ciencias matematicas de la UPNFM al 2012. Retrieved from http://vega.fis.cinvestav.mx/~zepeda/PROYECTOS/MCTP/MCTP/Centroamerica/Universidades/InformacionProfesoresMatematicasUPNFM.pdf

INIEES (2011). Investigación educativa en la UPNFM: 2006–2010 (1ª ed., pp. 1–486): UPNFM.

Martínez, S., & Edén, D. (2002). *Calidad de la educación en contextos multiculturales*. Comayagüela. Retrieved from www.oas.org/udse/seminario_mx/honduras.doc.

Matemáticas UNAH. (2008). *Historia de la Carrera*. Retrieved from https://sites.google.com/site/carreramatematicasunah/historia-de-la-carrera

Meléndez, M. (2012). *Sistematización de la Experiencia de Capacitación en el Área de Matemáticas Dirigida a Docentes de 12 Escuelas Normales: PROMETAM Fase II.* (Máster), UNFM. Retrieved from http://www.upnfm.edu.hn/bibliod/images/stories/tesnov/Formaciondeformadores/Maria%20Dolores%20Melendez.pdf

OHM (2008). Participantes de Honduras en la Historia de las Olimpiadas Internacionales. Retrieved from http://www.honmatsps.com/resultados_honduras_olimpiadas_internacionales.htm

Pérez, G. (1996). Historia de la educación superior de Honduras: Antecedentes históricos 1733–1847. Retrieved from http://www.guspepper.net/CAP1.html

Pérez, G. (1997). Historia de la educación Superior de Honduras: Ciento cincuenta años de vida universitaria, I. Retrieved from http://www.guspepper.net/CAP2.html

Portillo, A. (1998). La Educación Superior en Honduras 1733–1997 Vol. II. *Bosquejo Histórico de las Unidades Académicas*. Retrieved from http://www.guspepper.net/Vol2.htm

Portillo, A. (2003). La Educación Superior en Honduras 1949-2000 Vol. III. *Bosquejo Histórico de las Unidades Académicas*, pp. 1-239. Retrieved from http://www.guspepper.net/LA%20EDUCACION%20SUPERIOR%20EN%20HONDURAS%201733-1997%20Vol%20III.pdf

Portillo, A. (2011). La Educación Superior en Honduras 1941-2005. *Cronología de las Universidades creadas en el Período de 1941-2005*. Retrieved from http://www.guspepper.net/versi%C3%B3n%20final-3.pdf

Ramos, L. (2006). Una estrategia metodológica para desarrollar olimpiadas matemáticas en el nivel medio del sistema educativo hondureño. (Máster), UPNFM. Retrieved from http://www.upnfm.edu.hn/bibliod/images/stories/Tesis/luis_armando_ramos_palacios.pdf

Rivas, R. D. (1993). Pueblos indígenas y garífuna de Honduras: (una caracterización). Editorial Guaymuras.

Salgado, R. (2000). La educación superior en Honduras. Retrieved from http://www.ufg.edu.sv/ufg/theorethikos/art7.doc

Salgado, R. (2004a). La formación inicial, profesionalización y capacitación docente en Honduras: Transición hacia un nuevo sistema de formación docente. Honduras: IESALC.

Salgado, R. (2004b). La formación docente en la región: de las normales a las universidades, pp. 1-69. Retrieved from http://www.oei.es/docentes/articulos/formacion_docente_region_normales_universidada.pdf

Salgado, R. U. (2006). La formación de docentes en América Latina. Fondo Editorial Universidad Pedagógica Nacional Francisco Morazán.

Sekiya, T. (2011). Evaluación del Impacto sobre el Proyecto de Mejoramiento de la Enseñanza Técnica en el Área de Matemática en el nivel de educación básica en Honduras. *Paradigma*, 43-64.

Secretaría de Educación Pública-OEI (2001). Sistema Educativo Nacional de Honduras. Retrieved from http://www.oei.es/quipu/honduras/#sis

SEP (2003a). *Diseño Curricular Nacional para la Educación Básica*. Tegucigalpa. Retrieved from http://www.se.gob.hn/DNBC/ciclo1.pdf

SEP (2003b). *Diseño Curricular Nacional para la Educación Básica*. Tegucigalpa. Retrieved from http://www.se.gob.hn/DNBC/ciclo3.pdf

UNAH (2008). Plan de Estudios de Matemáticas. Retrieved from https://sites.google.com/site/carreramatematicasunah/plan-de-estudios-1

UNESCO (2011). Datos mundiales de Educación 2010/2011, pp. 1-44. Retrieved from http://www.ibe.unesco.org/fileadmin/user_upload/Publications/WDE/2010/pdf-versions/Honduras.pdf

UPNFM (2007). *Memoria 2007*. Tegucigalpa. Retrieved from http://www.upnfm.edu.hn/index.php?option=com_rubberdoc&view=doc&id=138&format=raw

UPNFM (2011). *Memoria 2011*. Tegucigalpa. Retrieved from http://www.upnfm.edu.hn/index.php?option=com_rubberdoc&view=doc&id=350&format=raw&Itemid=88

Zuniga, M. (1987). La Educación Superior en Honduras, pp. 1-98. Retrieved from http://unesdoc.unesco.org/images/0008/000847/084755so.pdf

About the Author

Marvin Roberto Mendoza is a professor at the National Autonomous University of Honduras. He is currently finishing his doctoral studies in mathematics education at the University of Los Lagos, located in Santiago de Chile. He has taught at the secondary and university levels in Panamá and Honduras. He has participated in national and international conferences on mathematics education. His research interests include the development of variational thinking, problem solving, and teacher training.

Chapter 11

MÉXICO: The History and Development of a Nation and Its Influence on the Development of Mathematics and Mathematics Education

Eduardo Mancera and Alicia Ávila

Abstract: Mathematics and mathematics education in México has developed significantly since the twentieth century. Prior to that, only the academic practices of Spain or France were reproduced. Currently, there is a robust development of both disciplines. To illustrate this process, we give a historical background, from the Aztec and Mayan times to the Spanish subjugation, going through the struggle for independence until the present.

Keywords: Mathematics education; history of mathematics education; curriculum development; history of mathematics education in México.

Introduction

The development of mathematics education in a country can only be understood in its political and social contexts because from these it is possible to identify opportunities for the emergence and burgeoning of a discipline such as mathematics, and another discipline, such as mathematics education. To understand the magnitude of this statement, it is appropriate to do a cursory journey across México's history and, simultaneously, along this path traced by the rise of a nation, make special reference to some of the problems associated with the emergence of the disciplines that concern us in this chapter.

It is worth noting how — although the development of the nation was not the result of the influence of mathematics or mathematics education — the role of these disciplines changed and found opportunities for their development within their respective communities. There are many characters, of which we will only mention some as points of reference.

Pre-Columbian Times

The archaeological traces of the peoples that lived in the region date to 2400 B.C., but the importance of the existing social groups reached a summit around 300 A.D. In comparison, around that time the Greek philosophers were making important discoveries about mathematics and its structure, like Diophantus' impressive results in algebra contained in his *Arithmetica* — a work eternally linked to Fermat's Last Theorem.

Although the natives of what now constitutes the United States of México lived in diverse communities of a great cultural, architectonic, and social richness, there were cultures that began to stand out around 50 A.D. In particular, we highlight the Mayas and Aztecs for their cultural, sociopolitical, and commercial control of vast expanses of land. Indubitably, the extent of the dominance of these communities has made it possible to learn more about their social, cultural, and religious organization, among other topics. The archaeological findings attest that Pre-Columbian mathematics focused on human activities, described by Bishop (1999) as a general trait of every culture: counting, locating, measuring, designing, playing, and explaining. However, only the Aztecs and Mayas are recognized for the creation of symbols for a numerical system (Cajori, 1993). This was an important aspect in the development of mathematical knowledge. It was also relevant for the advancement of other fields of study, including science, architecture, and commerce.

According to archaeological findings, the Aztecs as well as the Mayas had units of measurement for everyday use, and they computed the area of regular and irregular shapes. For instance, from studies on the Codex Vergara,[1] it was concluded (Williams & y Jorge, 2008) that the Aztecs had a symbolic representation for fractions connected to land surveys and tax calculations. They dealt with arithmetic through pictures of hearts, arms, and arrows related to fractions, to measure and survey plots of land. In this context, they developed at least five different algorithms to find the perimeter and area of a surface.

---

[1] This codex is located at the National Library of France and at the Santa María de Asunción at the National Library of México.

The mathematical knowledge of the Mayas is better known. The vigesimal number system had two versions: the commercial and the astronomic, with algorithms for arithmetic operations. As is often stressed, the most important aspect of Mayan numeration is the presence of zero.[2]

Certainly, the development of Mayan arithmetic allowed them to register and analyze multiple phenomena that led to outstanding astronomical findings, and from conjectures about its development, they identified advanced knowledge about the movement of the stars.

Mayan geometry was also very important since it is present in several activities: city planning and the shapes of their buildings, pottery, and textiles, among others. Every city was erected taking into consideration the position of the sun and the stars relative to the earth. The main plazas were located in the center of the cities, and were surrounded by houses that had a clear social hierarchy. The first rows were occupied by chiefs or by those with high positions, and the rest were distributed among the rest of the population.

A typical practice among both the Mayas and the Aztecs was that knowledge was reserved to certain elites, like priests and the leaders of political or social life. In this sense, mathematics was a rarely shared commodity.

Currently, there are many indigenous groups in the country, and although some Aztec and Mayan traditions have endured, given that the Spanish conquest claimed the vast territories under their control, other ethnic groups are now more prevalent.

At this time, the spring of Greece and Rome was over. Europe was still under the influence of the Dark Ages and the Renaissance was just beginning. Yet, what would be known as America was far behind Europe in terms of technological and scientific knowledge.

The Colonial Period

The height of the Aztecs and Mayas occurred between 1200 and 1500 A.D., but Mayan culture had a different development and covered a larger period. The exploratory incursions into the Americas in 1492, 1494, 1498, and 1502 marked the decline of the human groups that were present on this side of the world. The conquest of what would be known as México began in 1519. By 1521, the Aztecs had been essentially conquered, though further incursions

---

[2]There are several works about it, for example Lam, Magaña, and Oteyza (2010).

prolonged this process until 1525. This was not a time of intellectual pursuits, but of religious and cultural obedience, including slavery and other social ills.

In 1535, the Viceroyalty of New Spain was established, thus the social and cultural structure of the conquest determined the norms of the sixteenth century that lasted until the end of the seventeenth century. During this period, education was reserved for Spanish noblemen. When the native groups were educated, it mainly meant evangelizing them as a way of inducing them to cooperate in the conquest. With the passage of time, around the mid-seventeenth century, only the mestizos and *criollos* were able to further their academic education.

Education at all levels had a strong religious component, and at the basic level, it was geared toward eliminating the knowledge of the native groups, which was not totally accomplished, yet their rich mathematics knowledge was hidden from them. In this period, the mathematics of the conquerors was oriented toward trade and building urban centers, although they maintained an interest in astronomy (D'Ambrosio, 2001).

We can claim that the formal impetus for science and technology in México began in 1551 with the creation of the Royal and Pontifical University of México, which for a century led intellectual and religious development. Nevertheless, there was no special interest in mathematics and its teaching, other than as utilitarian knowledge, dedicated to basic quantitative and spatial relations. This was reiterated half a century later through the *Ordinance of the noblest teachers of the art of reading, writing, and counting (Ordenanza de los maestros del nobilísimo arte de leer, escribir y contar)* by Don Gaspar Zuñiga y Acevedo in 1600 (Gonzalbo, 2011).

The colonial period was not a favorable time for the development of mathematics or mathematics education. Only basic knowledge was transmitted from the conquistadors to the conquered, and the great wealth of the natives' knowledge continued to be ignored. It is worth noting that in Latin America, from colonial times, education has been state policy, through which ideological control was orchestrated by the training of cadres from the elites, where the conquered were essentially displaced — a trait still prevalent in different regions of the Americas (Ossenbach, 1993).

In the Royal and Pontifical University of México, physics and mathematics courses were offered, but from an Aristotelian perspective that did not give any importance to the use of quantitative methods for the comprehension of physical phenomena. This institution gave preference to the training of physicians and attorneys.

During the sixteenth and seventeenth centuries science suffered a major setback in México. Whereas in Europe various scientific disciplines were being developed, on this side of the world the only discipline using science was medicine.

Echoes of the Illustration: The 18th Century

The Illustration, which began in France between the seventeenth and eighteenth centuries, disseminated the ideas of liberty, science, and politics throughout Europe, yet they took a while to reach Mexican land. This influence reached its peak during the eighteenth century as a result of intellectuals and clerics who participated in various activities in México.

There were several religious orders in the country. Among them, the Jesuits had introduced many new ideas about the pursuit of knowledge, but their activities were considered inappropriate for the ruling class; hence, they were expelled in 1767. In this way, some self-taught *criollos* began to disseminate scientific knowledge. They were joined by some Spanish scientists, who translated and published some scientific works that bolstered research and teaching. They were able to keep this rhythm until the end of the eighteenth century.

In Europe also, from the mid-eighteenth century on, the Industrial Revolution defined the important mathematical terms that every citizen should learn to become a productive member of society. We still preserve that input: natural numbers and their operations, fractions and their operations, decimal numbers and their operations, and areas and perimeters of geometric figures, among others (Howson and Wilson, 1986).

The colonial period began its final decline toward the end of the eighteenth century. The resulting discontent propelled the organization of the people, in particular the middle class, to begin its struggle for independence.

First Half of the 19th Century: Independence

The struggle for independence took place between 1810 and 1821. The new country was referred to with different names: the United States of México, the Mexican Republic, and even Mexican Empire and *Anahuac* (which means "land surrounded by water" in Mexican).

During the colonial period, México was the term used to refer to the most important city of New Spain and to the lands it ruled. México is a word of pre-Hispanic origin that evokes the cities surrounded by lakes Tenochtitlan and Tlatelolco: *Mexi* means "moon" or "center of the maguey" and *co* means "where it is located". The word "Mexican" was used to refer exclusively to individuals who lived in the city of México or those who spoke *Náhuatl*, the Mexican language (Clavijero, 1964).

The social pact that gave birth to a nation consisting of diverse federative entities lasted 12 years. Historically, it was a time when certain public policies were implemented. Moreover, some great ideas were generated, that even though many were not realized, remained as part of the educational ideology for years to come.

Given the context during independence, not much attention was devoted to the advancement of science. Education was more open to the existing social groups, but it continued to be linked to the methods and contents of the colonial period, with a strong Spanish and French influence. As is often the case with belligerent conflicts, the war for independence halted the development of the Mexican educational system. While in Europe the institutions of higher education were developing at a rapid pace as of the seventeenth century, this struggle set back the scientific and technological development of the country by almost two centuries.

Certainly, the work of Mexican intellectuals also affected the political turmoil, civil wars, military pronouncements, and foreign invasions. There were neither conditions for peace nor the collegial atmosphere needed for the efforts of academicians to bear fruit. Nevertheless, the war allowed for certain progress in knowledge. For instance, shops were organized to tend to the needs of the war, and since there was much emphasis placed on combat tactics and military discipline, the making of gunpowder and the smelting of copper and iron were developed. Moreover, medicine transformed into a scientific enterprise given that tending to the wounded and preventing diseases were key elements to succeed in the war. Furthermore, paleontology, social science, and public administration were furthered. Also, science and scientific knowledge began to take into account the political context. For instance, in the spirit of the first constitution is found such principles as "to govern scientifically"(González, Gessure, and González, 2006).

When the war was over, some important developments could be considered, such as the creation in 1823 of the first Lancastrian normal school to train teachers. A year later, the Normal School for Mutual Instruction was established in Oaxaca. Other important aspects were addressed, like the need to develop cartography and to conduct a statistical survey of the human and natural resources. This is how in 1824 several knowledgeable men convened to conduct a land survey of México.

During this time, Mexican intellectuals like Manuel Payno and Guillermo Prieto founded journals for the dissemination of scientific, literary, judicial, and historic knowledge. Yet, in 1833, the Royal and Pontifical University of México closed; this reduced the discussions about science, and in particular, about mathematics.

The Federal Republic was replaced in 1835 by the Centralist Republic, which concentrated the decision making power in the capital.

Second Half of the 19th Century

The social turmoil and wars did not end, and the country faced intervention by the United States of America from 1846 to 1848. This entailed the loss of national territory and created the conditions for a new period known as the Reform, from 1857 until 1861. During this period new issues arose, like internal wars and parallel governments. These further restrained the development of mathematics and its teaching.

At the end of the Reform, there was another military intervention, the French, which lasted until 1866. Not until then did the country begin to tread a more constructive path in terms of creating an educational system. In fact, the liberal leanings of the Reform stamped an important characteristic on to the type of country that they wanted to create. They realized that education should have a different purpose and that they should pay more attention to teacher training and to the establishment of exemplary schools. Again, there were more ideas than actualizations, and several educational initiatives that were put on paper, never materialized.

The liberal government led by Benito Juárez proposed a "progressive" national project and sought the emancipation of society in the scientific, religious, and political fields (González, Gessure, and González, 2006). During this time the Law of Public Instruction was passed (1867); it had a "scientific and positivist" spirit and provided for the opening of elementary schools for boys and girls. It also broadly delineated the mathematical instruction that they would get: "the four arithmetic operations for integers, common fractions, decimals and denominators, and the decimal metric system" (*ibid.*). For the most part, mathematics teachers at the elementary level used French textbooks translated into Spanish. This is interesting since the French influence lasted long after the end of Spanish domination; that is, the European influx continued to be important (*ibid.*).

The Normal School for Teachers was inaugurated in México two years after the bill was approved by the Union Congress. Before the creation of this institution, more normal schools had been established throughout the Republic. The construction of elementary schools became an important program from 1874 onward. During this time México basically reproduced the teaching methods and contents that were considered essential in other countries.

In Europe, the efforts of mathematicians to increase research led to the creation of professional organizations to further this goal. For instance, the following professional societies were created: Moscow Mathematical Society (1864), London Mathematical Society (1865), *Societé Mathématique de France* (1872), *Circolo Matematico di Palermo* (1884), New York Mathematical Society (1888; in 1894 it became the American Mathematical Society), and the *Deutsche Mathematiker-Vereinigung* (1890). This movement helped to consolidate the profession of the mathematician, given that until then only some courses were offered at universities, especially in departments related to engineering. These societies began to have a leading role in various countries since they led several educational reforms and became part of a surge of opinions that would propel changes in plans and programs of study within Europe, and years later, in other countries such as México.

Around this time, but far removed from being related to advances in the mathematical community, in 1890 an educational proposal was presented in México that had some points worth noting. The roles of boys and girls were differentiated. Boys took nine classes, while girls took ten because the girls' program included needlework. At each level the precise content was given, along with teaching recommendations. For instance, for second grade, there were mental and objective calculations with digits in the four arithmetic operations with numbers from 1 to 1,000; teachers had to fix in the minds of their students the multiplication table up to 10 by daily exercises with the abacus; arithmetic and geometry appear as separate subjects (González, Gessure, and González).

During that time, the principles of mandatory, free, and secular elementary education were rescued from the Reform Laws of 1857. The training of teachers improved with the creation of two schools in México City: the Normal School for Male Professors (1887) and the Normal School for Female Professors (1890). Also, coed education was advanced and the national educational system was unified.

At the end of elementary school, one could enroll in a normal school. The study plan approved in 1892 for the Normal School for Male Professors included mathematics during the first three years (arithmetic, algebra, and geometry), as well as teaching methods. However, in the Normal School for Female Professors, only arithmetic and algebra were taught during the first years and geometry during the second. There was a clear distinction in terms of gender. In the case of females, the study of mathematics should be limited, replacing political economy with domestic economy, and adding gender-based chores and music education up to piano instruction or melody. Teaching less mathematics to future female teachers reflected several things, but mainly their expected domestic role.

Toward the end of the nineteenth century, Mexican professors began to write textbooks for teachers. However, some normal school graduates thought that a textbook would limit the teacher's role in the classroom, forcing him or her to follow sequences, contents and pedagogical recommendations.

The growth and influence of the mathematics community, mainly the European, were vast. They aimed at reforming university education, but for that they needed the support of the international mathematics community. Therefore, at the International Mathematical Congress of 1893, the distinguished mathematician Felix Klein (1849–1925), paraphrasing Karl Marx's Communist Manifesto, stated: "Mathematicians of the world, Unite!" This propelled the creation of a world organization, the *Internationale Mathematische Unterrichtskommission* (IMUK), which would group the existing organizations under an umbrella that came to have great influence in the scientific world, in which there was an absence of Mexican mathematicians, as well as mathematicians from other Latin American or Caribbean nations.

The Revolution

At the beginning of the twentieth century, the bellicose spirit of the country returned. The new uprising sought to thwart the re-election of President Porfirio Díaz and demanded a better distribution of wealth and land. This led to the Mexican Revolution in 1910.

Despite the state of social tension, the Secretariat of Public Instruction and Fine Arts was created in 1905, which allowed for the unification of the educational activities and tended to the national needs of the education sector.

The liberal educational tradition and the idea of elementary education as a means to foster the economic development of the country were still present, but the nation lacked the necessary infrastructure to implement these ideas. Hence, the construction of elementary schools was a priority across México. Construction had begun a few decades earlier, but most schools were built in the urban centers; hence, the need to build schools throughout the country was felt.

In the international sphere, the community of mathematicians moved forward decisively, and in 1908, the International Commission on Mathematical Instruction (ICMI) was created during the 4th International Congress of Mathematicians held in Rome and presided over by Felix Klein.

In contrast, México remained immersed in internal conflicts. The Mexican Revolution begun in 1910 — a century after independence — had a protagonist: President Porfirio Díaz. Díaz's policies were able to unify the population against him, demanding changes to the constitution and his exit from power. He was indeed a controversial character, and he is better known for his desire to remain in power than for the work he did for the country. However, in terms of education, some improvement was seen.

In 1910 a new plan of study for elementary education was proposed, which preserved some aspects of the former one, like gender difference. This decision affected the integration of mathematical content. For instance, arithmetic and geometry were grouped into one subject, but the allotted time for mathematics instruction was reduced to a seventh of what it used to be. To adjust for this, there was also a reduction in geometric content. The extra time available was used to train boys in military exercises and girls in domestic chores.

The agreement to reduce the amount of time devoted to mathematics was the result of debates between two groups with opposing views. One favored a scientific education and sought to increase the time allotted to mathematics instruction; the other favored a more practical and patriotic education. The latter came out victorious.

The plan of study kept the content fairly intact, with some minor variations. The curriculum now consisted of different types of numbers (natural numbers, decimals, and fractions) with the four arithmetic operations, identification of geometric shapes and solids, and the estimation of magnitudes. They were expected to deal thoroughly with fractions in primary education. It also included the graphical representation of equations of first and second order in the fifth grade.

These ideas were gradually making room for those of Pestalozzi (1746–1821). The footprint of these can be seen in the fight against the rote memory approach and in fostering students to approach mathematics with the knowledge they already had and that would serve as a starting point for teachers. This approach encouraged educators to use examples of everyday life to assist comprehension, problem-solving as a key element of instruction, and the manipulation of objects before explaining a concept. Hence, students were required to learn to count using pebbles or use the corners of a table before introducing the notion of an angle (González, Gessure, and González, 2006).

In 1912, the demand for elementary school teachers was increasing and this forced the government to take action. Hence, a night normal school was created to train teachers in the capital city who had not graduated from a teacher education program. This allowed teachers from the outskirts of México City to attend classes on weekends. This institution was an antecedent of the Oral Normal School, which in turn preceded the Federal Institute for Teacher Training (*Instituto Federal de Capacitación del Magisterio*) for in-service teachers.

In terms of the organizing of mathematicians, in 1911 the *Sociedad Matemática Española* was created; this was the first mathematical society in the Spanish-speaking world. In México, mathematics was still only studied at engineering schools, as a tool for that professional activity. Yet, mathematics as an international community came to the fore as an effect of World War I (1914–1918), since the belligerent countries understood the need to train scientists of the highest caliber. They in turn would support the advancement of science and technology by contributing to the war efforts (through arms manufacturing, creating more effective means of transportation, improving communication facilities, aviation, and medical support), which was only possible with a sound preparation in mathematics. Therefore, in the 1910s, many departments and schools in the mathematical sciences sprouted in Europe. Most of Latin America remained a mere observer of this process.

The 1920s and 1930s

In the two following decades, the country remained in an uneasy calm, with mobilizations and protests. However, the issues were somewhat resolved, which gave the country some stability needed to further education. That is how in 1921 the Secretariat of Public Education (SEP), now still active, was created. This became the center from which all educational programs were administered for every level. It controlled the normal schools, in addition to evaluation and curriculum development projects for basic education.

Beyond curricular transformation, there was a rising interest in education. Hence, the curriculum remained fairly stable while focusing on the initial training of teachers and on providing support for in-service teachers. In fact, in 1922 and 1928, the first rural normal schools were created, supported by local governments. In 1924, the Normal School for Male Professors transformed into the Normal School for Male Teachers, which helped in the unification of all normal schools. In that same year, a preschool, elementary and secondary schools, and professional cycles were added to this normal school.

The first institution to train teachers for secondary school was the Faculty of Advanced Studies in 1923, which in 1929 organized courses for foreign teachers. The following year, this service was offered to secondary school teachers from border countries.

In 1931 the International Mathematical Union (IMU), founded in 1920, was consolidated. Within it, the ideas to unify mathematics and work on its foundations were born. In México, some institutions began exchanges with IMU that would later help to further mathematics in the country.

Science started to become important, and for that, the training of scientists. The Faculty of Sciences at the National Autonomous University of México (UNAM) was opened in 1935, a year after the creation of the Physics Institute. In this context, the existing academic cooperation between the incipient university bodies permitted that a significant number of the faculty members of UNAM would teach courses and guide the mathematical training of the students at the Superior Normal School.

In 1936, the Institute for the Training of Secondary School Teachers was created, and with it began the first exchanges between universities and normal schools. The institutions maintained their autonomy, but academic cooperation was fostered.

From 1940 to 1969

During these three decades the country remained embroiled in political conflict, yet there was a favorable climate for education, in particular for curriculum development and the initial and continued education of teachers.

For instance, the Institute for the Training of Secondary School Teachers transformed into the Institute for the Advancement of Secondary School Teachers, which in 1941 was renamed as the Superior Normal School. This new institution created seven bachelor's degree programs: Spanish, Mathematics, English, Psychology, Pedagogy, Social Science, and Natural Science. Although the purpose of these programs was for teachers to become middle school teachers, many worked as high school teachers.

At the beginning of the 1940s, there were some Mexican nationals studying mathematics abroad, like Enrique Bustamante Llaca (1915–1977), who earned his doctorate in mathematics at Princeton University in 1941. Another distinguished personality during this period was Remigio Valdés Gámez, who was in charge of the First National Congress of Mathematics in 1942, held in Saltillo, Coahuila. In 1943, the Mexican Society of Mathematics (SMM) was created as a direct result of this congress. By then, substantial mathematical activity began to be seen in México, in which the Mathematics Institute at UNAM played a key role. Incidentally, Valdés Gámez was also instrumental in the subsequent creation of the National Association of Teachers of Mathematics (ANPM) in 1969.

In 1944 the Federal Institute for Teacher Training was created, which had a system of instruction by correspondence. Likewise, in-person courses were given throughout the country during the academic year. Each student had to cover six courses and then a professional examination, to receive the appropriate degree.

The General Directorate of Normal Education was created in 1947 to regulate and control the training of teachers in the country. In this context, normal schools benefited with the participation of the incipient Mexican mathematical community.

In the international sphere, IMU's activities were impaired by World War II. The differences between mathematicians from warring countries became apparent, and the organization had to resume work years later. This perhaps negatively impacted on the influence that mathematicians had on mathematics education, even though they had created ICMI. However, ICMI did not become a commission of IMU until 1952, when it assumed a leading and consistent role in mathematics education. As such, it sponsored activities and publications that promoted reflection, collaboration, and the dissemination of ideas about the theory and practice of mathematics education.

During the mid-twentieth century, the mathematics community was convinced of the need to solve several internal issues to unify it. In addition, these discussions led to ideas about the acquisition of mathematical knowledge. These ideas, inspired by the works of the Bourbaki group, were advanced by ICMI members during the Royaumont Seminar in 1959 sponsored by the Organisation for European Economic Co-operation (OEEC). This led to the New Math movement, although this took longer to be considered in México.

In 1959, while the principal ideas of the New Math were being discussed abroad, the Secretariat of Public Education took on the task of providing free textbooks to all elementary school students. It also proposed the creation of a bachelor's degree in mathematical physics, as well as the Graduate School and the Higher School of Physics and Mathematics (ESFM) within the National Polytechnic Institute (IPN). For this, a commission was created in 1960 composed of 20 scientists and engineers from IPN. They developed the plan for the Graduate School and the creation of the Center for Research and Advanced Studies of the National Polytechnic Institute (CINVESTAV-IPN), which opened in 1961. ESFM also opened in 1961.

In the 1960s, the concerns for the development of science and technology in México were repeatedly presented and there was a critical mass of professionals that could support the programs for the development of mathematics. Furthermore, participation of Mexican mathematicians had increased in IMU and ICMI, which opened the door to begin work in incorporating the New Math ideas in the country.

In 1960, ICMI helped to create the Inter-American Committee on Mathematics Education (IACME), which was the first regional group sponsored by this organization, with Marshall Stone as president. It had the intention of promoting the New Math in the western hemisphere. Yet, the New Math foundational ideas and their diffusion took about a decade to be introduced in México, with the assistance of mathematicians from UNAM and IPN, although there was internal disagreement among mathematicians about how closely to follow the recommendations proposed by IACME. The disagreements were eventually overcome in some countries, but not without lingering doubts about the work to be done.

IACME was also important in the region as its members began to promote the creation of teacher organizations and mathematics education societies to face the problems posed by the New Math, and even to advocate for the physical and moral defense of various mathematicians who were harassed in several Latin American countries by local political regimes. In this way, Valdés Gámez organized the National Congress of Mathematics

Education in 1967. Out of this meeting came the aforementioned ANPM, formally established in 1969.

There were no important curricular reforms during the 1940s or 1950s, but in the 1960s, although the contents remained essentially unchanged, the methodologies were transformed and there was more emphasis on lesson plan development. Moreover, the study plans of the 1960s with respect to arithmetic and geometry had as their goals the development of quantitative reasoning and the ability to make connections, use precise language, foster analysis and research, and validate mental discipline (SEP, 1961). The nature of these goals, essentially formative, was complemented by an interest in the everyday utility of the discipline (Ávila, 2006).

During this time schools tried to provide every student with a small board or a graph paper pad to work out the exercises. The abacus, meter, square, and the wooden protractor supported instruction. There were also boxes with different wooden shapes of linear, surface, and volume magnitudes.

It should be mentioned that the link to everyday life did not merely have a utilitarian goal. There was also a belief that the link with everyday experiences would help make sense (in cognitive terms) of mathematical ideas and would promote learning. The idea was that the school, as an artificial construct, inhibited interest, whereas ordinary dealings had "vital meaning" that would solve the problem of a lack of interest in students (Ramírez, 1945).

The 1970s

This decade witnessed the attempts at a radical transformation in the teaching of mathematics. However, the entrance of the New Math created tension at all educational levels. The incorporation of logic and set theory, besides the emphasis placed on mathematical structures, triggered many discussions. Yet, the mathematics community, supported by scholars from other fields (a reform was also taking place in the teaching of Spanish), emphasized these changes and the "specialists" imposed their point of view. It is worth noting that these changes were being implemented in México over a decade after they had started in other countries, at a time when they were being strongly questioned.

The modernizing wave inherited the central tenet from higher education (in regard to specialists), which claimed that "to teach mathematics, one has to know mathematics". With that type of slogan they began their homemade version of the New Math. They first began implementing changes at the middle levels (lower and upper secondary), and later at the elementary level. This second stage hinged on two points: (a) that the new curriculum for the middle levels demanded the acquisition of certain knowledge in order to succeed at it, and (b) children could begin learning from an early age the principles of the New Math (Dienes, 1966).

In the international context in elementary education, the ideas of the Hungarian mathematician Zoltán P. Dienes were a basic reference. His books were translated into many languages and introduced in many countries. Some of his ideas included "logic blocks", number problems in different bases, and a novel conception of the teaching of fractions and geometry. His main proposal consisted of children developing a sense of ownership for mathematical structures as well as abstracting common elements from diverse but isomorphic situations, through a process partially inspired by Piaget.

In México, Dienes' proposals were taken with extreme caution. Yet, although the abandonment of everyday applications of mathematics was not openly declared, it in fact happened and curricular changes were radical. First, three new areas were introduced: logic, probability, and statistics, which had never been thought to be part of elementary education. But in the known areas, there were also novelties. For instance, in geometry, symmetry made its debut and took up much space in the new curriculum. Moreover, there were elements of analysis, graphs, and the classifications of shapes. Triangles, up to then classified by angles and sides, were now also classified by the numbers of axes of symmetry; other polygons were classified by the number of rotational symmetries. Even perpendicularity was given a new flavor: "Two lines, each symmetric with respect to the other, are called perpendicular" (Filloy, et al., 2001, p. 104).

The Cartesian plane was among the newcomers. Its inclusion was justified by its mathematical relevance, since analytic geometry allowed for the transformation of geometric problems into algebraic ones and for the application of methods from both areas. In arithmetic, the changes were also significant. Numerical calculations, supposedly taught in a mechanical way focusing on "real life" problems, gave way to the distributive, associative, and commutative laws. To introduce the integers, a frog leaping on the number line was devised for the early grades. The international movement that gave rise to these ideas, however, was based on an overestimation of the effect that early contact with formal mathematics would have on students, and on an overestimation of the intellectual abilities of the average student.

In 1979, a new reform movement began to take place in France, deriding anything related to the New Math. The United States soon joined the tide. This movement came to be known as Back to Basics. In the mocking spirit of the era, a translation of Morris Kline's book *Why Johnny Can't Add* became very popular in México. Other than that, the debate that ensued around the reform movement is not well-documented, nor its results in terms of learning. Nevertheless, the new perspective that questioned the existence of the New Math in the international arena, would give way to new curricula with markedly distinct objectives.

The modernizing atmosphere created confusion in the management of school content and in the need for the training of teachers. This led to the creation of the National Pedagogical University (UPN) in 1978, which sought to develop studies related to education. However, the mission soon changed to giving an opportunity for teachers without the bachelor's degree to get one. The reason for this is that a normal school diploma was not considered a higher education degree.

During the mid-1970s, a group of mathematicians who had been involved in the 1972 reform (Eugenio Filloy, Carlos Imaz and Juan José Rivaud), created the Section on Educational Mathematics (*Sección de Matemática Educativa*, SME) at the CINVESTAV-IPN. This initiated the work of specialists in mathematics education (although they preferred the term "educational mathematics"). They created a master's degree program in the field and fostered international exchanges. This decade was very important in the beginning of formal activities in mathematics education in México.

The 1980s

The following decade was one of continuing development for the country. Inertia in the educational system led to a similar inertia in mathematics education. However, there were various reactions against the New Math and suggestions that the curriculum that had been based on the New Math needed to be changed.

In fact, for elementary education there was an effort to change the curriculum in lower primary grades, while at the secondary level, programs were being constantly adjusted, but principally related to the specific programming of the contents.

An intense period began, which was related to the dissemination and acceptance of constructivist ideas by both educational officials and teachers. Nevertheless, there were no changes in the curriculum. Attention was focused on teacher education: not only in the normal school system, but also for high schools and universities. Eugenio Filloy coordinated a national program that had as its objective the education of teachers for high schools and universities. As the program continued, the normal schools were included and the academic level of their teaching personnel increased. The program led to the creation of bachelor's degrees dedicated to mathematics teaching in various universities and a master's degree program that could be completed through part-time study. Participation in IACME, PME,[3] and ICMI increased.

Parallel to mathematics education programs offered in CINVESTAV, other options were created in UNAM in the Academic Unit for Professional Development and Graduate Studies of the College of Sciences and Humanities, and in the Mathematics Research Center in Guanajuato. However those two programs were closed after relatively short lives.

Other universities and normal schools began to offer master's degrees in mathematics education. They were supported by graduates of SME and specialists from other institutions. SME increased its influence not only with its master's degree, but it began to offer a doctoral program in collaboration with foreign universities that permitted some of its faculty members to obtain a doctorate in the teaching of mathematics.

The founders of SME further supported the creation of a journal called *Matemáticas y Enseñanza* (Mathematics and Teaching). After sporadic publication, it was managed for a year by the Center for Mathematical Research (CIMAT). In that context, there arose the idea of unifying efforts that had led to various short-lived publications into one journal that is still being published today, *Educación Matemática*.

The normal schools were also transformed during the 1980s. The course credits of normal school graduates had not been recognized as university credits, but instead were considered to be technical or semiprofessional courses. This meant that some normal school graduates had difficulty being admitted to the university system. The reform of the normal school system in the mid-1980s reclassified them as institutions of higher education and created a high school diploma in pedagogy that was needed for

---

[3]International Group for the Psychology of Mathematics Education.

admission to them. Research was expanded in the UPN, as was the preparation of specialists in mathematics education through a master's degree in education.

In 1989, another period of reform in basic education was begun with the Program for Modernizing Basic Education. The plans and programs of study gave special emphasis to the integration of all levels of basic education, including preschool, elementary and lower secondary, via problem-solving and attention to the development of mathematical abilities. At the lower secondary level, courses were to be organized around general themes: algebra, geometry, trigonometry, probability, and statistics. It was hoped that the past tendency to have fragmented courses composed of tidbits of each theme would be avoided.

A proposal was presented to teach mathematics via problem-solving to be tried in a sample of schools. The proposal emphasized putting emphasis on meaning rather than on abstract concepts and symbols without context. It eliminated from the curriculum such topics as logic, sets and structure as the center of attention. The support for the proposal was based on research in mathematics education, especially with respect to the development of mathematical abilities and curriculum development from ICMI Studies (Mancera, 1991).

This new curricular proposal was known as the Operational Test (*Prueba Operativa*, PO) because rather than being immediately applied to all schools, it was tested in a sample of over 500 public and private schools from throughout the country. Thus, the proposed curriculum and teaching strategies were discussed with all the teachers involved in the PO. In contrast to the 1970s, all free textbooks in this project were written by classroom teachers.

Moreover, not only teachers, but principals and supervisors were consulted for the first time. Similarly, there were discussions with specialists from the most representative public and private institutions in the country. This educational proposal was not implemented at every school in the country, but the participation of the various social sectors was intense and there was a diversity of reactions.

The 1990s

The PO was well-accepted in the schools where it was applied. From the results of the PO it was possible to define the definitive curriculum, the necessary components for initial and continuing teacher professional development, the contents of the free elementary textbooks, and the desired characteristics of the textbooks for the other grades (Mancera, 1991).

Without a further evaluation of the PO, it was decided at the SEP not to implement it at every school. Although the actions of the PO continued to function, the results were not discussed and the experiences that could have resulted from the wide-ranging participation of the teachers were not appreciated.

In May of 1992, an "Agreement on the Modernization of Basic Education" was signed. That document delineated the actions that would be taken with respect to reformulating content and educational materials, and the revalorization of teaching. To the considerations, programs, and other dispositions related to elementary education, aspects of teacher professional development were added. Teachers were to be "protagonists for educational transformation" (SEP, 1992, p. 12) and normal schools were important "because they were the ones to develop and prepare teachers for Basic Education" (SEP, 1992, p. 4).

Once again, mathematics was given prominence in the education that children should receive: "The foundations of a basic education are reading, writing and mathematics. These abilities need to be assimilated simply but rigorously, facilitating lifelong learning and providing the rational support needed for reflective thinking" (SEP, 1992, p. 10). Specifically with respect to mathematics, it was necessary to "reinforce the learning of mathematics throughout the school years, highlighting the development of the capacity to relate and calculate quantities with precision and to strengthen the knowledge of geometry and the ability to clearly pose problems and then solve them" (SEP, 1992, p. 11). This new approach considered problem-solving as a means and end of mathematics.

The curricular reform derived from the Agreement introduced in 1993 a reform in which — in ways distinct from the reform of 1972 — the emphasis was on what might be called the didactic relationship, that is, in the relationship among the students, the teacher and the mathematical knowledge. There is, however, one characteristic common in the two last reforms, that innovation is the result of a wave of international thinking. This thinking was based on evaluation studies conducted at the end of the 1970s that made public the persistence of difficulties that children have in finding solutions to problems that were not identical to other problems that they had studied (Perret, 1985, pp. 24–25). These results were repeated at least in the United States, Great Britain, France and Switzerland.

In México, the influence of the United States was present as well as that of France. It was officially declared that this new perspective for teaching mathematics was derived from research results in the previous decades from national as well as foreign institutes that called into question the prevalent methods of teaching mathematics (SEP, 1993). In fact, the results allowed a rethinking — initially in an academic and experimental environment — of the assumptions concerning the teaching and learning of mathematics. The results of various studies — later translated into principles, sequences and recommendations for teaching — constituted to a large extent the basis for the new Mexican programs and materials.

Putting students in touch with what they were to learn via problem situations was declared to be the nucleus of the Mexican educational reform in the 1990s. Although that intention was woven into the materials with elements from other perspectives that at times diluted the intent, it was emphasized and reiterated in different ways throughout the plans and programs. The basic idea was to introduce problem-solving from the beginning, letting students use their own resources because this would allow them to construct new knowledge that, later, would help them to find solutions to more complex problems.

It has been suggested by Mancera (2000) that "the focus on problem-solving refers to a variety of approaches from the simple incorporation of problems into classroom lessons to proposals that are very elaborate and supported by theories on cognitive development and information processing." The curricular materials do not explicitly define a strategy to help in solving mathematical problems. A problem is understood to be a situation in which the person trying to solve it may not have a specific strategy, although elements to imagine or create a solution may be there. A problem — in this assumed perspective — requires analysis, reflection, anticipation of solutions, and testing to generate solutions. The value of this process promotes knowledge.

Nevertheless, even though the reform was generally well-received, it was also criticized. For instance, Cambray-Núñez (2003) has pointed out that constructivism in the reform was not very clear. He says that this can be seen on the one hand because no official document circulated by the Secretariat of Education treats constructivism in any depth, and, on the other hand, the authors of the free textbooks — by their own admission — essentially had no knowledge of constructivism, which certainly made it difficult for them to write instructional materials with a constructivist perspective.

At any rate, the reform remained in place for about 15 years, and a few years after its implementation began, a teacher training program was introduced to help educators understand this new approach to teaching in terms of content and methodology.

The New Millennium

By the beginning of the new century there were already important developments in mathematics education in México. For instance, the number of publications had increased and more institutions had developed programs in this field.

The topics that are studied include the learning process in mathematics, notions and knowledge of teacher, the effects of manipulatives, communication and information technologies, and experimental teaching. Other topics include adult education and special education (at a superficial level). Teaching in indigenous schools remains an unmet challenge.

There was also work on the use of various media in the teaching of mathematics. This had to do with the investments that were being made in this kind of education. Although most of the investments were in programs that were not particularly successful, they did permit the development of education materials and experimentation.

Investment in education has increased, but continues to be insufficient and most of it is spent on fixed costs: salaries and benefits. The school infrastructure has not been significantly improved. Investments in computing equipment and electronic blackboards have not shown benefits.

The curricular reform of 1993 was modified in several stages: 2006 for secondary and 2008 for primary. It was reviewed several times until 2011. In part, the new curriculum and the revisions that it underwent were conducted by the same working groups that had been present in 1993. One of the reasons that this stands out is that there are no data to indicate whether or not the reforms led to positive results. There were declarative documents giving a rationale for the new reforms, but without sufficient data.

This a recurring story in the reforms at the school basic level in México. Certainly, the pressure created by international studies such as PISA and TIMSS favored the most recent changes; it is also not known whether these studies are enough to warrant curricular reforms. In general, the current thought is the need to improve the training of in-service teachers instead of carrying out reforms, but the decision-makers in México have their own agendas, and they do not always consider what should be done based on the available research.

With the intent of contributing to the understanding of the teaching and learning of mathematics, several organizations have sprouted, helping to organize academic gatherings. For instance, in 2007 the 12th IACME was held in México, as was International Congress on Mathematical Education (ICME) 11 in 2010. The ANPM has had almost 30 national congresses and SMM has also held congresses on the teaching of mathematics.

In the annual meetings of the National Council of Teachers of Mathematics (NCTM), under the heading "Bridge across the Americas", the topic of teaching mathematics from a bilingual and multicultural perspective and their effects on immigrant Mexicans in the United States has been addressed. These are pending research topics to be considered by mathematics education researchers based in México.

Virtual Educa and other programs have helped to promote opportunities for the continuing professional development of teachers through the virtual environment of *C@mpus de las Matemáticas*. Opportunities for courses, diplomas and graduate studies have grown as can be seen in the SEP catalogs on continuing professional development.[4]

Final Commentaries

There is no consensus on the direction that curricular reforms should take. Mathematicians have one position, mathematics educators another, and applied mathematicians yet another. Thus each sector emphasizes what is the closest to its own work. To help better understand this situation, more serious studies are needed on the effects of mathematics education at each educational level and on the initial and continued education of teachers.

If one considers the results of achievement tests, things do not seem to have changed much throughout decades of reforms. The results of México in international tests created pressure to change the way mathematics is taught. For this, action must be taken to allow for the convergence of the training of teachers and their continued education, with students' aptitudes, and the learning environment. We must also evaluate plans and programs

---

[4]SEP catalogs on continuing professional development are available at http://www.virtualeduca.org/campusmat/sep/

of study, and their implementation in the classroom, which requires time to show their strengths and weaknesses. This approach must also consider that teacher training requires not only mathematical content, but also how this knowledge is learned and used by students.

Regardless of the observations stated above, the growth of research and the interventions in the educational system promise a great capacity to confront the problems that might arise with curricular development, educational materials, the use of technology, and the training of teachers. Although the gestation period of this level of development has covered many years, the capacity to contribute to the advancement of the teaching and learning of mathematics among mathematics education researchers has not been enough (Ávila, 2013). The renovation and betterment of the educational system need new perspectives and understandings that are ever more critical of and challenging to the national educational reality.

Bibliography

Ávila, A. (2006). *Transformaciones y costumbres en la matemáticas escolar.* México. Paidós. Col. Educador.

Ávila, A. (2013). Sobre pasado, presente y futuro de la investigación en educación matemática en México, in A. Ávila, A. Carrasco, A. Gómez Galindo, T. Guerra, G. López-Bonilla, & J. L. Ramírez (Eds.), *Una década de investigación educativa en conocimientos disciplinares en México: Matemáticas, ciencias naturales, lenguaje y lenguas extranjeras* (2002–2011), pp. 111–126.

Ávila, A., & Mancera, E. (Eds.) (2003). El campo de la educación matemática, 1993–2001, in A. López (Ed.), *Saberes científicos, humanistas y tecnológicos: procesos de enseñanza y aprendizaje I*, pp. 39–352. México. COMIE-SEP-CESU.

Bishop, A. (1999). *Enculturación Matemática, La educación matemática desde una perspectiva cultural.* Barcelona: Ediciones Paidós Ibérica SA.

Cajori, F. (1993). *A History of Mathematical Notations.* New York: Dover Publications Inc.

Cambray-Núñez, R. (2003). *Reform process of the mathematics curriculum for basic education in México during 1992–2000.* Tesis de doctorado en educación. Muncie, IN: Ball State University.

Clavijero, F. X. (1964). *Historia Antigua de México,* Ed. Porrúa Libro II, México.

D'Ambrosio, U. (2001). La matemática en América Central y del Sur: Una visión panorámica, in A. Lizarzaburu, & G. Zapata (Eds.), *Pluriculturalidad y aprendizaje de la matemática en América Latina.* España. PROEBID Andes/DSE (Deutsche Stiftung für Internationale Entwicklung)/Ediciones Morata, pp. 49–87.

Dienes, Z. (1966). *Mathematics in primary education: Learning of mathematics by young children*. Hamburg: UNESCO Institute for Education.

Filloy, E., Rojano, T., Figueras, O., Ojeda, A., & Zubieta, G. (2001). *Matemática Educativa, Tercer grado. Libro del maestro*. México: McGraw-Hill Interamericana Editores.

Gonzalbo, P. (2011). *Ordenanza de los maestros del nobilísimo arte de leer, escribir, y contar el humanismo y la educación en la Nueva España*, SEP/el Caballito, México.

González, R. M., Gessure, F., & González, S. (2006). La enseñanza de las matemáticas en las escuelas primarias de México (Distrito Federal) durante el Porfiriato: programas de estudio, docentes y prácticas escolares. *Revista Educación Matemática, 18*(3), pp. 39–63.

Howson, A. G., & Wilson, B. J. (1986). *School Mathematics in the 1990s*. ICMI Studies Series, Cambridge University Press, England.

Lam, E., Magaña, L. F., & de Oteyza, E. (2010). *Puntos, rayas y caracoles*. México. Editorial Terracota.

Mancera, E. (1991). La modernización de la educación básica. El enfoque de la Modernización Educativa. *Revista Educación Matemática, 3*(3), pp. 4–9.

Mancera, E. (2000). *Saber matemáticas es saber resolver problemas*. México. Ed. Iberoamérica.

Ossenbach, G. (1993). Estado y Educación en América Latina a partir de su independencia (siglos XIX y XX). *Revista Iberoamericana de Educación*, 1 — Estado y Educación.

Perret, J. F. (1985). *Comprendre l'écriture des nombres*. Berna. Peter Lang.

Ramírez, R. (1945). *Técnica de la enseñanza*. México: Editorial Técnico-Educativa.

SEP (1961). *Programas de educación primaria aprobados por el Consejo Nacional Técnico de la Educación*. México: Secretariat of Public Education.

SEP (1992). Acuerdo nacional para la modernización de la educación básica. *Diario Oficial*, 4–14. Retrieved from http://www.sep.gob.mx/work/models/sep1/Resource/b490561c-5c33-4254-ad1c-aad33765928a/07104.pdf

SEP (1993). *Plan y programas de estudio. Educación básica primaria*. México. Secretaría de Educación Pública.

Waldegg, G. (Ed.) 1995. *Procesos de enseñanza y aprendizaje II*. Vol. 2. México. Fundación SNTE para la Cultura del Maestro Mexicano.

Williams, B. J., & y Jorge, M. D. C. J. (2008). Aztec arithmetic revisited: Land-area algorithms and Acolhua congruence arithmetic. *Science 320*(5872), pp. 72–77.

About the Authors

Eduardo Mancera obtained his bachelor's degree in physics and mathematics from the Higher School of Physics and Mathematics at the National Polytechnic Institute, and master's and doctoral degrees from the Center for Research and Advanced Studies of the National Polytechnic Institute (CINVESTAV-IPN). He is the author of numerous textbooks across all levels from elementary school to higher education. He has been a researcher and professor at several universities. He is currently the director of C@mpus de las Matemáticas de Virtual Educa and the first vice-president of IACME.

Alicia Ávila has been a professor at the National Pedagogical University (UPN) since 1988. She holds a doctorate in education from the National Autonomous University of México (UNAM) and has been a member of the National System of Researchers since 2005. Besides authoring multiple research articles and books, she is also the chief editor of the *Educación Matemática* journal. Her research interests are elementary mathematics education which focus on curricular reforms in México during the 20th century, mathematical literacy, and mathematics in adult education.

Chapter 12

PANAMÁ: Towards the First World through Mathematics

Euclides Samaniego, Nicolás A. Samaniego, and Benigna Fernández

Abstract: In this chapter we describe the history of mathematics education in Panamá, from colonial times to more contemporary developments. During this tour, we will highlight some of the aspects that characterize Panamanian mathematics education today and finally conclude with the greatest educational challenges for Panamá.

Keywords: Mathematics; development; sustainability; educational system.

Evolution of the Educational System in Panamá

The Panamanian educational system is strictly related to the different stages of the history of this country. The informal way of knowledge transmission that the first settlers of Panamá had remains the same due to the efforts of these indigenous peoples to keep their traditions. Teaching was mainly oral and practical.

The arrival of the European settlers set a new landmark in the history of this indigenous process. The Catholic Church not only dedicated great efforts to formal teaching, but also to evangelization, thus generating important effects on the colonial society and the original peoples (Céspedes, 1981). At this time the missionary teacher appears as a character ubiquitously present in the Mesoamerican territory.

When the Spanish settlers built the first cities in Panamá, a revolution in the educational system was observed during the sixteenth, seventeenth, and eighteenth centuries. There were two very well marked trends because of the ecclesiastic power over the Panamanian educational system. On the one hand, the church was in charge of the education of the clergy for the evangelization in the new land and, on the other, it was also dedicated to educating the indigenous people. This process involved education in the Christian faith, values, and ethics, the teaching of the language (Spanish was the official language in Panamá in the seventeenth century), handicraft techniques, and the social organization of settlers (Meduca, 2002).

There were two schools during the seventeenth century in Panamá City: Saint Augustine High School founded by the Order of Saint Augustine and Saint Javier High School, founded by the Order of the Jesuits.

In the nineteenth century, Panamá got its independence from Spain and became part of the so-called Gran Colombia. However, the social and economic aspects of the recently acquired territories were largely ignored by Gran Colombia (Winkler, 1994). Being part of Gran Colombia meant following the Neogranadine (Colombian) law constituted by political and social issues concerning education in the eighteenth century, but Gran Colombia never implemented the law in Panamá (Meduca, 2002). In this respect, to open a school in the Panamanian territory, a decree from the Vice-President in charge of the executive branch of Gran Colombia was necessary.

It was not until the 1870s that the Colombian legislation impacted educational issues in Panamá. Thus, the first Panamá Normal School for Boys was established in 1872. It was open for 15 years and was important for Panamá during the time it was part of Gran Colombia (Meduca, 2002).

The University of San Javier was founded in 1874 by the bishop Francisco de Luna Victoria y Castro (Céspedes, 1981). It is important to highlight that the sixteenth, seventeenth, and eighteenth centuries were the scene of a cultural confrontation that ended up in a new society, which culminated in the making of the Panamanian nation.

The first official documents showing data about the teaching of mathematics in Panamá discusses the Organic Law of Education N°1 issued in 1877 by the Legislative Assembly of Panamá (Orozco, 1996). It states that subjects like arithmetic, reading, writing, Spanish, composition and writing exercises, general concepts of geography and history, moral, measurements and weights legal system, practical use of a dictionary are to be taught in primary schools (Meduca-Prode, 2005).

By the end of the nineteenth century, many primary schools had been created across the country, as well as schools exclusively for girls or for boys. The first public high school was *Colegio Balboa*. Several important historical events happened: The Federal State of Panamá (1885), the inauguration of the transisthmian railroad, the French Canal construction (1880–1903), and the Thousand Days' War (1899–1902). The growing process of education stopped during the last decade before the separation of Panamá from Colombia, since schools were closed and the educational system was interrupted due to the loss of autonomy of the Isthmus and the Thousand Days' War (Meduca, 2002).

The First 50 Years as a Republic

After the separation movement in November of 1903, the Republic of Panamá had to define its own identity as a nation. One of the first steps in that direction was the inclusion of the principles in the political constitution as a guide of the educational system (Meduca, 2002). Thus, the Political Constitution of the Republic of Panamá was issued at the beginning of 1904 and it was published by the Public Organic Law 11 of March 23, 1904 (Céspedes, 1981). This law divided the educational system into elementary schools and high schools (in arts or philosophy, industrial, and professional).

As history evolved, the following events were the basis of the Panamanian educational system (Céspedes, 1981; Winkler, 1994; Meduca, 2002):

- The Normal School for Boys (*Escuela Normal de Varones*) and the Normal School for Teachers (*Escuela Normal de Institutores*) were created on April 15, 1904 as the first schools to train elementary school teachers.
- In 1904, two schools were founded. In May, the Music and Reciting School and in August the Higher School for Boys (a middle secondary school for boys).
- In 1906, Commercial School of Languages.
- In 1907, the School of Arts and Trades. It is still functioning as *Escuela de Artes y Oficios Melchor Lasso de la Vega*.
- In 1906, the School for the Indigenous People was created. That same year, 75 schools were active, 72 for boys and 3 for boys and girls. In 1908, there were 222 elementary schools.
- In 1907, the National Institute was created. This high school offered two options: a regular high school diploma and a diploma concentrating on commerce. It is still open and has a new organizational structure.
- In 1915, a decree temporarily approved the plans, programs, and requirements for elementary school, high school and schools to train teachers

for elementary schools. This is the beginning of the standardization of teaching in elementary schools in the country.
- In 1920, secondary education expanded as a result of the creation of normal schools in the countryside. Moreover, a second organic law was issued that reiterated the principle of free and compulsory education, and foresaw the creation of a university.
- In 1930, the Normal School of Santiago de Veraguas was created and the other normal schools disappeared. Today it is known as Normal School Juan Demóstenes Arosemena.
- In 1946, Organic Law of Education N°47 was passed.

Panamanian Educational System since 1950

The coming of the second half of the twentieth century demanded some significant curricular adjustments. That is why the Ministry of Education put into effect new plans and programs in 1953. They were structured by a declaration of objectives (Winkler, 1994). This new adaptation included the following subjects: crafts activities, home economics, recreational and artistic activities, science, physical education, social studies, Spanish and the indigenous languages, mathematics, ethics and religion, and English (Meduca, 2002).

To manage the necessity of upcoming changes of plans and programs, the Ministry of Education created the National Office of Educational Planning (Céspedes, 1981). The first results appeared when the adapted programs were put into effect in 1961 by a Revising Commission of Plans and Programs that same year (Zapata, 2008).

In order to get a broader dissemination of education, the Service of Education on radio and television was launched during the 1970s creating the Technological Center for Education for this purpose (Meduca, 2005). Besides this, a Program for Nutrition in Schools was established with the purpose of decreasing the dropout and failure rates. This would help to increase the nutritional level of the elementary school children and, at the same time, improve academic performance (Céspedes, 1981).

The educational reform of 1975 aimed at giving the National Educational System a new structure and orientation. It was a very important attempt to change the national education system (Orozco, 1996). However, due to many circumstances it was abolished in 1979. After this, the Coordinating Commission of National Education was created and was made up of the most representative sectors of the community of education (Céspedes, 1981). After its abolition in 1979, Panamá had to wait for 30 years for a real reform of the educational system. Today it is known as the curricular transformation, which started in the middle of 2009 (Meduca, 2002).

After a very long and deep process of evaluation, the reality of the national education system began. It lasted throughout the 1980s and became the starting point for a transformation (Meduca-Prode, 2005). After diagnosing the status of education, this commission worked on a proposal to reform the Organic Law of Education N°47 of 1946 (Meduca, 2002). It was approved in 1995 by Law 34 of July 6.

In 1983, the programs of education went through a process of revision in an effort to update them. This process of making structural changes to the national educational system would take a lot of time (Meduca, 2005).

The 1980s were a time distinguished by a deep social and political crisis that had a negative impact on education. Education progress was slowed down, as was the adaptation of positive international reforms to Panamanian realities.

At the beginning of the 1980s, a higher number of credit hours for subjects like biology, physics, chemistry and ethics was added to the study plan (Meduca, 2002).

A constant process of research on the reality and awareness of education to promote its importance in the Project for National Development marked the first half of the 1990s (Preal-Cospae, 2002).

These actions reached the authorities and the society as well. As a result, the Panamanian society recognized all the efforts towards making people aware of the strategic importance of education in overcoming the fundamental problems the country was facing in its development efforts (poverty, illiteracy, unemployment, family disintegration, drug addiction, violence, among others) (Meduca, 2002). The aspects mentioned above were the reference framework for the beginning of the structural process of transformation of the educational system, which started during the second half of the 1990s. The nature, scope and content of the so-called reforms are strategically created to be long term (Meduca, 2005). This development was expected to reach its higher intensity in the first decades of the twenty-first century as a way of visualizing all the efforts made into real changes to the educational system.

One of the most important achievements in this educational process was the approval of the Organic Law of Education N°34 of July 6, 1995 which entailed amendments to the Organic Law of Education N°47 of 1946. Thus, the legal framework was updated and the conditions to start the processes of change in education were made (Meduca, 2005).

The curricular transformation started with the elementary and middle schools in 110 pilot centers, beginning with grades PK, 1, 2, 3, and 7 (Meduca, 2007).

Similarly, significant work has been done on the design of the programs for elementary and middle school (grades 4, 5, 6, 8, and 9), and for adults and young people.

In addition, 350,000 textbooks and 37,000 guides for Spanish, mathematics, science, and social studies were distributed free of charge. Moreover, significant progress was made in providing computer equipment for educational use (116 elementary schools and all the high schools) and internet access (180 elementary schools and high schools).

There has been a wide range of academic program options for secondary school (high school programs in ports management, tourism, environmental and science education, and computing). There has also been a consistent teacher training program. Today 100% of multi-grade teachers (teachers who teach two grades at the same time) have been trained, whereas only 84% of the regular ones have accomplished this.

Other steps have been taken. They include the establishment of a national system of assessment, programs to help parents and children economically to attend schools in areas of extreme poverty (6,000 scholarships have been given), and health care for the homeless.

Mathematics in Panamá

In 1501, the Spaniards arrived in Panamá subjugating the native peoples, and through evangelization, the educational process started. The Franciscans, followed by the Augustinians and the Dominicans, initially carried this out. However, from 1575 until their expulsion in 1767, the Jesuits guided education in the isthmus. In spite of the difficulties of that epoch, education had a formal character by creating schools and going beyond evangelization. The Spanish language and the basic arithmetic operations were taught.[1] They also founded the first university[2] in 1749, the Royal and Pontifical

---

[1] Studying arithmetic included guarisma rules which are quantities expressed in Arabic numbers, but with the symbols in Spanish.
[2] Universities of Santo Domingo, Lima and México.

University of San Javier offering bachelor's, master's and doctoral degrees in philosophy and theology.

With the expulsion of the Jesuits, the *Colegio de Panamá*, founded by the Dominicans in 1715, became the leader in training priests; only Latin grammar was taught there. More academic options were offered, like teaching trades to poor people.[3] In 1785, the Franciscan Order founded the *Colegio Propaganda Fidae* to train priests. In 1795, the Jesuits opened the *Colegio Seminario*, which had disappeared in 1671 after an assault by the pirate Henry Morgan. In 1803, the administration of that institution was under the Dominicans' responsibility with precise instructions for teaching Spanish grammar. By the end of the eighteenth century, there was a Latin Department and several elementary schools.

Learning by memorization from the theoretical models was introduced by the scholastics in colonial times. The fashionable method was repeating and repeating answers to questions until they were learned by memorization.

Panamá was culturally isolated by 1805. Mariano Arosemena[4] said, "This region is so far away from dealing with enlightened men since only books like *The Ingenious Gentleman Don Quixote of La Mancha*, *La Voz de la Naturaleza*, *La Medicina Doméstica*,..., were seen and some other similar ones in Spanish and Latin. The primers, first reading books, and handbooks of arithmetic came from other parts of the Americas where there is a press...".[5] Only an elite group read these books, because most of the population did not know how to read.

The independence of Panamá occurred on November 28, 1821 and it voluntarily united with Colombia. In 1855, it became the "Sovereign Federal State", but as a part of Gran Colombia allowing the army to have autonomy in dealing with its internal affairs.

The Colombian state created the School of the Isthmus (*Colegio del Istmo*) by decree in 1824, which became a university in 1841. It taught scientific subjects such as arithmetic, practical and tropical agriculture, mechanics, farming, pharmacy and rudiments of surgery. The Lancastrian method was followed at the time. It had its origins in the poor areas of

---

[3] Miserably poor people: orphans, elderly people, sick people and widows without any money.
[4] Mariano Arosemena (1794–1868) Panamanian neoclassic writer, journalist and politician.
[5] The press was introduced in Panamá in 1821.

London in which the teacher could attend to a large group of students with the help of class assistants (teacher aides) who were outstanding students who worked with small groups. This study plan was radically different from the colonial pattern. The technical-secular tradition had overtaken Panamá.

Mass education or popular education began in 1836. A diploma in arts or philosophy, like the one Justo Arosemena[6] got at *Colegio de San Bartolomé* in Bogotá in 1833, allowed students to enter a university. To be enrolled in the law program, prospective students were required to pass literature classes, which were made up of Spanish and Latin together, Greek, English, French, literature, fine arts, elocution, and poetry. Additionally, they needed classes in philosophy that consisted of mathematics, geography, physics, logic, ideology and metaphysics, as well as ethics and natural law.

During the second half of the nineteenth century, the number of newspapers increased; there was the beginning of a new literary movement, and a few civic-spirited leaders who believed in public education added their voices. Colombian educational practices came into force during this period following the decree of the National Executive Power in 1870 that President Eustorgio Salgar passed with the purpose of developing laws for public education. This decree indicated the subjects that should be taught both in elementary schools and in higher education. This classification was not adopted in Panamá until after its independence in 1903. The Organic Law N°1 of 1877 issued by the Legislative Assembly of the Sovereign State of Panamá, established that "the teaching in elementary schools included subjects like: reading, writing, moral, arithmetic, weights and measurements, legal system, Spanish language elements, practical use of the Spanish dictionary, composition and recitation exercises, general features of Panamanian geography, and history."

Once the Isthmus adopted the Colombian Constitution in 1886, it lost autonomy and went on to become a Department; therefore, a period of deterioration of public education began. However, governors of the department of Panamá made a great effort towards giving a boost to teaching. Manuel José Hurtado promoted the creation of the first Normal School

---

[6]Justo Arosemena (1817–1896): politician, lawyer, sociologist, essayist, professor, and historian. He struggled for autonomy in Panamá and is considered the father of the Panamanian nationality.

in Panamá in 1872. It is worth pointing out that Manuel José Hurtado (1821–1887) was an engineer and teacher dedicated to commerce and his profession, and he represented the Department of the Isthmus before the Nueva Granada government on several occasions. He was a natural and exact sciences teacher. Because of his generosity and efforts for popular education, he has been called Father of Public Education in the Isthmus. Therefore, Hurtado's birthday, December 1, was declared Teacher's Day. Also, a high honor for Panamanian teachers is to be awarded the Order Manuel José Hurtado.

The Panamá Normal School had noticeable impacts on Isthmus education and life. German teachers were hired to train teachers according to the Pestalozzi system. Among other objectives, this system was introduced to "improve the mechanical teaching of arithmetic and grammar".[7] Oswald Wirsing ran this school and the Panamanian teacher Manuel Valentín Bravo was the vice-principal. During the fifteen years that this school was functioning, it graduated teachers who contributed to motivating others to become teachers and they held important positions before and after 1903.

It is important to understand the fact that teachers had a national meeting in 1883 and took the following topics into consideration: teaching system by Pestalozzi, assistants for schools, classification of schools, savings bank, newspapers for public education, orthography and others.

Publication of the magazine *La Escuela Normal* was launched, which as the name indicates, should have been an important help for teachers. Years later, the *Revista de Instrucción Pública* (a magazine of public education) was published. By the end of 1896, Governor Ricardo Arango, who was supported by eminent personalities, not only created and equipped schools, but also restored the Normal School for Girls (*Escuela Normal de Institutoras*) from which devoted female teachers graduated.

The Thousand Days' War, at the beginning of the twentieth century (1899–1902), led to the closing of schools and the educational process was completely paralyzed. When the war ended, the next to last governor of Panamá, Facundo Mutis Durán, issued Resolution N°68 of March 31, 1903, in which he proclaimed that to re-establish the public order, "it was vitally

---

[7] "Pestalozziano system was characterized by the child's mind developed through hands-on activities, experiences through games, drawing exercises and discussions to improve the level of reading and writing, taking into account the spontaneous activities of each" (Triana, 1846).

necessary to give preference to elementary public education, which was totally abandoned during the war. To do so, the restoration of the teacher training schools for both sexes was urgent in order to graduate the right teachers as soon as possible."

At the beginning of the twentieth century, elementary school education barely existed in Panamá, there was no high school education, and almost 100% of the population was illiterate.

Panamá's separation from Colombia occurred on November 3, 1903. On January 15, 1904, the Provisional Board of Government informed the National Convention that they considered that "Concerning public education; abandoned during three years of war, everything needs to go through considerable reforms. It is your duty to organize it on a scientific basis, according to modern systems and procedures" (Arosemena, 1949).

The new republic was in need of qualified personnel for the different services of the State and the educational system started to be organized despite having few trained teachers. The Organic Law for Public Education N°11 of March 23, 1904 is important for the legal and constitutional principles adopted. One of the most outstanding aspects refers to teacher preparation: "The teacher training schools will prepare qualified teachers and they will do their best in order to have students acquire enough knowledge, not only moral and intellectual, but also regarding fundamental principles about agriculture, industry, and commerce. Besides, it will prepare practical teachers, which are not just teachers but erudite teachers. It will also require teachers and principals to be certified by a board of examiners prior to being appointed to a school." This law allowed the Executive to hire teachers and technicians from abroad to work in educational areas, and empowered the Executive to found several schools, with special attention on normal schools. On April 15, 1904, a decree created and organized them. Among the most outstanding ones were the Normal School for Boys and the Normal School for Girls. In three years, they began graduating teachers, thus contributing significantly to the development of Panamanian education.[8]

In 1907, Law 11 ordered the creation of a "National Institute in which some professions will be taught and preparatory instruction for others will be given". In this institute, students received a high school education in arts and philosophy. "The professional studies will be focused on law, surveying

---

[8]They included Octavio Méndez Pereira, José Daniel Crespo, Cirilo J. Martínez, Alejandro Tapia, and others.

and topography, agronomy, dentistry, nursing auxiliary, commerce and languages, statistics, and post office services."

It was evident that teaching in elementary schools was not efficient due to untrained teachers. Actually, a public education and statistics inspector said, "I am convinced that concerning teaching it is not well oriented. Education is not integral as the contemporary pedagogy canons demand; importance is given to definitions, details, and repetition of many ideas. Education must have a scientific character, as elementary school only teaches general culture in such a short time; the learning process is exclusively directed to intelligence. In other words, as the aim to be achieved is not the gradual and harmonic development of the mind faculties, and as the active method is ignored, everything about knowledge is just given to the child. Consequently, he or she is not allowed to observe or to think or to judge or to discover."

Decree N°2 of 1910, the Organic Law of Elementary School Education, established the following: "Teaching has to be essentially practical, adequate to the social conditions of the country in general and particularly to the necessities where each school works." It also says, "From the third grade on, the teacher will call attention to the nature of alcohol and narcotics, and on the effects they cause on human beings." Moreover, "Candidates who want to join the educational system without a certificate of competence should undergo a test," and that, "The final tests at elementary schools look for stimulating students, investigating how they have improved during the academic year and evaluating the pedagogic conditions teachers have."

The foundations of the national educational system were laid in the first six years of the republic. In the following decade, importance was given to pedagogic and technical aspects of education as well as to the systematization of the administration and the number of elementary schools continued to increase. Teachers and highly qualified technicians were incorporated into the education system and the first teachers who studied abroad with a scholarship joined the system. The German teachers[9] who were hired in 1911 to be part of the educational system had a contractual obligation to teach "according to the German modern methods". During the second decade of the republic, the graduates from national schools also started joining the academic staffs of elementary schools. The gradual increase of trained

---

[9]Richard Neumann, Otto and Eugenio Lutz, and George Goetz stand out.

teachers and a more efficient inspection service undoubtedly contributed to the improvement of teaching.

Plans, programs and regulations for elementary schools, high schools and teacher training schools were temporarily approved by decree in April in 1915. These programs showed the objectives and characteristic of each subject and methodological instructions about how to teach them were given. Similarly, programs for the different subjects in teacher training schools and high schools were designed.

It is important to highlight the work of Frederick E. Libby, who led the General Inspection of Elementary School between 1915 and 1923. He was a specialist in school administration. One of his achievements was to organize the functioning of elementary school, supported by the inspectors. Libby established a practice where sixth grade students had to take final tests that were prepared and graded by him and the Secretary of the Inspection, Cirilo J. Martínez. This practice gave the General Inspection a way to observe how efficient the teaching was and to evaluate the teacher's work.

Another outstanding person was Richard Neumann who introduced the Herbart teaching methodology of formal steps to Panamá. With this, the programs had an introduction, objectives, educational philosophy and a section about how to teach the subject. Many teachers' classes were planned with this method. Two of Neumann's disciples, Guillermo Méndez Pereira and Luis Tapia, prepared the *Teacher's Assistant* in two volumes, which was a guide for many teachers for many generations.

Under Libby's administration, many adolescents were trained for teaching even if they had not even finished elementary school. Having a sixth grade diploma at that time meant the holder had a preparation that was not common.

Another action taken to improve the preparation of teachers was the creation, in 1913, of a specialized course of mathematics for three years at the National Institute, under the direction of Eugenio Lutz. A cadre of competent mathematics teachers was formed in this course.

The Office of the Public Education Secretary presented a detailed study of the situation of education in the country in 1920. It suggested that mathematics and Spanish had an "exaggerated dominant" place in the current program. Moreover, in this report it was also said that there was a lack of proper methodology and of the necessary command of the subjects. There were excellent teachers, but they were few. The tendency to use memorization prevailed in schools.

Memorizing facts in class and being ready to reproduce them was judged by teachers as the first responsibility of the student. The teaching of reading and arithmetic was inefficient.

Teachers keep their teaching in an environment of dogmatism where only their voices were present and there is no opportunity for students to have a general and continued participation in class. The teaching needs to be socialized, in other words, allow students to participate in the activities of the school and make them feel that they are members of a group. The school must extend its range to the community (Cantón, 1955).

In the 1920s pedagogical ideas of the active school and the new approach of education that Octavio Méndez Pereira presented in his *Memoirs of 1924 and 1926* prevailed. He cried out for an educational reform showing his disagreement with the orientation and character of Panamanian schools and offered what he called "the ideals" that education must follow. New programs for elementary schools, high schools, teacher training schools, and vocational schools were designed, but their orientation and contents did not reveal any real change regarding the purpose of education.

The University of Panamá was inaugurated on October 7, 1935 and started classes the following day with a registration of 175 students. It offered programs in education, commerce, natural science, pharmacy, pre engineering and law. This institution has been involved in preparing teachers since its founding.

In 1947, the Commission of Study of National Education, directed by Otilia Arosemena de Tejeira, finished a study about the status of education in which, besides teachers, economists, architects, engineers and sociologists participated for the first time. The fundamental objectives of education were set out. The study revealed deficiencies in the educational system and proposed corresponding recommendations.

Among others, it underlined that the school should look for developing basic skills in students. In this regard it suggested, "They should have the ability to apply number relations in the interpretation and solution of current problems of life, the ability to use the scientific method in the solution of problems by observing systematically and by interpreting facts."

The study plans were modified in 1926, 1953, 1957, 1961, and 1975. For example, in 1953, the modification was based on the statement of goals, objectives of education and basic skills related to these objectives. The program of studies was comprised of the following subjects: arts and crafts activities, recreational and artistic activities, natural science and hygiene, physical education, social studies, national language, mathematics, ethics and religion, and English.

These programs were a significant reaction against "intellectualism", "verbalism" and other faults that were attributed to Panamanian schools. Similarly, they insisted on the importance of research and experimentation.

Based on criticisms, the programs of 1953 were restructured in 1957 and reformed in 1961. The 1961 reform increased the hours assigned to Spanish, mathematics, social studies, natural science and hygiene. At the lower secondary level, modifications to the plan occurred in 1926, 1935, 1945, and 1954. High school programs were revised in the early 1960s and updated in 1976.

Not only have the normal schools prepared teachers, but since its founding the university has also prepared teachers. The academic structure of the University of Panamá included colleges such as surveying, economics and political science, law, pharmacy, mathematics, arts and philosophy. Its orientation was essentially professional and intended to attend to demands for highly qualified personnel to satisfy the essential tasks of the state and the private sector of that moment. Although there was an educational area, it was in 1937 when the School of Education was born. It was part of the School of Philosophy, Arts and Education until 1985. It was on July 13, 1994 when the School of Education was officially named the School of Sciences of Education. Before 1970, teachers who were prepared to teach in elementary schools at the university were given a teaching degree in pedagogy. Today they receive a bachelor's degree in Education as do future teachers for other educational levels.

The School of Natural and Applied Sciences of the University of Panamá offers a bachelor's degree in Mathematics and, since 2001, a bachelor's in Mathematics Teaching.

However, it is the School of Education that offers the pedagogy courses. In this way, everyone who wants to study to become a high school teacher must be registered for three semesters in undergraduate courses in the specialty chosen.

The Ministry of Education also provides teacher preparation at the Normal School Juan Demóstenes Arosemena. Nowadays, many universities have some master's and doctoral courses in education and in mathematics.

As has been noted, mathematics has been taught since colonial times. It continues to have an important place in the study plans. In spite of efforts made to prepare teachers, and the efforts of teachers to educate students, the difficulties in learning seem to have multiplied, not only today but in the past and not exclusively in Panamá but also in other countries.

The difficulties that this subject causes in Panamá are reflected in several aspects:

(1) The frequent use of private mathematics tutors.
(2) The entrance examination at universities shows that the biggest difficulties are in mathematics.
(3) The low quantity of students who start the degree program in mathematics and the few who finish it.
(4) Even with entrance examinations, or preliminary courses, there are difficulties in this subject. That is the reason why many drop out or face problems trying to stay in the program. This is the case at the Technological University of Panamá (UTP) and others.
(5) The demand for specialists in this subject is seen when vacancies are occupied by non-specialists in the field, who have some credits that allow them to teach at the high school level, like engineers and others who have graduated from the Technological University of Panamá whose academic program had some mathematics courses.
(6) The loss of vocation of the mathematics teacher.

Despite difficulties, the panoramic view for the future of mathematics is encouraging. Talent has been stimulated through activities that keep young people away from indifference and apathy, and promote hard work to face the learning of the challenging mathematics. The following is worth noting:

(1) Panamanian Mathematical Olympiad

The Panamanian Mathematical Olympiad (Olimpiada Panameña de Matemática, OPM) is a competition where students from seventh to twelfth grades from all schools participate voluntarily. The Ministry of Education, the University of Panamá and the Autonomous University of Chiriquí convene it.

It has the following objectives:

- To stimulate interest of young people in mathematics.
- To motivate young people to develop skills in solving mathematics problems.
- To identify students who show relevant mathematical abilities.
- To promote the exchange of information and experiences among students, teachers, and researchers.

(2) **National Network of Mathematics**
The mission of the National Network of Mathematics is to awaken student interest in understanding and learning how useful mathematics is for all the aspects of life, thus contributing to the formation of citizens. Its vision is to educate students with mathematical competencies, which allow them to face the technological, scientific, and social changes by making informed decisions, developing thinking skills from a critical and analytic perspective, being very respectful of dialogue and a culture of peace. Among the courses of action are the following: Train and follow up on teachers, organize meetings for teachers of different levels, establish agreements and commitments, and strengthen the organizational structure.

(3) **Other Activities**
Other activities have been developed, like the Mathematics Camp 2013, organized by the National Secretariat of Science, Innovation and Technology (SENACYT) together with the Panamanian Mathematical Olympiad Foundation and the University of Panamá; the Project for the Popularization of Science (by SENACYT), and several congresses.

Education in Panamá: Five Goals to be Improved

Considering the status of education, and taking into consideration the primary objectives and defined goals in the National Initiatives towards Improvement of Education (1944–2009),[10] general goals, objectives, baselines, specific goals, indicators and projections of its fulfillment for years 2016 and 2021, are presented. The general goals agreed upon at working sessions held with the participation of experts on education are:

(1) All girls and boys aged four (4) to five (5) years old will be in school.
(2) Students will complete high school.
(3) Good quality of education throughout the country.
(4) Good quality of education for all.
(5) More investment in education.

---

[10] Originally published by the Economic and Social Development Foundation of Panamá (FUNDESPA) as text of the document "Public private alliance: towards state policy in education", done in collaboration with COSPAE, CONEP, Cámara de Comercio, Industrias y Agricultura de Panamá, Unión Nacional de Centros Particulares and PREAL, in June 2009.

From the five general goals presented, goal 4 will be dealt with since it focuses on the teaching of mathematics.

Five objectives have been presented for this fourth goal:

(1) To improve the results on tests of all students.
(2) To decrease the percentage of grade repetition.
(3) To eliminate the dropout levels in elementary school, middle school, and high school.
(4) To increase the number of qualified teachers.
(5) To increase the amount of hours for basic subjects.

Having as a reference the baseline of year 2010, Panamanian students were below the Latin America average in mathematics, reading and science (SERCE-LLECE 2008, elementary school).[11] SERCE-LLECE tests given to third graders showed that 16% of these students of elementary school were below level I in Mathematics, 50% in level I, 25% in level II, 7% in level III and 2% in level IV.

In the mathematics tests, third graders from level I recognize facts and basic concepts of the geometrical and numeral domains and the information processing of SERCE's conceptual framework. In level II, they recognize facts, concepts, characteristics and direct and explicit relationships in the different conceptual domains of SERCE. They solve simple problems in familiar contexts, which involve recognition and use of one basic operation (addition, subtraction or multiplication). In level III, students recognize concepts, relationships and characteristics that are not explicit in the different conceptual domains. They solve simple problems that involve the recognition and use of the four basic operations. In level IV, students solve complex problems in the different conceptual domains with strategies based on the use of data, characteristics and relationships that are not explicit (SERCE, 2008).

In the SERCE-LLECE tests given to sixth graders in elementary schools, the results indicated that 4% of these students were below level I in mathematics, 27% in level I, 49% in level II, 18% in level III, and 2% in level IV.

In the mathematics tests, sixth graders recognize facts, concepts, relationships and characteristics in the different conceptual domains of SERCE.

---

[11] SERCE-LLECE 2008. Second Comparative and Explanatory Regional Study — Laboratorio Latinoamericano de Evaluación de la Calidad de la Educación (Latin American Laboratory of Assessment of the Quality of Education).

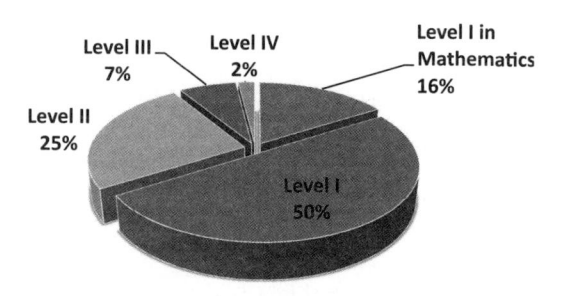

Figure 1. SERCE-LLECE tests given to students in the third grade.

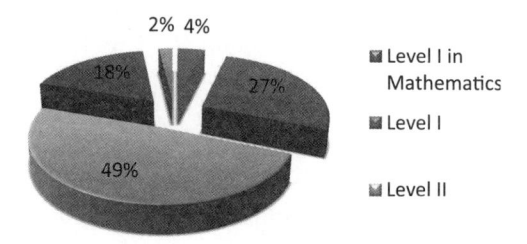

Figure 2. SERCE-LLECE tests given to students in the sixth grade.

They solve simple problems of additive structures in the numeral domain. Students in level II recognize facts, concepts, characteristics and relationships of SERCE's different concepts. They solve problems that require simple strategies with relevant explicit information involving one or two of the four basic operations in the conceptual domains of SERCE. Students in level III, solve problems in the SERCE's conceptual domains involving the use of concepts, relationships and characteristics of higher cognitive level. They can interpret information from different representations. In level IV, students solve complex problems of SERCE's conceptual domains with no explicit information requiring the use of relationship and connections among different concepts (SERCE, 2008).

SERCE's specific goal is to apply census-based tests of achievement to students every two years. These tests include learning assessment for elementary schools, which are consistent with international standards and supervised by independents in 2016.

According to these specific goals, no third grade student of elementary school will be under level I in the mathematics tests, 51% in level I,

30% in level II, 12% in level III and 7% in level IV in the SERCE's test by 2016.

According to these specific goals, no third grade student of elementary school will be under level I in the mathematics tests, 41% in level I, 30% in level II, 17% in level III and 12% in level IV in the SERCE's test by 2021.

According to these specific objectives, no sixth grade student of elementary school will be under level I in the mathematics tests, 16% in level I, 54% in level II, 23% in level III and 7% in level IV in the SERCE's test by 2016.

Based on these specific objectives, no sixth grade student of elementary school will be under level I in the mathematics tests, 21% in level I, 44% in level II, 20% in level III and 15% in level IV in the SERCE's test by 2021.

Indicators

Following up to the fourth general goal of education, about good quality of education for all, the indicators to be taken into consideration are: (a) number of national tests of assessment of the learning given and (b) the result of the mathematics tests.

Fulfillment of Projections According to the Trends for Years 2016 and 2021

These projections indicate that whether or not the tendency is maintained, Panamanian students will continue to be left behind in relation to the Latin American level.

Higher Education

Most of the students at the university level have been registered in a bachelor's degree program, although an important increase in the number of applicants for master's degree is observed; almost doubling in only four years (see Table 1). One of the causes of this tendency is that some universities have substituted the final graduation work for subjects for a master's degree.

Another relevant change during this period is that the number of technical degree programs at universities has decreased. Evidently, private education has gained ground. Thirty-five out of forty universities in the

Table 1. Registration at universities in the Republic of Panamá, by academic level.

Registration	2001	2007	2008	2009	2010	2010–2011
Total	**117,864**	**132,660**	**134,290**	**135,209**	**139,116**	**21,252**
Official	97,253	95,704	92,680	88,999	88,534	−8,719
Technician	16,476	8,847	9,978	9,767	9,715	−6,761
Bachelor	96,812	114,384	114,160	115,261	117,700	20,888
Postgraduate	2,162	2,936	3,222	2,870	3,612	1,450
Master's degree	2,387	6,362	6,742	7,121	7,893	5,506
Doctorate	27	104	188	190	196	169
Private	20,611	36,956	41,610	46,210	50,582	29,971
% Private	17.5	27.9	31.0	34.2	36.4	19

Source: Instituto Nacional de Estadística y Censo

country are private. They have more than doubled their registration in 2010 with respect to 2007. In 2001, 17.5% of students were attending these universities, and in 2010 it had increased to 36%. Elements that stimulate students to register at these private universities are shorter degree programs, less rigorous admission requirements, flexible schedules (day, night, weekends), professional training (distance learning and blended learning), as well as financial incentives given by private enterprise and a diversity of options to pay with credit.

Panamanian Students are Focused on the Science of Education

By 2010, almost one third of students who completed university studies were education students prepared as teachers. This situation is similar to 2007 as seen in Table 2

The demand for bachelor's degrees has increased since it is a requirement to teach, an exigency closely related to the accreditation system approved in 2007.

The degree programs where most of the private and public universities graduated students in 2010 were similar to year 2007 focusing on areas like commerce and administration with 23% and medicine 7%.

Table 2. Graduates from public and private universities.

Sector	2007 Number	2007 %	2010 Number	2010 %
Science of Education and Teacher Education	6,582	32.6	7,100	32.5
Arts	251	1.2	334	1.5
Humanities	347	1.7	402	1.8
Behavioral Social Sciences	377	1.9	413	1.9
Journalism and Information	294	1.5	348	1.6
Business and Administration	4,847	24.0	5,012	23.0
Laws	819	4.1	888	4.1
Life Sciences	86	0.4	134	0.6
Physical Sciences	118	0.6	166	0.8
Mathematics and Statistics	396	2.0	443	2.0
Computing	1,158	5.7	1,226	5.6
Engineering and Engineering Trades	965	4.8	1,028	4.7
Manufacturing and Processing	633	3.1	676	3.1
Architecture and Construction	636	3.2	690	3.2
Agriculture, Forestry and Fishing	133	0.7	145	0.7
Veterinary	26	0.1	3	0.0
Medicine	1,452	7.2	1,521	7.0
Social Services	280	1.4	327	1.5
Personal Services	478	2.4	519	2.4
Transport Services	83	0.4	126	0.6
Environmental Protection	135	0.7	169	0.8
Security	86	0.4	151	0.7
TOTAL	20,182	100.0	21,821	100.0

Great Challenges for Panamá

More than thirty years have passed without any improvement in infrastructure. Nowadays, 201 projects of major repairs and 3,505 of minor repairs are in progress. These projects reach the amount of 15 million dollars. They include construction and equipping of laboratories for 57 high schools and the construction of eight schools of excellence.

The Ministry of Education reports that 91.2% of its teachers are formally qualified. The results of assessments of the educational system show there are deficiencies in it, and it implies the need for training in educational issues at a higher level.

Regulations in educational matters allow teachers to start working very young, 18 years old approximately. It allows them to get a permanent job in the public system of education or a chair at universities by having a few years of service. Job stability gained at the beginning of their educational work does not require them to pursue professional development because of current laws that protect teachers who have this status.

Therefore, low levels of assessment results are obtained in mathematics, less than regional averages. Almost one third of third graders are in the basic level, but by sixth grade they do not even reach the basic level.

These results should be used as tools to make important adaptations in educating teachers and in teaching methodologies and strategies. There are encouraging signs with the recent implementation of competency-based curriculum, a measure that can cause a significant impact on students.

The Panamá Canal and its Influence on the Teaching of Mathematics

From the period of the French Canal, the Panamá Canal and its current expansion have influenced the curriculum in terms of training. The Zonians, residents of the Canal Zone, had their own schools and universities, many of which still work with American plans. The struggle for full sovereignty did not affect their educational systems.

Today, the revenues from the Panamá Canal are partly geared toward education, since the upkeep of the Canal requires an educated and able cadre. In fact, the creation of the Technological University of Panamá (UTP) was a direct result of the power transition of the Canal to Panamanian authorities. UTP serves the nation, among other things, by providing trained professionals for the able administration and maintenance of the Canal.

Final Considerations

Education in Panamá has made important advances in the last decades especially in accessing and covering all the teaching levels.

Significant achievements include that elementary schools are available for all students, dropout rates have been reduced as more children stay more permanently in the system. There is the beginning of more attention and education for preschool children, as well as improvements in gender equality in elementary schools teaching. Literacy levels, average number of years of work, and the number of teachers holding higher level degrees have all increased.

Regarding quality, international assessments and tests place Panamánian students under levels reached by other countries in the region. Nearly 50% do not have basic competencies in language, mathematics and science, considered fundamental to be incorporated into social and productive life in an active way.

Improving the quality of education is an essential challenge implying not only better student learning results, but also giving students the knowledge, abilities, attitudes, competencies and skills in order to adequately incorporate them into today's competitive world.

Bibliography

Arosemena, M. (1949). *Historical notes.* Publications Ministry of Education.
Bernal, J. B. (1994). *Planning, management and financing of higher education.* Panamá: Universidad del Istmo. pp. 31–32.
Cantón, A. (1955). *Development of pedagogical ideas in Panamá from 1903 to 1926.* Panamá: National Printer.
Céspedes, F. (1981). *Education in Panamá. Panamanian Culture Library.* Panamá. Mariano Arosemena (1794–1868), neoclassical writer, journalist and politician Panamá.
MEDUCA (2002). *National education system of panamá,* Panamá Ministry of Education and Organization of American States. Retrieved from http://www.oei.es/quipu/panama/index.html
MEDUCA (2005). *Education Statistics 2005.* Department of Statistics of the Ministry of Education of Panama. Online document available at http://www.oei.es/quipu/panama/index.html and http://www.oei.es/quipu/panama/estadisticas_media2005.xls
MEDUCA-Prode (2005). *Profile of the new teacher Panamanian.* Online document available at http://www.oei.es/webdocente/Panama.htm and http://www.meduca.gob.pa
Ministry of Education (MEDUCA) (2007). Development Project/IDB. Preparation of the reform of secondary education in Panamá. Performance Study of Secondary Education in Panamá. Panamá.
Ministry of Education (MEDUCA) (2005). *Strategic plan 2005–2009,* Panamá, April 2005.
National Council of Education (CONACED) (2006). *A document for action in Panamá's educational system.* First Report to the President of the Republic, Panamá.

National Council of Education (CONACED) (2008). *A document for action in Panamá's educational system.* First Report to the President of the Republic, Panamá.

NOMISMA. Report. Survey. UNIPAN-BID. Phase III. (1998). Quoted by University of Panamá. Great Congress. p. 175.

Orozco, L. E. (1996). Financing and management of higher education institutions in Latin America. In *CRESALC-UNESCO. Basis for the Transformation of Higher Education in Latin America and the Caribbean.* Caracas, p. 39.

PREAL-COSPAE (2002). *The challenge is to progress: Educational progress report of Panamá.* Retrieved from http://www.oei.es/quipu/panama/preal_panama2002.pdf

SENACYT (1999). *National strategic plan for the development of science, technology and innovation 1998–2000.* Panamá.

Thompson, A. (1992). Teachers' beliefs and conceptions: A synthesis of the research, in D. A. Grows (Ed.), *Handbook of Research on Mathematics Teaching and Learning: A Project of the National Council of Teachers of Mathematics,* New York, Macmillan, vol. 326, pp. 127–146.

Triana, J. M. (1846). *Manual that should have the teachers in schools to teach Spanish Grammar.* Bogotá, J. A. Cualla.

Trejos, M., Lebrija, A., Oliveros, O., Gutiérrez, J., Gomez, R., Elisha E., & Flores, R. (2006). *Mathematics for all our commitment,* Research report, SENACYT, University of Panamá.

Vasquez, S. (2006). *Statistics 2005–2006 forecast. Problems in Mathematics.* Retrieved from http://mensual.prensa.com/mensual/contenido/2006/08/21/hoy/vivir/708553.html

Wilson, S., & Cooney, T. (2002). Mathematics teaching change and development, in G. Leder, E. Pehkonen & G. Törner (Eds.), *Beliefs: A Hidden Variable in Mathematics Education?* Dordrecht: Kluwer Academic Publishers, pp. 127–147.

Winkler, D. (1994). *Higher Education in Latin America. Efficiency and Equity Issues.* Washington: World Bank, p. 17.

Zapata, R. (2008). *Evaluation of the race curricular Teaching of Mathematics,* Department of Mathematics, University of Panamá, Master Thesis, Central Institute of Management and Supervision of Education (ICASE). University of Panamá.

About the Authors

Euclides Samaniego teaches in the area of artificial intelligence in the Faculty of Computer System Engineering at the Technological University of Panamá. Dr. Samaniego is the author of six technical books and several articles related to science, technology, and informatics. He is a doctoral advisor and the Ministry of Education of the Republic of Panamá awarded him the Honor of Merit for his valuable contributions in the development of the Curriculum Transformation of Media Education and motivation for continued growth process towards the pursuit of educational excellence.

Nicolás A. Samaniego teaches in the area of computer graphics in the Faculty of Computer System Engineering at the Technological University of Panamá. Dr. Samaniego is a doctoral advisor. The Ministry of Education of the Republic of Panamá awarded him the Honor of Merit for his valuable contributions in the development of the Curriculum Transformation of Media Education and motivation for continued growth process towards the pursuit of educational excellence.

Benigna Elena Fernández de Guardia is a professor of mathematics in the Faculty of Computer System Engineering at the Technological University of Panamá. She has also taught at the middle school level and at the University of Panamá. She has degrees in mathematics, engineering, and mathematics instruction and has been a speaker at several national and international conferences.

Chapter 13

PARAGUAY: A Review of the History of Mathematics and Mathematics Education

Gabriela Gómez Pasquali

Abstract: This article describes the history of mathematics and its teaching in Paraguay, from the founding of its first universities until the present. It tells how, in spite of isolated individual efforts, the country could not adjust to the rhythm of mathematical development in the region, with the result of falling behind. It also highlights the recent but promising creation of graduate programs, as well as an incipient but cohesive mathematical research activity.

Keywords: History of mathematics education; mathematics in Paraguay; history of mathematics; mathematics education; Paraguayan Mathematical Society.

Beginnings of Mathematics Teaching in Paraguay (1926)

College of Physical Sciences and Mathematics, and the arrival of the Russians in Paraguay — 1926

The first steps in mathematics higher education were taken in 1926, as a result of the creation of the College of Physical Sciences and Mathematics at the National University of Asunción (UNA). Its first Dean was Sergio Bobrowsky. A very important milestone in the development of mathematics in Paraguay was the arrival of Russian mathematicians (Sergio Syspanov, Krivoshein, Sergio Coradi, etc.) to the College of Physical Sciences and Mathematics (nowadays called the School of Engineering) after the Bolshevik revolution. Roberto Sánchez Palacios, Luis A. Paleari, Roque Saldívar,

Luis L. Volta, Juan Cámeron, among others were trained by the Russians. These teachers, in turn, transferred what they had learned to the next generations of teachers, including Francisco Pujol, Ángel Secchia, and many others.

National Institute of Physics and Chemistry (MOE) — 1957

In 1957, the National Institute of Physics and Chemistry was created and affiliated with the former Ministry of Education and Culture, which worked out of the National School of the Capital (*Colegio Nacional de la Capital*). This institute functioned until 1959.

Between 1957 and 1958, the National University of La Plata in Argentina held a refresher training program for Latin American university professors. From Paraguay, Juan H. Domínguez, Juan C. Lebrón, Luis L. Volta, and Ángel P. Secchia participated in courses in algebra, physics, and nuclear physics. At the time, there was a bachelor's degree (*licenciatura*) in mathematics in the School of Philosophy at the National University of Asunción. This school had started as a School of the Humanities, but later came to be known as the School of Philosophy (J. Von Lücken, personal communication, September, 2013).

Paraguayans on Specialization Trips — 1960

In 1960, the Regional Center of Mathematics in Latin America was created at the University of Buenos Aires, with the aim of strengthening the mathematical knowledge of the youth in Latin America. From Paraguay, José Luis Benza and H. Feliciángeli participated. At the same time, this regional center had highly respected mathematicians, such as don Julio Rey Pastor (Spanish), Jean Dieudonnée (French), Ostrowski (Russian), Antonio Monteiro (Portuguese), Santaló (Spanish), Mischa Cotlar (Ukranian), Manuel Sadoski (Argentinian), González Domínguez (Argentinian), and other remarkable mathematicians from the United States and Europe. Later, other Paraguayans such as Rose Marie Estigarribia, Lina Wehrle, and Alcides Vergara made important contributions to the center (J. L. Benza, personal communication, November, 2013).

Institute of Sciences of the UNA (1962) and First Generation of Graduates (1965)

In April of 1962 the Institute of Sciences of the UNA was created, by an agreement of cooperation between the UNA and UNESCO. It was intended to train students in "Natural Sciences and Exact Sciences". The first class of mathematicians to graduate from this institution was in 1965.

At this time, the study of mathematics was strongly supported by the United Nations Development Program and UNESCO. With the sponsorship of these agencies, Ernesto García Camarero came to Paraguay. The presence of this remarkable Spanish scientist gave a great impetus to mathematics in Paraguay.

An advance was made to students, Fernando Kropf, Edgardo Brown and Luis Talavera, who were given scholarships to attend the Balseiro Institute of Physics in Bariloche (Argentina), and to Luis Fernando Meyer who went to the United States to continue his studies in engineering.

When these students completed their studies, they returned to the Institute of Sciences. This was a significant step for the country in the development of the basic sciences, and especially mathematics.

Ernesto García was hired to organize the Institute. He was the driving force behind the introduction of what is now known as modern mathematics (group theory, ring theory, linear algebra, etc.). In the past, only elementary mathematics was taught; this was the first time that this mathematics was taught in Paraguay. Without a doubt, García Camarero was a catalyst for the evolution of local mathematical activities. Two Paraguayans who had graduated from the University of Buenos Aires were brought to the National Institute of Sciences: José Luis Benza and Horacio Feliciángeli. They later became engineers. They were joined at the institute by two physicians who had graduated from Bariloche, Argentina: Fernando Braun Moreno and Fernando Kropf. These were the first four Paraguayans with a strong training in modern mathematics.

This period was characterized by the intention to popularize mathematics by developing refresher courses in schools, both in the capital and in the countryside. Notable people such as Santaló and Faba from the University of Buenos Aires and La Plata arrived in Paraguay to develop specialized classes intended for teachers and students of mathematics. From the beginning, women played an important role in the development of different areas of mathematical education, notably, Feliza Buzó and Carmen Gómez, as well as students of the Institute of Sciences, many of whom received scholarships abroad (H. Feliciángeli, personal communication, November, 2013).

Creation of the Paraguayan Mathematical Society and Other Relevant Activities at the End of the 1960s

In the mid and late 1960s, a strong emphasis on modernizing mathematics took place, causing excitement throughout the discipline. Conferences, seminars, and meetings abounded. Perhaps the most important activity was laying the groundwork for the creation of the Paraguayan Mathematical Society (SMP) and the seed for the current National Computing Center. The Paraguayan Mathematical Society was founded on October 18, 1964 and was active for 15 years, and then was ultimately reactivated in 2005.

There were many activities developed by the SMP: a series of lectures about mathematical topics (Mathematics Week), the publication of a monthly newsletter, a database of the mathematics books present in the country at the time (at private and public libraries), the publication of a magazine titled "Communications" (which allowed for an exchange among similar organizations from abroad), the establishment of the Society's library, and the organization of three meetings in Paraguay with the Argentinian Mathematical Union.

In February 1964, the Institute of Sciences — with support and funding from UNESCO — organized a training course for secondary school teachers. This had a significant impact on mathematics education because of the interest of the participants and the material that was presented. The course was six weeks long, eight hours per day from Monday to Friday (lectures and recitation). It covered the six volumes of the *Intermediate Mathematics* by Julio Rey Pastor. This course was definitely an important contribution to the spread of mathematics at the secondary school level.

The Student Union of the Institute of Sciences organized the Science Fair, where classic experiments were conducted. Among these were Foucault's pendulum that proves the earth's rotation; Buffon's needle, an experiment that allows for the calculation of π in a statistical way; a binary adder, the arithmetic unit of a computer which was built with electromagnetic relays, and other experiments that were presented in Paraguay for the first time.

The departments of Mathematics and Natural Sciences of the Institute of Sciences organized a "Seminar of Mathematics Applied to Development" that was very important for the development of computer science in the country, to be discussed later.

Teachers and students contributed as guides in the scientific exposition organized by the French Embassy and the *"Palais de la Decouvert"* of Paris, which brought some of its expositions to Paraguay.

The first workshops for secondary school teachers in the interior were advanced by José Von Lücken, a graduate of the first class of mathematicians at the Institute of Sciences and who, during his distinguished professional career, was a major influence in the mathematical training of numerous professionals in diverse academic areas. Although there was no assessment system, it is accepted that mathematics education improved as a result of this initiative.

The Birth of Computing in Paraguay: Creation of the National Computing Center — 1967

The experience with computer technology present in Italy and Argentina allowed for the possibility of having a computing center in Paraguay. A variety of institutions, such as the Department of Mathematics of the University, the Paraguayan Mathematical Society, and some other organizations that were not part of the University at the time, came together to promote the creation of this new center. The contributions by the engineer Luis F. Meyer, who supported this proposal from the beginning, must be highlighted.

The seminar "Mathematics Applied to Development", organized towards the end of 1965, was well received, not only by the University but also by the public sector. This included Technical Planning Secretariat, Municipality of Asunción; ANTELCO (currently COPACO, the national telephone company); CORPOSANA (currently ESSAP, Corporation of Sanitary Works, the state company responsible for water distribution); and the National Bank for Development, among others. One of the issues put forth during the event was whether Paraguay needed a Computing Center; the most important conclusion of the seminar was that this was essential and should happen without delay. The urgent need to have a Computing

Center in the country became evident (not only for the University, but also for many other institutions), and Luis F. Meyer was put in charge of developing this project that resulted in the creation of the National Computing Center (CNC) in 1967, with Meyer as its first Director. The Development Bank and CORPOSANA worked simultaneously with the same goals.

Creation of the Institute of Basic Sciences (ICB) — 1969

In 1969, UNA decided to pay special attention to those schools that prepared professionals in fields related to science and technology. The Honorable University Higher Council created the Institute of Basic Sciences (ICB) during its session of June 27, 1969 with the aim of collecting in one location all the services needed for the teaching of the Basic Sciences at the University. In March of 1972, it moved to its current location at the UNA.

A bachelor's degree in Mathematics did not last long in the School of Philosophy as it became part of the Institute of Sciences of UNA in the late 1970s. Also, for some years the Catholic University offered a bachelor's degree in Mathematics.

In 1972, a group of mathematicians noticed the need for the creation of graduate programs and wanted to develop academic courses in advanced mathematics. However there was no agreement among the potential beneficiaries. The aim was to offer Masters', doctorates, and specializations. Although there was agreement to promote this initiative, it was not possible to reach a consensus among the relevant authorities.

On June 21, 1990 the ICB was transformed into the School of Exact and Natural Sciences (FACEN).

Most of those with a bachelor's degree in Mathematics work as teachers in secondary schools and universities. In response to this, FACEN opened the first program in Mathematics Education in 2011 — a combined bachelor's degree. Two years later they were expected to have their first graduates.

It is important to note that the ICB (founded in 1969) and the FACEN (since 1990) graduate new undergraduate students in mathematics every year.

What has not been developed until recently (from 2006 onwards) is mathematical research since most graduates of mathematics work as teachers in secondary schools and universities.

Founding of OMAPA — 1989

A group of enthusiastic teachers and students created the non-profit civil organization OMAPA in 1989. At first this acronym stood for Paraguayan Mathematical Olympiad (Olimpiada Matemática Paraguaya) and in 2007 the name was changed to Multidisciplinary Organization for the Support of Teachers and Students (*Organización Multidisciplinaria de Apoyo a Profesores y Alumnos*). The founding of OMAPA revitalized the world of mathematics at the elementary and secondary levels. Its appearance was key for the imposition of a structure in mathematics education in Paraguay. Throughout its 25 years of existence, OMAPA has been able to create a community of teachers who are convinced that problem solving is a pedagogical tool that must be used in the classroom on a daily basis in the teaching-learning process.

A community of students who share their passion for mathematics now thrives all across the country, especially when departmental or national meetings are held. At these meetings young students get to know each other, become friends and talk about mathematics. The support of parents, who notice that their children receive excellent academic training because of these activities, has been instrumental in the success of this initiative.

The biggest impact of the Mathematical Olympiads in the teaching of mathematics has been to present problem-solving as a vital part of the teaching of mathematics and to explicitly establish that doing drill exercises alone is not the same as doing mathematics. The problems are disseminated to teachers and students through workshops and annual collections of problems carefully targeted to the Paraguayan reality. These problems help promote mathematical thinking.

Among its regular activities in support of teachers, and as part of the 25th Iberoamerican Mathematical Olympiad, OMAPA organized in September of 2010 the 22nd Iberoamerican Symposium on the Teaching of Mathematics. This three-day symposium attracted national and international teachers, with speakers and researchers hailing from Brazil, Costa Rica, France, Puerto Rico, Venezuela, and Paraguay. Approximately 800 Paraguayan teachers and 50 teachers from abroad participated.

In the beginning, OMAPA attracted mainly private schools, but gradually it was able to include public schools. The slow rate of growth at the beginning stands in contrast to the rapid acceleration it has had since 2006, thanks in part to the Ministry of Education, the National Council of Science and Technology (*Consejo Nacional de Ciencia y Tecnología*,

CONACYT) and Itaipú. In 2013 more than 2,500 elementary and secondary schools (out of approximately 6,000 in the country) participated in the National Mathematical Olympiad, in versions for both the youngest students and older ones. Moreover, about 3,500 teachers from across the country participated in the workshops on the methodology of problem solving.

The most significant contribution of OMAPA in the evolution of mathematics in Paraguay is the Program of "Scientific Initiation with an Emphasis on Mathematics for Talented Students". Outstanding students are invited to participate in intensive courses that train them for international competitions such as Southern Cone, Rioplatense, Iberoamerican, and the International Mathematical Olympiads. During the first years of this program, it had only about 40 participating students, mostly from the capital area. In 2013 it reached over 200 participants from 13 of the 17 departments of Paraguay. Twenty graduates of this program are currently doing undergraduate or graduate work related to mathematics in renowned universities of the United States (such as MIT, Columbia, Stanford and Cornell), as well as universities in Spain, Italy, Brazil, Argentina, India, and other countries. Almost all of them have merit-based scholarships that they received because of their outstanding performance in international competitions (R. Berganza, personal communication, September, 2013).

Committee of Mathematical Education of Paraguay (CEMPA) — 1996

A group of teachers, after their first international experience at a Congress in Brazil in 1996, decided to create the Committee of Mathematical Education of Paraguay (CEMPA). The goal was to promote the continuous training of teachers of mathematics through participation in national and international conferences and the development of workshops and courses targeted towards the specialization of teachers in this area (N. Centurión, personal communication, October, 2013).

According to CEMPA, the unregulated appearance of teacher training institutes and the reduction of the years of training required of teachers were the reasons for the failure of teacher training programs. The disappearance of the training of so-called regular teachers in 1963 and the reduction in length of training from 7 years to 2, along with several attempts at educational reform, resulted in a lack of excellent teachers. In particular, these teacher training institutes, instead of teaching the pedagogy of mathematics, were more interested in teaching the minimal requirements.

In light of this situation, CEMPA, composed of 15 teachers from public and private schools, was charged with promoting the training of unionized elementary and secondary school teachers (M. A. Ayala, personal communication, October, 2013).

In 2007 CEMPA organized a National Congress of Mathematics Education that gathered teachers of all levels from across the country. This two-day congress had about 500 participants from Paraguay and abroad. In 2009, CEMPA, as part of the Iberoamerican Federation for Mathematics Education (FICEM), hosted the eighth Meeting of Mathematics Education in the Southern Cone (M. E. Ovelar, & A. Demestri, personal communication, October, 2013).

Late Creation of the National Council of Science and Technology (CONACYT) — 1997

While around the world and in the region there was an emphasis on research since the beginning of the twentieth century, in Paraguay this did not happen until the end of the millennium, specifically in 1997, when the first institution to advance research was founded. Law 1028 of Science and Technology formally provides the first mechanisms for a national system for the development of science, technology and innovation, and creates the National Council of Science and Technology (CONACYT) as its governing body.

With neither a budget nor officers, CONACYT started its work creating and implementing public policies favoring research, with a weak though increasing support of local researchers. Gradually, it began to foster awareness in governmental agencies of the urgent necessity of giving support to scientific research and technology.

As a result, it is very likely that CONACYT will receive public funds beginning in 2014 and for a minimum of 10 years. The Paraguayan scientific community (not only the mathematics community) is anxiously waiting for this to happen, in the hope that this would propel the advancement of science and technology in the country.

Basis for the Beginning of Mathematical Research — 2004

Although Paraguay has had a steady flow of mathematics graduates at the bachelor's degree level since 1965, mathematical research had not been considered necessary until the beginning of the current millennium. With a small budget and a misguided idea of the discipline, mathematics was falling behind in terms of research progress, while other disciplines such as medicine, chemistry, agricultural and environmental sciences were enjoying a sustained growth.

The urgent need to improve this situation, the hopes of many people, and a singular mission, imply that three events shaped the beginning of a new path in the process of building the mathematical agenda in the country.

The visit of Uruguayan Roberto Markarian, who conducted an evaluation of mathematics in Paraguay, as requested by UNESCO, and the first contacts with institutions from abroad were viewed as a new beginning for the mathematical and scientific community.

The first contact with French mathematician Michel Jambu was made by Gabriela Gómez Pasquali, while in Spain. His visit to Paraguay, from the International Center of Pure and Applied Mathematics (CIMPA), was an important step that would open the doors to a new stage of mathematics in Paraguay.

At the same time, the young professional Christian Schaerer, who was the first Paraguayan to earn a doctorate from the Pure and Applied Mathematics Institute (IMPA) of Brazil, met Marcelo Viana (Vice Director of the institution, and one of the most important mathematicians from Latin America). Viana mentioned to Schaerer, "You have no idea how long we have been waiting for a Paraguayan; we want to organize a Mathematics Summer School[1] in your country."

These events, and after so many years, made it possible to attract the attention of the international mathematics community. This interaction occurred almost spontaneously, with the intention of advancing mathematical research in the country.

---

[1] An EMALCA (*Escuela de Matemática de América Latina y el Caribe*). These are one- or two-week seminars on current research topics. They are sponsored by UMALCA (Mathematical Union of Latin America and the Caribbean).

As a strategy of intervention, several immediate actions were executed in order to bring quick results for sparking mathematical research in Paraguay. Working with Miguel Ángel Volpe (a professor at the Engineering School) and other professionals, the search for the first people to send to IMPA began. Two students from OMAPA, Carlos Sauer and Ingo Dick, and later the engineering students Max Duarte and Darío Alviso, would be the first ones sent to summer courses.

Meanwhile, with the help of IMPA, the first contact with the Mathematical Union of Latin American and the Caribbean (UMALCA) was made, which allowed for the organization of the first Latin American and Caribbean Mathematics School (EMALCA) in Paraguay.

Antonio Cano strongly supported hosting mathematical schools and decisively promoted this initiative. A professor and council member in the School of Engineering, Cano worked to strengthen the first steps in the process. From the capital to the countryside, students and teachers joined this initiative. There was excitement about the possibility of having a summer school in Paraguay.

There was a natural participation of people from IMPA, authorities of FACEN, the Engineering and Polytechnic Schools of UNA, CONACYT, and other institutions such as the Paraguayan Mathematical Society. After an evaluation of the efforts underway, they decided to intervene with more specific actions. As a result, training courses were created along with specialized study-abroad tours for students, producing a sustained flow of students doing graduate studies abroad.

History is made from a series of events and people who set the process in motion. The contact with a Paraguayan scientist living in Paris, Juan Carlos Rolón, was crucial. Rolón's decision to return to Paraguay in 2006 was a major step in strengthening scientific research, particularly mathematics research, in the country.

The role of OMAPA in nourishing the system, through the Program of Scientific Initiation with Emphasis in Mathematics, was crucial. It became an important source of the strengthening of this new process.

An EMALCA every two years, mathematics meetings, and three-month schools interspersed among these other events, were among the initiatives that complemented these first steps. These events allowed scientists to visit Paraguay periodically. The EMALCAs remain crucial for the organization of the scientific community in Paraguay.

With these schools, UMALCA proved to the national authorities that the organization of the mathematical and scientific process within Paraguay was important for the international community, and that the moment had come to develop it in a systematic way. In this way, the authorities felt the importance for strengthening and doing science in Paraguay. UMALCA offered the international political support needed to consolidate the process.

More than 200 students have gone through these schools. Many of them have traveled abroad and returned to the country, or are about to return, and are in possession of graduate degrees in the mathematical sciences and engineering (C. Schaerer, personal communication, September, 2013).

Boom of the New Organization of Civil Society — 2011 to 2013

In 2011 the Paraguayan Pure and Applied Mathematics Society (SPMPA) was founded as part of FACEN. SPMPA started its activities the same year, with the successful organization of the First Paraguay Congress of Pure and Applied Mathematics. With the participation of mathematicians from Argentina, Brazil, and Perú, it had a strong appeal for and was well received by local mathematicians.

In rural cities such as Encarnación, Ciudad del Este, and Concepción, several collectives of people with bachelors' degrees in mathematics were created with the aim of strengthening the discipline. The Paraguayan Mathematical Society is associated with them and helps them to consolidate and work among themselves.

In 2013, the Mathematical Research Center (CIMA) was launched, aiming to strengthen the incipient research in this area in Paraguay and to promote the international alliances with prominent international centers. Several possible alliances are under study, including ongoing negotiations with centers in Chile and France.

Masters, Doctorates and Recent Advances

One of the most relevant accomplishments that have come as a result of the international contacts made at the beginning of the millennium was the opening of a Master's degree program in Mathematics in 2006, conducted by FACEN. Two graduating classes have produced about 30 master's degree students in pure mathematics, with a master's degree in applied mathematics in the process of seeing its first graduating class. This program was developed with the support of the international academic community and the participation of distinguished Spanish, Argentine, and Brazilian mathematicians.

UMALCA, through the EMALCAs, has helped to structure the master's degree program in mathematics (FACEN) and consolidate it with the doctoral program in computer science from the Polytechnic School at UNA. In 2011, Diego Pinto Roa became the first doctor in the mathematical sciences in the history of Paraguay to graduate from a local university, specializing in optimization.

There are five students expected to graduate between 2013 and 2014. Moreover, about 15 students have earned a master's degree in mathematics or computer science, either in Paraguay or abroad.

We finally can say that mathematics research has begun in earnest with the establishment of the Scientific and Applied Computing Laboratory at the Polytechnic School of the UNA, under the direction of Christian Schaerer.

Also, important results, such as the relations established through research agreements with universities from abroad, and the exportation of developed models from the capital of the country to disadvantaged areas of Paraguay, have been seen in this new era. As a result, the demand in the sciences has grown, and to satisfy this demand, the Paraguayan government has created the Law of National Funds for Public Investment and Development (FONACIDE). By means of this law, Paraguay will assign 30% of the funds that it receives from the export of electricity to Brazil (approximately 30% of $274,000,000) to the Excellence in Education Fund.

Challenges and Perspectives

Scientists, teachers, students, and Paraguayan institutions dream of a scientific process that will reach all parts of the country and will enjoy a rapid and sustained growth.

The search for public and private support for the creation of training centers in mathematics, as well as initiatives such as programs for talented students and the Mathematical Olympiads, are challenges that ought to be taken up by Paraguayan society. Repatriating those who will be trained in the sciences abroad and giving them the possibility of reaching their full potential as professionals — with stable jobs, good wages, and being fully incorporated into the system — are other goals that will have to be considered in the near future.

Supporting CONACYT is a priority to strengthen the work of scientists. New undergraduate and graduate programs, with the training of young professional scientists who will be incorporated into the system, should complement this process.

Although a promising future awaits us, the work in Paraguay is complex. The challenge is to strengthen teachers and schools, as well as to create spaces for teachers, scientists and other professionals, where they can interact and work in research centers and universities.

Acknowledgements

Anselmo Ramos and Verónica Rojas deserve special credit; the former for conducting interviews and the latter for her editorial assistance. Moreover, the interviewees who provided much of the information herein presented — due to the lack of primary and secondary sources — also merit praise: José Von Lücken, José Luis Benza, Horacio Feliciángeli, Christian Schaerer, María Angélica Ayala, María Estela Ovelar, Nélida Centurión, Rodolfo Berganza and Avelina Demestri.

Bibliography

Consejo Nacional de Ciencia y Tecnología (2012). *Fortaleciendo el capital humano para la ciencia y la innovación. 2011–2012 Reporte de avances de programas de apoyo a la ciencia, tecnología e innovación en Paraguay*. Asunción: Author.

Consejo Nacional de Ciencia y Tecnología (2013). *Estadísticas e Indicadores de Ciencia y Tecnología del Paraguay 2011*. Retrieved from http://secit.conacyt.gov.py/estad%C3%ADsticas-e-indicadores-de-ciencia-y-tecnolog%C3%ADa-del-paraguay-2011-0

Consejo Nacional de Ciencia y Tecnología (2013). *CONACYT: rector y articulador de políticas públicas en ciencia y tecnología*. Retrieved from http://secit.conacyt.gov.py/conacyt-rector-y-articulador-de-pol%C3%ADticas-p%C3%BAblicas-en-ciencia-y-tecnolog%C3%ADa

Facultad de Ciencias Exactas y Naturales, Universidad Nacional de Asunción (1999). *Seguimiento de egresados*. Asunción: Author.

Facultad de Ciencias Exactas y Naturales, Universidad Nacional de Asunción (2013). *Historia*. Retrieved from http://www.facen.una.py/es/institution/history

Facultad de Ingeniería, Universidad Nacional de Asunción (2013). *Historia.* Retrieved from http://www.ing.una.py/?page_id=220

Ley 1.028 General de Ciencia y Tecnología (1997). Retrieved from http://paraguay.justia.com/nacionales/leyes/ley-1028-jan-31-1997/gdoc/

Ley 4.578 que crea el Fondo Nacional de Inversión Pública y Desarrollo (FONACIDE) y el Fondo para la Excelencia en la Educación y la Investigación (2012). Retrieved from http://es.scribd.com/doc/108489287/Ley-4758-Que-crea-el-Fondo-Nacional-de-Inversion-Publica-y-Desarrollo-FONACIDE-y-el-fondo-para-la-excelencia-de-la-educacion-y-la-investigacion

Molinas, J., Elías, R., & Vera, M. (2004). *Estudio y Análisis del Sector Educativo en Paraguay, informe final.* Asunción: Instituto Desarrollo; JICA.

OEA (1966). *Informe sobre las tareas realizadas por el Prof. Ernesto García Camarero.* Retrieved from http://elgranerocomun.net/IMG/pdf/EGC_OEA_INFORME_abril_1966.pdf

Sociedad Matemática Paraguaya (2013). *Historia.* Retrieved from http://www.cc.pol.una.py/SMP/historia.html

Universidad Nacional de Asunción (2013). *Breve reseña de la Universidad. Cronología de la UNA.* Retrieved from http://www.una.py/index.php/la-universidad/resena-historica

About the Author

Gabriela Gómez Pasquali is a member of the board of directors of the Paraguayan Mathematical Society, and was its president from 2000 to 2004. She is dedicated to the improvement of education in Paraguay, particularly mathematics. In 1989, along with a group of colleagues, she founded OMAPA, an organization that administers teacher workshops, as well as math Olympiad training and other activities for the mathematically talented across the country. She has been editor and co-author of 10 textbooks for Editorial Santillana. She is an agricultural engineer with a master's degree in environmental science and public policy.

Chapter 14

PERÚ: A Look at the History of Mathematics and Mathematics Education

César Carranza Saravia and Uldarico Malaspina Jurado

Abstract: In this chapter we present fundamental aspects of the history of mathematics in Perú — as an academic activity — begun at the National University of San Marcos. We highlight the importance of providing an adequate mathematical training for primary school teachers, and describe several initiatives that have been developed to address that, and to improve mathematics education in general, parallel to the development of mathematics in our country.

Keywords: Mathematics in Perú; history of mathematics; mathematics education.

Mathematics as an Academic Activity in Perú: A Historical Review

The first Peruvian professional who worked as a mathematician was Francisco Ruiz Lozano (1607–1677). Born in Lima, he was appointed to the first university chair of mathematics at the National University of San Marcos (*Universidad Nacional Mayor de San Marcos*, UNMSM). His treatise on comets was the first book written in Perú on such a topic. He also determined the geographic coordinates of the most important locations on the Peruvian coast.

Cosme Bueno (1711–1789) was a Spaniard who had lived in Perú from his youth. He was a medical doctor who in addition to medicine, taught mathematics, optics, and dioptrics. In addition to being a medical doctor, he became the Chief Cosmographer. He is considered to have introduced mathematical activity as part of the academic life in Perú. Two of his students, Gabriel Moreno (1735–1809) and Joaquín Gregorio Paredes (1778–1839), would in turn succeed him as university chair of mathematics. Later, José Domingo Choquehuanca (1789–1858), known for his praise of Simón Bolívar, wrote his *Essay on the Complete Economic and Political Statistics of the Province of Azángaro in the Department of Puno in the Peruvian Republic for the Five-Year Period from 1825 to 1829, Inclusive*, which was published in 1833. He was followed by José María Córdova y Urrutia (1806–1850), a civil servant, who in 1839 published *Historical, Geographic, Industrial, and Commercial Statistics of the People Who Live in the Provinces of the Department of Lima*.

Mathematics as a regular academic activity began officially at UNMSM in 1857, when the Faculty of Sciences was created by Supreme Decree. José Granda and José Joaquín Capelo were the first two faculty members in mathematics. Granda (1835–1911) was born in Camaná, Arequipa and graduated from the Central School of Arts and Manufacturing in Paris. He was named to a professorship of transcendental mathematics in 1866 and was the first to graduate with the Doctor of Sciences degree. Capelo (1852–1925), born in Lima, was a civil engineer who studied mathematics and later joined the faculty in the School of Sciences.

Federico Villarreal (1850–1923), a student of Granda and Capelo, was born in Túcume, Lambayeque. He entered UNMSM in 1877 to study pure mathematics. He graduated with a bachelor's degree in mathematical sciences in 1879 and a doctorate in 1881. His dissertation was entitled "Classification of third order curves". For this, Villarreal was awarded a gold medal that Granda had donated to the School of Science. Moreover, his dissertation was the first one to receive the grade of outstanding according to the new Instructional Regulations.

Parallel to what was happening in Lima, there were interesting events in Arequipa. Miguel Garaycochea (1815–1861) attended the *Colegio San Francisco* and obtained a bachelor's degree in International Law from the University of St. Augustine that won a prize for the quality of his graduation examination. He received a doctorate from the same university in 1838. Federico Villarreal wrote the following about Garaycochea's doctorate:

> The Honorable Departmental Council, on August 27, 1832, had granted a doctorate to R.P.F.[1] Juan Calienes, and on October 26, 1838 he left it at the discretion of the faculty for it to grant it to

---

[1] Reverendo Padre Franciscano.

one of the most impoverished and diligent students who needed its support to be able to study and who had been a student at *Colegio San Francisco*. Having registered the results of the examinations, they found that the most outstanding student whose poverty had been verified was Don Miguel W. Garaycochea, who was unanimously granted the doctorate (Carranza, 2007).

Garaycochea worked on a theory of polynomials with binomial calculus. He moved to La Libertad in 1846, where he became professor of mathematics at the University of Trujillo. Also in Trujillo, Juan de Dios Salazar published various educational works.

Years later, Godofredo García (1888–1970), born in Lima, was the favorite student of Villarreal. He graduated with a bachelor's in mathematical sciences with the thesis "Singular points of curves in the plane". In 1912 his doctoral dissertation was entitled "Resistance of columns of reinforced concrete". Federico Villarreal and Godofredo García are the two mathematicians that close out this era. At the time Perú was almost totally isolated from what was happening internationally in mathematics, but they were able to discover on their own some of the methods and theorems that were already known in Europe.

Federico Villarreal represents almost a half century of mathematics in Perú. He obtained his doctorate in mathematics from the UNMSM and also graduated as a civil engineer from the School of Engineering. He rediscovered the formula for raising a polynomial to an integer power and created a classification system for third degree curves. He published more than 500 works, the majority of which were applications to technology.

García, like his professor Villarreal, graduated with a doctorate in mathematics from UNMSM and as a civil engineer from the School of Engineering. His publications were numerous and varied: eight in real and complex analysis, 80 in theoretical mechanics, more than ten in astronomy, five in biomathematics, and more on resistance in columns.

In the 1930s, the panorama of Peruvian mathematics began to change. It was opened to international trends in two ways: the arrival of foreign books and journals for those who learned the corresponding languages, and the influence of Alfred Rosenblatt (1880–1947), a distinguished mathematician of international stature, who emigrated to Perú because of the climate of tension and intolerance caused by the Nazis. Rosenblatt was born in Krakow, Poland. He received a Ph.D. at the Jagiellonian University presenting a dissertation entitled "Integer functions with complex variables".

He arrived in Perú in 1936 and lived in Lima until his death in 1947. He worked intensely alongside Godofredo García at UNMSM, where he published, mostly in the *Journal of Sciences* (*Revista de Ciencias*), more than 130 articles in Polish, German, French, Italian and Spanish on real and complex analysis, geometry, topology, differential equations, rational and celestial mechanics, and hydrodynamics. It can be said that he initiated the diffusion of new trends from European mathematics in Perú.

In 1938, García, accompanied by Rosenblatt and other mathematicians from the UNMSM, promoted the creation of the Academy for Exact, Physical and Natural Sciences in Lima. It was indeed created on August 6, 1938 and García was elected as its first President. The creation of this academy was very significant in the scientific development of the country.

In that same era, Rosenblatt discovered an excellent student, José Tola. Tola (1916–1996) was born in Lima. From a young age, he was a student leader and later became a leader among young professors. With Rosenblatt as his advisor, he graduated from UNMSM on November 13, 1941 with a doctorate in mathematics. He presented a dissertation entitled "The equivalences of forms of continuity of operations by sequences and ranges in topological spaces". There is an unproven conjecture that the results of this dissertation were obtained simultaneously by Henri Cartan in Paris as an advisee of Maurice Frechet. Considered to be Rosenblatt's best student, in 1945 Tola was named Professor and Director of the Institute School of Physical and Mathematical Sciences in the Faculty of Sciences at UNMSM.

In the 1950s, Tola began to form groups of students. Those in the first group were José Reátegui, Gerardo Ramos, César Carranza, Oscar Valdivia, Leonor Laguna, Roberto Velásquez, Juan Guerra, and Jorge Sotomayor. They were interested in the study of topology, differential geometry, measure and integration, functional analysis, algebra and differential equations. In the second group were Alberto Vidal, Holger Valqui, Víctor Latorre and Ernesto López. Their main interest was in theoretical and experimental physics. From 1953 to 1961, Tola helped both groups to travel abroad for graduate studies in France, Spain, Germany, United States, México, Argentina, Brazil or Belgium. All of them returned and were able to spread new trends in topology, differential geometry, measure and integration, functional analysis, algebra, differential equations, theoretical physics and experimental physics.

Since 1961: Near the end of 1961 an administrative crisis in UNMSM led to the closing of the Institute School of Physical and Mathematical

Sciences. José Tola, who was an adjunct professor at the National University of Engineering (*Universidad Nacional de Ingeniería*, UNI), convinced Mario Samamé, at the time President of the UNI, of the need for an Institute of Mathematics. Samamé had received a doctorate in mathematics from UNMSM and enthusiastically accepted the idea. In March of 1962 the Institute of Mathematics (IMUNI) was created. It was dedicated to research and the preparation of mathematical leaders for the country. Tola and a small group of his students (Reátegui, Ramos, Carranza, Guerra, and Sotomayor) staffed IMUNI and began a new stage for mathematics in Perú. It is important to mention that at no time did Tola or his students abandon UNMSM. They stayed on as adjunct professors and enjoyed the affection of the students. Later, many of their students became professors or university administrators, and successfully canvassed for Tola and his students to be honored as Professors Emeriti.

From 1962 to 1968 there was a steady development of mathematics in UNI and UNMSM. Specific areas that were developed included linear and multi-linear algebra, group cohomology, real and complex analysis, functional analysis, algebraic and differential topology, qualitative theory of differential equations, dynamical systems, and partial differential equations. Common seminars were organized for students from both universities. A specialized library was organized at the IMUNI with recent books and a very important collection of journals. The first group of new doctoral graduates returned from Brazil, the United States, and France. Two dozen students left for graduate studies abroad and various visiting professors arrived. Marshall Stone (USA), Leopoldo Nachbin (Brazil), Luís Santaló (Argentina), and Carlos Imaz (México) shared pure mathematics with mathematics education. Warren Ambrose (USA), Paul Dedecker (Belgium), Laurent Schwartz (France and honorary professor of UNI), Francois Treves (France), Shiing Shen Chern (China), José Luís de Miguel (Spain), Mischa Cotlar (Russo-Argentinian and honorary professor of UNMSM), and Héctor Fatorini, Juan Carlos Merlo and Kelly Kestelman (Argentina) engaged in very intense mathematical activities.

IMUNI began its academic activities in 1962 in an off-campus location at the corner of Larrabure and Unanue Streets in the Jesús María district, and was there until the end of 1963. During that period, its members (Tola and his students Reátegui, Ramos, Carranza and Sotomayor) began to write articles related to their specialties and to publish them in the *Mathematical Communications* (*Notas Matemáticas*) journal, edited by IMUNI. They also gave seminars on algebra (Tola and Guerra), topology (Reátegui), analysis (Ramos and Carranza) and ordinary partial differential equations (Sotomayor). It is interesting to recall the seminar on analysis offered in the years 1962 and 1963 in which two UNI students stood out: César Camacho from mining engineering and Neantro Saavedra from mechanical engineering. Holger Valqui, a mathematician and theoretical physicist who had recently joined the faculty at UNI offered a seminar on the theory of elasticity. By the end of 1963 the Minister of Education, Francisco Miró Quesada, rewarded the efforts of IMUNI with a contribution of a million soles[2] to construct a new building for IMUNI on the UNI campus. With that contribution and the continued support of Mario Samamé, the new building was inaugurated in December of 1963.

In its seven years of existence (1962 to 1968), IMUNI contributed significantly to the advancement of mathematicians in Perú. Some of them are highlighted below.

César Camacho received his doctorate at the University of California, Berkeley, with renowned mathematician Stephen Smale as his advisor. Camacho became a leader in the area of complex dynamical systems. He is currently the Director of the Institute of Pure and Applied Mathematics (IMPA) in Rio de Janeiro. IMPA is considered to be one of the most important institutes in the world for dynamical systems. Camacho has received the most important award in Brazil and an award from the Third World Academy of Sciences (now called the World Academy of Sciences). He is an Honorary Professor at UNMSM, UNI, and the Pontifical Catholic University of Perú (PUCP).

Neantro Saavedra received his doctorate from the University of Paris. The famous mathematician Alexandre Grothendieck was his advisor. He is currently a professor in Japan.

Maynard Kong, a distinguished professor at PUCP, received his doctorate from the University of Chicago under the advisement of the renowned mathematician Richard Swan. Kong passed away in 2013.

Julio Ruiz and Teresa Tzukazan received doctorates at the University of Minnesota and currently work at the University of Río Grande do Sul in Brazil.

It is also important to recall some of Tola's students from UNMSM who participated actively in the seminars at IMUNI. Mario Piscoya, Pedro Espinoza, Luis Romero, and Carlos Chávez received doctorates from the

---

[2]Approximately $37,000.

University of Buenos Aires, and are professors at UNMSM. Pedro Contreras and Agripino García received doctorates from the Center of Advanced Studies at the National Polytechnic Institute in México and are professors at UNMSM. Michel Helfgott, with a doctorate from Montana State University, is a professor at UNMSM and at the State University of New York. Rolando Mosquera received his doctorate at the University of Paris 6 and is also a professor at UNMSM.

At the end of 1968, a political crisis led to the disappearance of IMUNI. José Tola was not caught off guard. As Vice-President of PUCP, he had participated in the creation of the Department of Sciences and of a bachelor's degree program in mathematics. Thus, in March of 1969, Tola, three of his colleagues (Carranza, Sotomayor, and Kong), and ten students moved to PUCP to initiate a third stage in Peruvian mathematics. They created a master's in mathematics in 1971 and contributed significantly to the preparation of university mathematics professors for the entire country (65% of their students were not from Lima and usually returned to their birthplaces to work at local universities). Today, PUCP offers a doctorate in mathematics, which along with the doctoral programs at UNMSM and UNI, is significantly promoting research in mathematics.

The Peruvian Mathematical Society (SMP) and the Institute of Mathematics and Related Sciences (IMCA)

The Peruvian Mathematical Society was created in 1957 by a group of mathematicians from UNMSM. It was José Tola who mustered the group. Francisco Miró Quesada, Rafael Dávila, Flavio Vega, Hernando Vásquez, José Ampuero, José Reátegui, Oscar Valdivia, César Carranza, and Roberto Velásquez were among those who joined. The session was held in an auditorium in the Faculty of Sciences at UNMSM. Tola was elected to be the first President; as such, he set the scope of work for the new institution. The original focus was on the professional development of university professors, and then of elementary and secondary school teachers. At the end of his term in 1961, Tola was replaced by Rafael Dávila. During Dávila's term, the SMP became inactive. In 1978, through the efforts of UNMSM, UNI, PUCP, University St. Anthony of Abad of Cusco (USAAC), and National University of Trujillo (UNT), the Fourth Latin American Mathematics School (IV ELAM) was held in Lima, organized by a commission of mathematicians from the sponsoring universities: Edgar Vera and Luis Huamán for

UNMSM; Julio Ruiz and Gerardo Ramos for UNI; César Carranza and Uldarico Malaspina, for PUCP; Alejandro Ortíz for UNT; and Abel Arce for USAAC. This was an example of what Peruvian mathematicians could accomplish when they joined forces. In the final assembly, it was decided that the organizing commission would be responsible for reactivating the SMP. The work of the commission continued until 1983. It prepared a program of activities that included writing statutes and holding national colloquia. The colloquia would follow the Brazilian model of week-long short courses offered at three levels, with notes previously disseminated. They also planned periodic lectures at various universities. Additionally, they proposed the publication of an annual informational newsletter, exchanges of national professors, visits from foreign professors, and agreements with other mathematical societies from throughout the world. It was also agreed that the governing body of the SMP, called the National Managing Council, would have seven members from three universities: three from UNMSM, two from UNI, and two from PUCP. The First Colloquium was held in 1983 and Luis Romero Grados, a professor at UNMSM, was elected as its first President; his term lasted from 1983 to 1984.

During the National Mathematics Colloquium in September of 1987 held in Cusco, an agreement was signed between the Peruvian Mathematics Society and the Brazilian Mathematics Society. Each of those groups was represented by its President: César Carranza of PUCP in the case of Perú and for Brazil it was César Camacho, a Peruvian mathematician working in Brazil who had been a student at UNI. This agreement was leveraged to get support from the International Center for Theoretical Physics in Trieste, directed by the Nobel laureate in physics, Abdus Salam. With that support from 1988 to 1996, 24 foreign mathematicians of international stature were able to visit Perú: three from Argentina, 15 from Brazil, two from Chile, two from Spain, one from Uruguay, and one from Venezuela. In groups of two or three each year, they gave month-long pre-doctoral courses: 20 at PUCP, three at UNMSM, and one at UNI. The participants were students or graduates of master's degree programs in mathematics from throughout the country and from countries in the Andean sub-region: Bolivia, Colombia, Ecuador, and Venezuela. A consequence of this effort was the creation at UNI in 1998 of the Institute of Mathematics and Related Sciences (IMCA). IMCA in some sense has replaced the defunct IMUNI. Thanks to the efforts of César Camacho and the support from the Institute of Pure and Applied Mathematics of Rio de Janeiro (IMPA) — as well as to an agreement with PUCP to support IMCA, both academically and financially — IMCA nourished a new vision for the future of mathematics in our country. With the collaboration of professionals from these institutions, both public and private, IMCA has become a recognized center for mathematics research in the country; in 2003 it began to offer a doctorate in mathematics.

Research in mathematics is advancing in Perú, particularly in optimization, dynamical systems, algebra and geometry, probability, statistics and differential equations. The main centers of research are PUCP, UNI, IMCA, and UNMSM. A sample of the international journals in which Peruvian mathematicians have published include *Bayesian Analysis*, *Journal of Differential Equations*, *Advances in Mathematics*, *Journal of Differential Geometry*, *Journal of Statistical Physics*, *Annales de l'Institut Fourier*, *Asterisque*, *Optimization*, *Journal of Algebra* and *Comptes Rendus Mathematique*.

It is important to highlight that Mathematical Olympiads — with the continued support of SMP, PUCP, and IMCA — have contributed in significant ways to the development of mathematics in our country. Perú has been participating in Olympiads since 1985. The Olympiads Commission of the SMP, presided by Uldarico Malaspina, has been in charge of selecting and accompanying the Peruvian delegations to various international and regional Mathematical Olympiads since 1989. The work of the Commission has been strengthened by the incorporation of various ex-Olympians. The Peruvian student delegations have won 48 gold, 90 silver, and 89 bronze medals in the Mathematical Olympiad of the Latin American Southern Cone, the Iberoamerican Mathematical Olympiad (OIM), and the International Mathematical Olympiad (IMO). Three of those gold medals were won at the IMO where nearly 100 countries participate (IMO 2008, 2010, 2011). Nevertheless, perhaps of more importance is the academic support that IMCA and PUCP provide to the young medal winners and others with notable mathematical abilities. Many of those students have received scholarships to study mathematics at the master's degree level either at IMCA or IMPA in Brazil. Several ex-Olympians have completed doctorates and are now professors at PUCP. A significant anecdotal note is that some of them presented brilliant doctoral dissertations in mathematics before finishing their undergraduate degrees in Perú.

It should be mentioned that the Olympiads Commission of the SMP has been academically supporting the Ministry of Education since 2004 by organizing the National School Mathematical Olympiads (ONEM) that are held annually and to which all secondary school students are invited. More than a million students participate in the first of four phases of this competition. The Commission is in charge of developing the tests for all four phases and for scoring tests in the final phase. The final phase is held in Lima and solutions must be developed with adequate justifications. During this final phase, the Commission also offers a workshop on problem posing and problem solving for team coaches.

The Teaching of Mathematics

In 1958, the International Congress of Mathematicians, held in Edinburgh, agreed to recommend reform of the teaching of mathematics. This was a consequence of the work of an important movement began in 1939 by a group of French mathematicians (Jean Dieudonné, André Weil, Claude Chevalley, Henri Cartan, Gustave Choquet, and others) and a Polish mathematician (Samuel Eilenberg), who using the pseudonym Bourbaki had published, in 1957, a monumental work entitled *Elements of Mathematics*. This recommendation and the meetings promoted by the International Commission on Mathematical Instruction (ICMI), in particular in Aarhus, Denmark in 1960, greatly influenced various international organizations to take a closer look at the teaching of secondary school mathematics. Thus, mathematics and its teaching began to receive priority attention in the form of seminars, professional development for teachers, graduate scholarships, and related research.

In July of 1960, the National Science Foundation (NSF), which was funding Summer Institutes for mathematics teachers at various universities in the United States, invited Francisco Miró Quesada to give presentations at those Institutes. Miró Quesada was one of the most renowned professors of the philosophy of mathematics at UNMSM. He traveled to the United States and visited eight of the Summer Institutes. He was able to have long conversations with the participants about their professional challenges and ideas about education. He was also able to observe their reactions to the New Mathematics. On his return to Perú, he discussed with José Tola the possibility of having the same kind of courses for Peruvian secondary school mathematics teachers. Throughout his career, Tola had advocated that any kind of educational reform should be based on the leadership of qualified personnel to direct it and well-trained professional developers to execute it. He immediately accepted the idea of Miró Quesada, and with the collaboration of two young professors (Reátegui and Ramos), they planned the first Summer Institute.

This First Summer Institute was held in 1961 at the National School of Nuestra Señora de Guadalupe, with the sponsorship of the Ministry of Education and the financial support of the Peruvian North American Cooperative Education Service. NSF provided technical assistance by sending Wade Ellis and Max Kramer. The Institute offered the following courses: Geometry (J. Tola), Logic (F. Miró Quesada), Algebra (G. Ramos), and Functions (J. Reátegui). The courses were held in the morning and over 130 teachers attended. The enthusiasm generated was so great, that as a result the National Association of Teachers of Mathematics was created. Francisco Armando Barrionuevo was its first President. This new association suggested the creation of a permanent organization that would address the professional development of teachers. On May 30, 1961, Tola and Abel Fernández met with Miró Quesada, Ramos and Reátegui, and agreed to create a non-profit scientific organization called the Institute for the Advancement of the Teaching of Mathematics (IPEM). IPEM was destined to hold professional development seminars for mathematics teachers of all levels, disseminate new concepts and pedagogical ideas related to the teaching of mathematics, support the publication of textbooks, and collaborate with organizations and national authorities in their efforts to support the study and teaching of mathematics. The first managing council of the IPEM was Tola as director, Miró Quesada as assistant director, and Ramos and Reátegui as members.

In 1961, with the sponsorship of ICMI and participation of invited representatives from 23 countries, the First Inter-American Conference on Mathematics Education (I IACME) was held in Bogotá. Gustave Choquet, spokesman for the Bourbaki group, proclaimed for Latin America: "It is necessary to take a new look at teaching at all levels, primary, secondary, technical, university, in light of the discovery of the great structures" Choquet (1962).

This first IACME and the exhortation of Choquet, were a landmark for reforms in the teaching of mathematics, particularly in Perú. Tola, who participated in the first IACME representing Perú, did not want to attempt educational reform in mathematics education unless the teachers received the necessary professional development. Therefore, IPEM's most important activity was the professional development of secondary school mathematics teachers, taking advantage of the return of Carranza and Guerra to Perú.

In 1962, IPEM organized the Second Summer Institute for secondary school mathematics teachers, again with the sponsorship of the Ministry of Education and financial support from NSF. Four courses were offered at the Chorillos Military School: Functions (Tola), Logic and Set Theory (Miró Quesada), Algebra (Carranza), and Geometry (Mariano García and Francisco Garriga, from the University of Puerto Rico). The Ministry of Education printed textbooks that were written specifically for the course by professors teaching them. The educational system was made up of lectures followed by practice sessions that were directed and graded by a group of young mathematicians (C. Castro, J. Guerra, E. León and R. Rivas). This system seemed to give the courses a true university level. Eighty teachers participated, 13 of them from Argentina, Bolivia, Chile, Ecuador and Central America.

In 1963, the Ministry of Education again sponsored the Third Summer Institute, this time at UNI. It had financial support from the Pan-American Union and the Ford Foundation. From an academic point of view, it followed the model used in the first two, but this time it was divided into two levels. In the first level were courses on Functions I (Reátegui), Logic and Sets (Miró Quesada), and Algebra I (Carranza). In the second level: Functions II (Tola), Algebra II (Ramos) and Geometry (H. Merklen, from University of Montevideo). Again there was collaboration with an important group of auxiliary professors (Castro, Guerra, Isla, León, Sotomayor, and Valqui).

In October of 1963, Miró Quesada requested leave from his position as Assistant Director of IPEM because he had been named Minister of Public Education in the first government of Fernando Belaúnde. This marked an outstanding stage in the development of mathematics in Perú for three reasons:

(1) preparation of mathematicians for the public universities (UNMSM and UNI);
(2) preparation of leaders for the teaching of secondary school mathematics (creation in UNI of the Regional School for Mathematics, which functioned from 1966 to 1968); and
(3) professional development for secondary school mathematics teachers (continuation of the Summer Institutes).

In 1964, the Fourth Summer Institute was held in the Higher Normal School "Enrique Guzmán y Valle" in Cantuta. Miró Quesada, as Minister of Education, had arranged for the total remodeling of the dormitories where the participants stayed. There was also substantial support from Stanford University (USA), NSF, the Pan-American Union, and UNI. Participants included 66 Peruvian teachers, and 61 from Argentina, Brazil, Bolivia, Colombia, Chile, Costa Rica, Cuba, Ecuador, El Salvador, Guatemala,

Honduras, México, Nicaragua, Panamá, Paraguay, the Dominican Republic, Uruguay and Venezuela. The foreign participants were selected by an International Commission that met in Lima in December of 1963. The Peruvian participants, 75% from outside of Lima, were selected by IPEM. The Institute required the complete attention of the participants who were lodged in Cantuta. Again, there were two levels of courses. In the first level the courses were: Algebra I (Carranza), Sets and Logic (Castro), and Function I (Reátegui). In the second they were: Algebra II (Ramos); Geometry (Merklen), and Functions II (Tola). Additionally, the following short courses were offered: Introduction to Geometry in the First Year of Secondary Education (Merklen), Introduction to Numerical Analysis and Computers (P. Willstätter), and a Seminar in Pedagogy (I. Sussman from Santa Clara University, California). There were textbooks specifically prepared and printed for each of the courses. The Seminar on Pedagogy discussed the teaching of secondary mathematics. It used the textbooks of the School Mathematics Study Group (SMSG) as the basis for the discussion and the participants were given copies of the SMSG textbooks. As in past courses, there was collaboration with auxiliary professors Mauro Chumpitaz, Emilio Isla, Eduardo León, Salvador Montes, Rosa Rivas, Miguel Tantaleán, Gloria Sánchez, and Holger Valqui.

At the end of the Institute there was a meeting to evaluate the results and make plans for the future of mathematics education in Latin America. Participating in that meeting were José Babini (Argentina), Djairo Figueiredo (Brazil), Edward Begle, Howard Fehr, Philip Hemilly and Rousell Phelps (USA) and José Tola (Perú). It can be said that this meeting and the Fourth Summer Institute were significant for reforms in the teaching of mathematics in Latin American, particularly in Argentina, Brazil, Chile, Colombia, México and Uruguay, where there was government support for the reforms.

The Higher Normal School "Enrique Guzmán y Valle" in Cantuta was again the site for the Fifth Summer Institute. Moreover, there was again support from the Ministry of Education, UNI, and NSF, and new support from the Peruvian Institute for the Development of Education. An International Commission that met in Lima in 1965 selected 86 Peruvians (66% from outside of Lima) and 11 participants from abroad. The courses functioned as in the past. For level one the courses were Functions I (Sotomayor); Algebra I (Carranza), and Sets and Logic (Miró Quesada). In the second level there were Functions II (Tola), Algebra II (Ramos), and Geometry (Merklen). Additionally, P. Willstätter taught a short course on

Introduction to the Study of Computers that used an IBM-1620 that was in a mathematics laboratory of UNI's Faculty of Mechanics and Electricity. As usual, there were printed materials for each course. Every week there was a required Seminar on Pedagogy to discuss secondary mathematics teaching that used the SMSG textbooks. The collaborating auxiliary professors were Carlos Cabrera, Eduardo León, Salvador Montes, Alejandro Ortíz, Gloria Sánchez and Roberto Velásquez.

The Sixth Summer Institute moved to UNI's Faculty of Physical and Mathematical Sciences and was held in 1966. It was still sponsored by the Ministry of Public Education, UNI and the Peruvian Institute for Educational Development. The Director was César Carranza. The participants were required to take Algebra I (Carranza) and Functions I (A. Ortíz), as well as a Seminar on the Teaching of Geometry (C. Cabrera and N. Cáceres). All the participants attended the lectures, the exercise sessions, and the evaluations. For this Insitute it should be indicated that:

(1) This was the first time that all the academic and financial support for the Institute came from Peruvian institutions. The majority of the 28 scholarships, provided by the Peruvian Institute for Educational Development, were awarded to the auxiliary professors.
(2) Most of the auxiliary professors of this Institute were young Peruvian secondary school teachers who had attended past Institutes. They were dedicated completely to mathematics and had received professional development in Perú and abroad. Thus, a group of secondary school teachers was strengthened that was highly qualified and very enthusiastic about contributing decisively to the development of mathematics teaching in the country.

In 1966, IPEM organized the Second Inter-American Conference on Mathematical Education (II IACME). A new Inter-American Committee on Mathematical Education (CIAEM) was elected with Marshall Stone as President and José Tola as a member of the Executive Committee.

UNI was again the site for the Seventh Summer Institute in 1967. This Institute had continued support from the Ministry of Education and UNI, and new support from the Ford Foundation. César Carranza and Gerardo Ramos were the Directors and Pedro E. Castillo served as Coordinator. There were 79 participants and this time they were divided into two levels: 59 in the first level (Level A) and 20 in the second (Level B). For the first level the courses were Algebra I (Walter Torres with Carlos Véliz directing the practicum), Analysis I (Alberto Fujimori with Edith Mellado directing the practicum), and a Seminar on the Teaching of Geometry (Víctor Agapito and Pedro Castillo). In Level B the courses were Analysis II (Alberto Cáceres with Yuri Haraguchi directing the practicum), Geometry (Boris Ayub with Carlos Gutiérrez directing the practicum), and a Seminar on the Teaching of Geometry (Rubén Muñoz). The courses were supplemented by a series of lectures presented by renowned national and international mathematicians.

The Eighth Summer Institute (1968) was held at UNI, the Ninth (1969) at the Peruvian University Cayetano Heredia, and the Tenth (1970) and Eleventh (1971) at PUCP. All of them had support from the Ministry of Education and other institutions dedicated to educational development. Beginning in 1972, professional development courses were offered by universities in Lima and the Peruvian Mathematical Society (SMP) in collaboration with provincial universities. Usually, they did not have support from the Ministry of Education, but continued to work diligently to provide professional development to mathematics teachers. Since its first colloquium in 1983 the SMP, in addition to its activities related to research in pure mathematics, has offered annual academic events that include short courses on mathematics for secondary school teachers.

Since 1997, the Peruvian Mathematics Education Society (SOPEMAT[3]), whose members are mathematics teachers from various levels, has been organizing National Congresses on Mathematics Education (CONEM) that contribute to the professional development of teachers and provide an opportunity to share experiences with national and foreign researchers. CONEM V in 2012 had keynote addresses from Luisa Ruiz Higueras and Luis Rico Romero from Spain, and Eduardo Mancera from México. The current President of SOPEMAT is Olimpia Castro Mora.

The Research Institutes on Mathematical Education (IREM) were created in France in 1968 and have made significant contributions to mathematics education. In the late 1990s, they began promoting the establishment of IREMs outside of France. Uldarico Malaspina, with the support of PUCP, promoted the establishment of a local branch. In August 2000, an agreement of the University Council created the IREM-PUCP[4] and designated Uldarico Malaspina as its Director. Since then, and in close

---

[3]SOPEMAT http://www.sopemat.org/
[4]IREM-PUCP http://irem.pucp.edu.pe/

collaboration with PUCP's master's degree program in the Teaching of Mathematics, it has offered colloquia on the teaching of mathematics to provide teachers of various levels opportunities to extend and strengthen their knowledge of mathematics, integrated with a scientific perspective on mathematics education.

The Sixth Colloquium on the Teaching of Mathematics was held at PUCP in 2012. Invited international presenters included Michèle Artigue and Raymond Duval from France, María Trigueros from México and Walter Castro from Colombia. The researchers in mathematics education who have participated in those colloquia and in the Sixth Iberoamerican Congress on CABRI — also organized by IREM-PUCP in 2012 — have developed workshops and research seminars for professors and students of the master's degree program in the Teaching of Mathematics. They have also had valuable support from Juan D. Godino and Vicenç Font from Spain, Patricia Camarena from México, Jean Marie Laborde and Colette Laborde from France and María José Ferreira da Silva from Brazil.

The master's degree program in the Teaching of Mathematics of the PUCP has promoted research in mathematics education in Perú in close collaboration with IREM-PUCP. The shared areas of research are: information technology and mathematics education, mathematics teaching, history in mathematics education, and mathematics in the curriculum and teacher training. Theses and research papers by the associated professors have been presented in international forums of significant prestige, such as Joint Meeting of the International Group for the Psychology of Mathematics Education (PME), International Congress on Mathematical Education (ICME), *Reunión Latinoamericana de Matemática Educativa* (RELME), *Simposio de la Sociedad Española de Investigación en Educación Matemática* (SEIEM), CIAEM and *Congreso Iberoamericano de Educación Matemática* (CIBEM). Their articles have also been published in such journals as *Educational Studies in Mathematics, Revista Latinoamericana de Investigación en Matemática Educativa, UNION-Revista Iberoamericana de Educación Matemática, Matematicalia* and *UNO-Revista de Didáctica de la Matemática*.

Final Considerations

Living in a society that is increasingly influenced by technology makes a strong mathematical preparation of future professionals, technicians, and citizens in general even more important. Their contributions in such a society need to be active, decisive, and creative. Converting information into knowledge requires creativity, logical reasoning, a capacity to identify and solve problems, a capacity for self-directed learning, and the management of information with scientific criteria.

Priority must be given to the teaching of mathematics from the earliest grades, as it is at that level that positive or negative attitudes are developed toward the various fields of knowledge, and in particular towards mathematics. Therefore, providing a good foundation in mathematics and its teaching for future in-service elementary school teachers is urgent. Certainly, attention must also be given to university mathematics teaching. At all levels, not only is it important to develop fundamental content, but also to stimulate active learning and mathematical thinking. It is also necessary to make connections to previous content, other disciplines, and to the contexts in which we live.

This historical review is incomplete, but presents fundamental aspects of mathematics and mathematics education in Perú. We hope that it will provoke the reflections of all of us who are involved in mathematics and its teaching. It is also our desire that it will be a stimulus to follow the example of our predecessors and continue to enhance the quality of mathematics and mathematics education in our countries, particularly in Perú.

Bibliography

Carranza, C. (2002). José Tola Pasquel y las universidades de San Marcos, Ingeniería y la Católica. Discurso póstumo pronunciado en Homenaje a José Tola en la Pontificia Universidad Católica del Perú.

Carranza, C. (2007). Historia de la matemática peruana. En Ciclo de conferencias sobre Matemática y Física Educativa, UNMSM, Lima. Retrieved from http://industrial.unmsm.edu.pe/olimpiadas_matematica/hmp.pdf

Choquet, G. (1962). The new mathematics and teaching, in F. H. Fehr (Ed.), *Mathematical education in the Americas: A report of the First Inter-American Conference on Mathematical Education*, pp. 71–78. New York: Teachers College Columbia University Bureau of Publications.

Malaspina, U. (2013). La enseñanza de las matemáticas y el estímulo a la creatividad. *UNO-Revista de Didáctica de las Matemáticas, 63*, 41–49.

Samamé, M. (1990). Godofredo García Díaz, in *Hacer Ciencia en el Perú. Biografías de ocho científicos*. Lima: Sociedad Peruana de Historia de la Ciencia y la Tecnología.

Velásquez, R. (1990). Alfred Rosenblatt en el Perú, in *Hacer Ciencia en el Perú. Biografías de ocho científicos*. Lima: Sociedad Peruana de Historia de la Ciencia y la Tecnología.

Velásquez, R. (1995). *Matemática en el Perú del siglo XIX*. Ayacucho: Universidad San Cristóbal de Huamanga.

Watanabe, L. (2004). *Federico Villarreal: Matemático e Ingeniero*. Lima: Ediciones COPÉ.

About the Authors

César Carranza Saravia has a Ph.D. in mathematics and is an emeritus professor at the National University of San Marcos, at the Pontifical Catholic University of Perú, and at the National University of Engineering. He was a member of the Inter-American Committee on Mathematics Education (CIAEM) from 1972 to 1979 and President of the Peruvian Mathematical Society for three terms. A major force in propagating mathematics and mathematics education in Perú, he is also a member of the National Academy of Sciences of Perú.

Uldarico Malaspina Jurado has a Dr.Sc., is a full professor at the Pontifical Catholic University of Perú and director of the Research Institute on Mathematics Education (IREM-PUCP). He is the President of the Olympiad Commission for the Peruvian Mathematical Society, a member of the National Academy of Sciences of Perú, and author of multiple publications on mathematics education published in journals like *Educational Studies in Mathematics*, *RELIME*, *UNIÓN*, *Matematicalia*, and *UNO*.

Chapter 15

PUERTO RICO: The Forging of a National Identity in Mathematics Education

Héctor Rosario, Daniel McGee, Jorge M. López,
Ana H. Quintero, and Omar A. Hernández

Abstract: This article reviews the history — through different political periods — and current status of mathematics education in Puerto Rico. It chronicles the struggles that have informed educational policies in the forging of a national identity in the field — policies that will hopefully foster the development of a society that values mathematical ideas.

Keywords: History of mathematics education; mathematics in Puerto Rico; history of mathematics; mathematics education; national identity.

The Colonial Experience

Puerto Rican identity has been shaped by the fact that our development as a nation has often answered to priorities and plans engineered from abroad. Given that Puerto Rico has been a colony of Spain (1493–1898) and of the United States since 1898, the nature of the state in Puerto Rico is intertwined with the Territorial Clause of Article Four of the United States Constitution. This clause assures that the United States Congress has final authority over Puerto Rico and its people. This has influenced our educational policies, which are still frequently "formulated to serve the needs of the imperial enterprise" (Navarro-Rivera, 2009).

Puerto Rico's Indigenous Taíno People

Puerto Rico's eminent historian of education, José Juan Osuna, warned that "in writing on education in Puerto Rico, the temptation is to neglect the origins; to think of education as beginning at a later time when we have a record of school laws and of the provision made by the government" (Osuna, 1949). Ironically, Osuna himself neglected the origins of Puerto Rico by relegating the Taínos to be mere beneficiaries of the Spanish rulers' goodwill, since "no other colonizing nation of the time was concerned with their welfare as much as Spain" (Osuna, 1949). He further claimed that "we find this concern manifested in an order to the effect that the [Taíno] children in each village should get together twice a day in a house next to the church in order that the chaplain might teach them to read and write, to make the sign of the cross, and to learn the prayers of the church" (Osuna, 1949). Regrettably, Osuna and others confused education with religious indoctrination and its political goals.

Osuna considered Puerto Rico to be "a Spanish colony in blood, in religion, in customs, and in traditions" (Osuna, 1949). Nevertheless, the Taíno Genome Project contradicts the blood assumption since the genetic composition of Puerto Ricans is, on average, 15% Taíno, 21% African, and 64% European (Via, et al., 2011). While European ancestry is dominant, 15% is higher than the 1/8 blood quantum required for membership among 18 Native American tribes.

The Taínos had a tributary system in place when the Spaniards arrived in 1493. It required that each *yucayeque* (community) pay a tribute in agricultural products to each regional *cacique* (chief) (Moscoso, 2003). The determination of the amount of the tribute and its relationship to the size of the village is a significant mathematical problem. In the absence of written records, its solution has been lost to history. We also know they navigated at night between the various Caribbean islands. Night navigation required some astronomical knowledge; however, the mathematical techniques they might have used again have been lost. While Puerto Ricans cannot speak of an indigenous mathematical legacy, as descendants of Incas or Aztecs can, it is clear that an operational competence existed well before the arrival of the Spanish conquistadors. Nevertheless, Puerto Ricans by and large do not identify with their indigenous ancestry, and there is no official recognition whereby someone may claim Taíno ancestry or any special rights thereof.

Colony of Spain (1493–1898)

Much of education under Spain was delegated to the Catholic Church (Navarro-Rivera, 2009), with brief intermittent periods after Spain's 1820 mutiny at Cádiz, and the subsequent *Trienio Liberal*, when the civil government undertook the provision of public education. As with most colonial possessions, the effects of the educational revolution and the development of public education arrived late in Puerto Rico.

Under Spanish rule, the curriculum and teaching strategies were very similar to those used in Spain and the rest of Hispanic America, except where there was a strong presence of Jesuit missionaries. In general, education was dedicated to religious instruction, with basic teaching of reading and writing, along with some notions of arithmetic and algebra (Coll y Toste, 1909). Teaching and learning were based on memorization. Humanistic studies were superficial and scientific studies almost non-existent. During this period, the first *catedrático* (professor) of mathematics in Puerto Rico, José Besabé, arrived in 1822. He resigned from his post after a few months of service and was replaced by Santiago Pérez. Both professors received their education in Spain.

The Jesuit *Seminario-Colegio*, founded by the *Compañía de Jesús*, functioned from 1858 until 1886 (Gómez-Diez, 2000). The Jesuits started a secondary school in 1862 where two years of mathematics instruction was offered. Besides arithmetic and algebra, the curriculum included geometry and trigonometry based on J. M. Fernández y Cardín's textbooks. Due to irreconcilable political differences with the civil government, the Jesuits were forced to close the school in 1886 and abandoned Puerto Rico (Gómez-Diez, 2000).

Colony of the United States (1898–Present)

As a result of the 1898 Spanish–Cuban–American War and the Treaty of Paris, the United States attained authority over Puerto Rico. "As a colony of the US, the major focus in the education of Puerto Rican children was the English language" and "the American way of life" (NEA, 2006). "Indeed, these were mandates of the US government" (NEA, 2006). Former presidents of the United States, including John Quincy Adams, Thomas Jefferson, and James Monroe, expressed a wish to annex Cuba and Puerto Rico. Charles Eliot, president of Harvard University, echoed those ambitions in 1898 when he stated, "I am inclined to the belief that we shall be able to do Cuba and Porto Rico some good; though to do so we shall have to better very much our previous and existing practices in dealing with inferior peoples" (Navarro-Rivera, 2009).

The new ruling power, the United States of America, "viewed state education as the most effective and efficient entity through which to undertake the colonization of Puerto Rico" (Navarro-Rivera, 2009). In 1899 the *Junta Insular de Educación* was established to centralize the administration of education in the island (Programa, 2003). "Thus, contrary to its tradition of decentralized governance, after the 1898 war the United States chose to govern its new colonies in the same centralized manner it governed the affairs of its Native American societies" (Navarro-Rivera, 2009). Confirmation of this policy is evident in the fact that 60 Puerto Ricans were sent to the Carlisle Indian Industrial School in Pennsylvania between 1900 until its closure in 1918. The historian Osuna was among the seven Puerto Rican students who graduated from the school.

According to a Carlisle brochure from the 1920s, "a thorough training [was] given in arithmetic, but no instruction [was] given in the higher mathematics" (Maryland, 2006). In a letter by Sun Elk, a Native American student at Carlisle, "There was much arithmetic. It was lessons; how to add and take away, and much strange business like you have crossword puzzles only with numbers. The teachers were very solemn and made a great fuss if we did not get the puzzles right" (Maryland, 2006). We can conclude that the sixty Puerto Ricans at Carlisle received similar training.

The Foraker Act was passed by the US Congress in 1900, recognizing Puerto Rican citizenship; the Jones Act of 1917 conferred US citizenship on Puerto Ricans, seven years before the Indian Citizenship Act of 1924. The Foraker Act also created the position of Commissioner of Education. In his foreword to Osuna's thesis book, José Padín, who held the post from 1930 to 1937, chose a conciliating tone arguing that "American 'imperialism' [endnotes by Padín] is inescapably infected with the spirit of democracy and an unquenchable zeal to treat peoples, even dependent peoples, as ends in themselves rather than as resources to be exploited for the benefit of the dominant power" (Osuna, 1949).

The first Commissioner of Education, Martin G. Brumbaugh, relocated Puerto Rico's Normal School to Río Piedras from Fajardo in 1901 (Osuna, 1949). The Normal School became the first department of the University of Puerto Rico (UPR) upon its founding in 1903. In 1911, the College of Agriculture and Mechanic Arts was established in Mayagüez (Osuna, 1949). These two campuses became — and continue to be — the foci of intellectual life and scientific and mathematical research on the island.

"Until 1947, the President of the United States appointed the Governor, the Attorney General, the Commissioner of Education, the Treasurer, and the Justices of the Supreme Court of Puerto Rico. As far as education was concerned, 'the rules of the game', regarding the language to be used, whether English or Spanish, the methodology and the course content were decided by the Presidential Commissioners, who usually were American educators with little knowledge of Puerto Rico, its culture and traditions" (García, 1964). In 1948 Puerto Rico had its first elected governor, Luis Muñoz Marín. Despite intentions, an educational system dominated by appointed officials acting in concert with local collaborators, fostered an intellectual dependency that hindered creativity in formulating educational policies. This problem persists today as Puerto Rico continues to debate its relationship with the United States.

Mathematics Education in the 20th Century

In *Marco Curricular* (2003), the Department of Education provided a synopsis of the historical development of the mathematics curriculum in Puerto Rico during the twentieth century. It portrays a questionably-smooth transition in the advancement of mathematics instruction that includes consideration of curricular developments but leaves out other important initiatives, like teachers' development. It is worth noting that *Marco Curricular* attempted to imitate some frameworks for mathematics

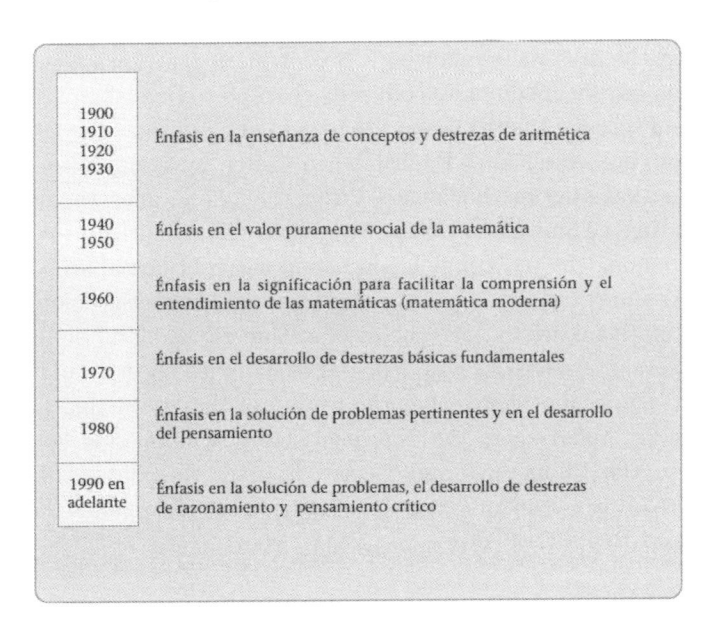

1900 1910 1920 1930	Énfasis en la enseñanza de conceptos y destrezas de aritmética
1940 1950	Énfasis en el valor puramente social de la matemática
1960	Énfasis en la significación para facilitar la comprensión y el entendimiento de las matemáticas (matemática moderna)
1970	Énfasis en el desarrollo de destrezas básicas fundamentales
1980	Énfasis en la solución de problemas pertinentes y en el desarrollo del pensamiento
1990 en adelante	Énfasis en la solución de problemas, el desarrollo de destrezas de razonamiento y pensamiento crítico

Figure 1. Curricular developments in the 20th century (Programa, 2003).

education in some US states (most notably California), especially since it was a late-comer in the list of framework documents for mathematics education. Some viewed this document as representative of the "state of the art" in the normative documents that were to guide mathematics education in Puerto Rico and the United States.

The Years 1900–1930: Emphasis on the Teaching of Arithmetic Concepts and Skills

López Yustos (1997) provided an overview of the structure of the education system around 1915. He informs us that elementary education was divided into urban and rural. The former consisted of eight grades, which granted a diploma upon completion. The latter consisted of only the first four grades, in addition to courses in agriculture and other farm-related trades taught to boys and girls. High school education was imparted in the regular high schools of San Juan, Ponce, Mayagüez, Arecibo, and Humacao, which offered complete programs. In addition, there were more than a dozen continuation schools that offered one or two years. These were established in some towns to delay the transfer of the young people to urban educative centers to complete their secondary education.

During the first years following US control, instruction was conducted in Spanish, but the medium was officially changed to English in 1904. In the majority of schools, Puerto Rican teachers used Spanish as the language of instruction, but switched to English when visited by a supervisor.

The mathematics textbook used during the first decades was *Arithmetic* by David Eugene Smith and George Wentworth. The Smith and Wentworth text was among the most popular in use in the continental schools of the period. Arithmetic skills were necessary to administer the colony, but by reducing mathematics to the teaching of arithmetic and later on in calculus for engineers, mathematics was deprived of its innate beauty and charm.

Marco Curricular claims that the results of the "Columbia University Commission", referring to the Teachers College Commission led by Paul Monroe in the Philippines and Puerto Rico in the mid-1920s, showed that Puerto Rican students "mastered" arithmetic (Programa, 2003). The Commission Report stated:

> In our charts the comparison of Porto Rico and continental children in arithmetic is conspicuous because of the unusually

high attainment of the pupils of the island. In arithmetic skills in the second, third, fourth, and fifth grades children of Porto Rico actually exceed those of continental schools. The attainment, for example, in the third grade in Porto Rico practically equals that in the fourth grade of continental schools (Survey, 1926).

The same document, however, states that students' mathematical achievement "from the eighth grade to the fourth-year high school represents a total gain in computational ability of less than half a year!" (Survey, 1926). That is, children were exceptionally apt in performing mathematical activities, yet opportunities for their advancement were not provided. This furthers the thesis of the pernicious effects of colonial policies. Perhaps the followers of Eliot in Puerto Rico saw advances in mathematics as appropriate for the intellectual elite at Harvard and Columbia, but not for "inferior peoples".

The Years 1930–1960

"The year 1932–1933 was a year of great activity in Puerto Rican education. The Department of Education prepared manuals for the use of the teachers and issued observations to supervisors on how to analyze the methodological needs of the teachers" (García, 1964). At the secondary level, "great pains were being taken to 'humanize' the teaching of mathematics, linking its meaning to the needs of life and initiating the student in the notions of business" (García, 1964). This emphasis was a reflection of the progressive education movement pioneered by Teachers College Columbia University professors, including members of the 1926 commission.

Marco Curricular claims there was another commission led by Columbia University professors in 1949 that recommended using "activities related to the community" (Programa, 2003). This too was a reflection of events in continental school mathematics following World War II that attempted to salvage what remained of progressive, socially relevant mathematics.

The mathematical gains reported in the 1926 report were somehow lost by 1940, possibly as a result of progressive educational policies. In June 1940, Commissioner of Education José Gallardo highlighted that "the teaching of arithmetic was so deficient in the elementary schools, that the teachers in the secondary schools had to devote a great deal of time to explaining the fundamentals of arithmetic so that students could understand the mathematics of the grades" (García, 1964). García cites texts and policies remaining from the "progressive" period:

> In Grades 7 through 9, the text used was *Mathematics and Life*. It consisted of three books, one for each year. The first one dealt with problems of the home, like salaries, occupations, etc. The second had to do with problems of the community, such as business, bank systems, and taxes. The students of the ninth grade would use the third book, which dealt with mathematical problems at the national level: natural resources, distribution of production, tax systems, and so on. In his annual report for the year 1943–1944, the Commissioner of Education pointed out that all efforts were being made to teach mathematics in the high school from the point of view of the needs and interests of the students. The decade of 1940–1950 highlighted a process of "socialization" in the teaching of mathematics and a nationalistic flavor in the whole educational process, crowned with the wise decision in 1949 that all instruction was to be imparted in Spanish and that the teaching of English would be intensified as a second language.

This prompted Erasto Rivera Tosado and Pedro A. Cebollero in 1948 to write the first mathematics textbook in Spanish under US domination, *Aritmética Social*, which became the standard for elementary levels. "The scope, as far as methodology and course content were concerned, did not differ much then from that of the previous decade" (García, 1964).

An important event for teachers' development during this time was the 1950 creation of the National Science Foundation (NSF). "The Foundation's programs reached Puerto Rico in 1957, when a grant was made to Inter-American University for a Summer Mathematics Institute for high school teachers of science and mathematics. The institute included three formal courses: Foundations and Philosophy of Mathematics, The Teaching and History of Mathematics, and The Theory of Numbers and Mathematical Recreations" and was directed by Mariano García (García, 1964).

Soon after this institute, the Soviets successfully launched Sputnik 1. "This created in the US a sense of urgency in the necessity of improving the teaching of Science and Mathematics and as a consequence multiple NSF-sponsored activities for teachers and students were developed" by universities in Puerto Rico (García, 1964).

The Years 1960–1990: Emphasis on Meaning to Facilitate the Comprehension and Understanding of Mathematics

The first elected governor, Luis Muñoz Marín, took office in 1949. At that time only half of school-age children attended school. During the first decade of the Muñoz administration the main educational effort was on universal school attendance. Between 1944 and 1962, public spending on education quadrupled in real terms, resulting in a massive expansion of schools, teacher hiring and the purchasing of books, materials, and equipment. In 1954 all children of school age started first grade. Once universal attendance was attained for elementary school children, Governor Muñoz declared the 1960s the "Decade of Education", switching the focus from quantity to quality.

Action on teachers' development was also motivated by the results of a study reported in 1959–1960, which showed that of 7,294 elementary school teachers on the island, 99.4% had not taken any courses in mathematics beyond their own high school preparation in this field. "Since August, 1961, the College of Education of the University of Puerto Rico is requiring six credits in mathematics of all students. This means that even those students who become elementary school teachers after pursuing only two years of college will have some training in college mathematics. The special course designed for this purpose was prepared by the Department of Mathematics in collaboration with the College of Education" (García, 1964).

From 1960 to 1968, Angel G. Quintero Alfaro, first as Under Secretary and then as Secretary of Education, developed a series of innovative projects designed to improve teaching from an activity in which students were passive recipients of knowledge, to an active endeavor more pertinent to the students' reality. As part of his reform, new educational materials were developed and teachers received professional development on how to use them. In the case of mathematics, new materials being developed in the United States were adopted since they differed from the basic skills approach (Quintero Alfaro, 1972). Following the lead of Edward G. Begle's School Mathematics Study Group, the New Math was introduced with an emphasis placed on mathematical terminology (mainly set theoretic), and its symbolism, and its uses. Given the difficulties that students had with this approach, soon the materials were changed to those developed in the newly created Curricular Centers, where teachers in collaboration with university professors created educational materials and helped to implement new teaching strategies in the classroom. Eugene Francis, a professor at the University of Puerto Rico at Mayagüez (UPRM), excelled in this endeavor starting a series of initiatives, both with teachers and with students, that have had a profound impact on the development of the teaching of mathematics.

The highlight during this era was the Department of Public Instruction's realization of the importance of providing for the academic needs of the mathematically and scientifically talented. As a result, the Department founded the specialized school *Centro Residencial de Oportunidades Educativas de Mayagüez* (CROEM) in 1967. CROEM is still considered one of the best public secondary schools on the island, and has an academic relationship with the Mayagüez Campus of the University of Puerto Rico. CROEM is part of three specialized mathematics and science schools, which include two later additions: University Gardens (grades 9–12) and Brígida Álvarez (grades 7–9).

In 1967, the Mathematics Commission of the College Entrance Examination Board created an algebra and trigonometry course, which *Marco Curricular* claims to be a "college-level" course. In 1968, courses in probability, statistics, and analytic geometry were introduced at the secondary level, as well as algebra courses for ninth-graders or talented eight-graders (Programa, 2003).

During the 1970s, a remedial curriculum was developed for students with "limitations" in the learning of mathematics at the elementary level. The *Proyecto Calendario Escolar Continuo*, which divided the year into five "quinmestres" was put in effect. Together with the quinmestres, a series of "Back to Basics" curricular materials were created that divided the curriculum into units to address the individual differences of students, with an emphasis on basic skills (Programa, 2003). Each student could then complete the various units at his or her own pace.

Marco Curricular claims that "studies conducted toward the end of the 70s revealed the risk" of students having only some mechanical facility with basic skills but no understanding of them or knowledge of their applications (Programa, 2003). The Mathematics Program of the Department of Education of Puerto Rico sponsored several projects to develop materials and methodologies that promoted discriminating learning in mathematics. At the University of Puerto Rico, with funding from NSF, Manuel Gómez established a Resource Center for Science and Engineering that promoted methodologies for teaching and assessment emphasizing critical thinking in mathematics. A component of the Center worked with secondary teachers and talented science and mathematics students. Eugene Francis, in Mayagüez, and Ana Helvia Quintero, in Río Piedras, jointly coordinated this effort. They promoted the development of materials for the teaching and learning of mathematics written by UPR professors that emphasized problem solving and critical thinking. The materials used in the Summer Camps for talented youth and in the professional development for teachers are still used today.

The 1990s and Onwards: Emphasis on Problem Solving, the Development of Reasoning Skills, and Critical Thinking

The reform movement in Puerto Rico was bolstered by a $10 million NSF grant in support of the Puerto Rico Statewide Systemic Initiative (PRSSI) and by the National Council of Teachers of Mathematics (NCTM) 1989 Standards and the 2000 Principles and Standards for School Mathematics. *Marco Curricular* states that it has been "nourished" by NCTM's documents. They formed the framework for the 1996, 2000, and 2007 versions of the Puerto Rico standards. The 2007 *Estándares de Contenido y Expectativas de Grado* developed under the direction of Jorge López at the University of Puerto Rico in Río Piedras, are improvements on the highly criticized 2000 version; however, it remains to be seen how the recent Common Core State Standards in the United States will impact the local standards.

Centros Regionales de Adiestramiento en Instrucción Matemática (CRAIM), a division of the Mathematics Department of the University of Puerto Rico in Río Piedras directed by Jorge López, devoted a significant effort to teacher training and the development of didactic materials. CRAIM was instrumental in the translation and adaptation of some of the units of the *Mathematics in Context* series for middle school, developed by Thomas Romberg, *et al.* of the University of Wisconsin at Madison. The units were translated into Spanish and adapted for Puerto Rico. The materials were piloted in a fifth grade of the Jesús Sanabria Cruz public school in Yabucoa with very dramatic results (all students except two ranked in the upper 10% in mathematics understanding according to the standardized tests administered by the Department of Education). CRAIM also developed a complete set of materials for elementary schools as a joint project with Koeno Gravemeijer of the Freudenthal Institute of the University of Utrecht in The Netherlands. Moreover, CRAIM developed materials and activities for talented students, thus contributing to the development of an initiative originated by Eugene Francis related to mathematical olympiads.

The Department of Education (DEPR) has had a notable ambivalence with respect to the methods used to train students for competitions in mathematics. The best students were typically from private schools while the funds for student training in this area were controlled by the DEPR, which naturally is reluctant to invest the monies in the training of students who are not in public schools. This situation has been partly responsible for the often mediocre showing of Puerto Rican students in international competitions. Local competitions, however, were organized very successfully by *Programa de Matemática* and Eugene Francis at their start, and later taken over by CRAIM and by Luis Cáceres in Mayagüez. Finally, CRAIM developed the series *Tesoros de la Matemática* for the DEPR, which consisted of interesting topics in mathematics written informally by the best mathematicians of Puerto Rico for the consumption of students and teachers of mathematics.

Education of Teachers

As noted earlier, the University of Puerto Rico was founded in 1903 from a Normal School. In 1901, this Normal School required only a 6th grade diploma for admission. It offered a two-year program to prepare elementary school teachers, and a four-year program to prepare school principals (López Yustos, 1997). The training of school principals was vital because, at the time, the function was being carried out by unprepared English teachers from the United States whose major qualification was that they could communicate with the Commissioner of Instruction. By 1920 the university continued to offer these programs but had expanded to offer two-year programs for teachers of home economics, two-year programs for high school teachers, and a one-year program that trained teachers for rural schools. There were 117 students in the program for rural teachers and 110 in the program for regular teachers. These programs aimed to correct the extreme shortage of teachers. In 1959, the UPR had campuses in Río Piedras, Mayagüez and San Juan (currently the Medical Center) that emphasized the humanities, engineering, and health sciences, respectively. The UPR system had 14,869 students that year. Of those, 11,654 were registered at the Río Piedras campus, out of which 3,441 were in the School of Education (López Yustos, 1997).

While the number of mathematics educators has risen consistently over the last century, a very restrictive education system, resembling that of North America — but with weak mathematics requirements from the Puerto Rico Department of Education — have combined to create the general perception that teachers do not have sufficient background to teach mathematics content. "During the year 1957–1958, 75.5% of junior high school mathematics teachers in public schools of Puerto Rico did not have a single credit in college mathematics, and 32.7% of senior high school teachers had not taken a single course at college level" (García, 1964). Professional development programs for mathematics teachers were urgently needed. The NSF-sponsored Institute of Mathematics, created by the Inter-American University in 1957, served 35 high school teachers. Besides the three aforementioned courses it offered, it hosted a conference series led by Dr. Derrick Lehmer, chairman of the Department of Mathematics of the University of California at Berkeley. This was the beginning of professional development programs for teachers supported by federal organizations. While initially most teacher training was done by CRAIM and the Resource Center in Río Piedras, professional development programs for mathematics teachers are now island-wide, reaching thousands of mathematics teachers and receiving both state and federal funding.

Luis Cáceres and Arturo Portnoy, from UPR-Mayagüez, have pioneered several educational projects for both teachers and talented students. Their teacher development program, *Alianza para el Fortalecimiento en el Aprendizaje de las Matemáticas* (AFAMaC), is of special interest. In 2004, Leida Negrón, then director of the Mathematics Program in the Department of Education, approached Cáceres with the idea of designing a professional development model that would comply with some new requirements of the NSF Mathematics and Science Partnership. In 2005 he created a program with three main points:

(1) Focus on content — Based on the simple premise that one cannot teach something one does not understand, AFAMaC has focused on insuring that the participating teachers gain an in-depth understanding of the mathematical and scientific concepts needed to be an effective teacher.
(2) Substantial contact hours — Realizing that teachers need the support of and exposure to mathematical professionals, AFAMaC offers 160 contact hours of workshops during an academic year, 80 of which take place in the summer.

(3) Sustainability — Professional development ought to be continued for several years and the results must be evident in the academic achievement of the participating teachers' students.

From 2004 to 2009 another major initiative, the *Alianza para el Aprendizaje de Ciencias y Matemáticas* (AlACiMa) developed alongside AFAMaC. The result of an NSF $35 million grant, AlACiMa had considerably more funding sufficient to impact education across Puerto Rico. They engaged 155 K-12 schools by offering professional development to teachers and creating 28 professional resource centers (AlACiMa, 2010). An interesting component of AlACiMa is that they "adopted the concept of learning communities" as a core principle. Participants at all levels were encouraged constantly, through workshops and other activities, to reflect upon their own learning (AlACiMa, 2010). "While the main thrust of AlACiMa was the professional development of in-service mathematics and science teachers," AlACiMa also included interventions "to strengthen the preparation of pre-service teachers" (AlACiMa, 2010). The funding for the program was provided by an NSF grant. Once that grant ended, the program has continued with alternate sources of funding — basically federals funds from the Puerto Rico Department of Education. It should be noted that grant termination is the foremost problem with government-funded initiatives.

Comparison of Mathematics Achievement in Puerto Rico and the United States

The National Assessment of Educational Progress (NAEP) consists of a series of standardized tests with assessments in mathematics and other subject areas. An English language version is administered nationally in the United States and a Spanish language version is administered in Puerto Rico. In 2003 and 2005, it served as a common metric for the US and Puerto Rico by reporting three achievement levels for students: basic, proficient and advanced. Table 1 (Baxter, *et al.*, 2007), presents results in mathematics for fourth and eighth graders in public schools of the United States, of Puerto Rico and of Mississippi (the state with the lowest scores) for 2005. The results in 2003 were similar.

In 2007, the last year for which results are available for Puerto Rico, comparisons using the above metric between the US and Puerto Rico were no longer made. Comparisons of the raw scores by subject area for fourth

Table 1. Comparison of NAEP results in 2005 for the US, Puerto Rico and Mississippi.

	Puerto Rico	United States	Mississippi
Percent of fourth grade students with basic level or better	11%	79%	69%
Percent of fourth grade students with proficient level or better	0%	35%	19%
Percent of eighth grade students with basic level or better	6%	68%	52%
Percent of eighth grade students with proficient level or better	0%	28%	14%

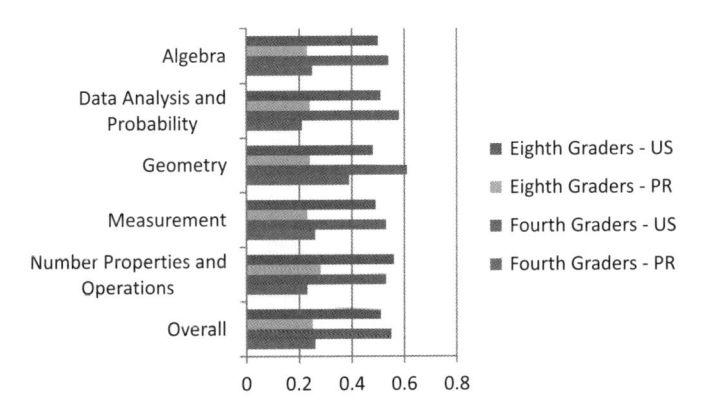

Figure 2. Comparison of NAEP results in 2007 for fourth and eighth graders in the US and Puerto Rico.

and eighth graders in the United States and Puerto Rico are shown in Figure 2 (Dion, Kuang & Dresher, 2008). These data speak to the worrisome state of mathematics education in Puerto Rico and the lack of preparedness of many secondary school students to enter STEM fields at the university level.

Some Geopolitical Aspects

In the early 1940s, a new law was passed that created the Superior Teaching Council that went on to become the Council on Higher Education: the governing body for all institutes of higher education in Puerto

Rico. In 1947, the initiative "Operation Bootstraps" advanced the right of universal education. Over the course of the next several decades, this resulted in new private and public university campuses in order to meet the greater demand for teachers and other professionals. In 1965, a law was passed that significantly expanded the powers to the Council on Higher Education, which included the power to "adopt and to promulgate the norms for the accreditation of private higher education" (López Yustos, 1997). Accreditation of the Council on Higher Education was necessary to obtain federal funds. Federal support has become indispensable to the private universities of Puerto Rico. As a result, antagonism developed between private and public universities as private institutions felt that their initiatives would only be approved if they resembled those already in place at public universities.

This antagonism between private and public universities, as well as the role of government in the UPR system, continues to be felt today. Puerto Rico's political background has resulted in divided political parties and turbulent power transitions which can affect mathematics education. Despite consistent recommendations from accreditation agencies that the University of Puerto Rico be shielded from politics, when a new governor of Puerto Rico is elected, this commonly results in the replacement of the president of the UPR system, followed by all campus rectors and faculty deans. In cases where the board of regents does not accommodate the government agenda, the legislative branch has passed laws that realign the board of regents in order to pursue its political needs. A recent case occurred in 2010, when the New Progressive Party increased the number of members from 13 to 17 to force a majority. This was bound to change after the general election of 2012, when the New Progressive Party lost to the Popular Democratic Party. Hence, immediately upon assuming power in 2013, legislation was passed that dissolved the former board of regents replacing it with a new "government board". The new board again has 13 members; however, its members now answer to the ruling Popular Democratic Party (Banuchi, 2013).

An interesting case study of how government intervention can impact mathematics education involves changing the admission standards of the UPR in 1993. The initial admissions standards were a balance between standardized test scores and grades. As this formula often favored better prepared students from private schools, public universities felt that they were handling the majority of Puerto Rican students from low income families and wanted more financial support from the government and political pressure mounted to make the UPR system more accessible to students from lower income families. As a result, the admission standards were adjusted to emphasize grades more than standardized test scores. To assess how this has impacted secondary school and university mathematics, two professors from the Department of Mathematical Sciences at the University of Puerto Rico at Mayagüez (UPRM) compiled data from the UPRM which is being published for the first time in this article. Mayagüez is the major engineering campus of the UPR system; hence changes in the mathematics preparedness of secondary school students in Puerto Rico should be quite evident and easily observed at this campus.

Figure 3 shows the rise in the admission index as a result of this change in the formula. Figure 4 shows mathematics achievement, as measured by the Puerto Rico College Board Mathematics Achievement test, since this change was implemented. The admission index would show that the students are becoming better prepared while the math achievement scores indicate that they are becoming less prepared. One cause for this contradiction could be grade inflation. As grades rather than knowledge have become the

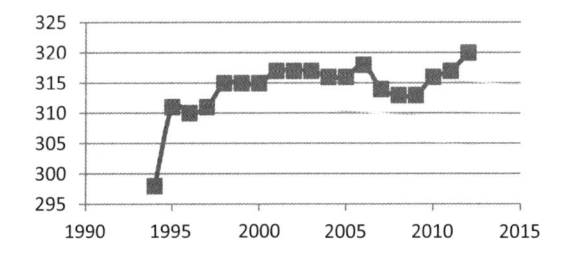

Figure 3. Average admission index for entering freshmen at the UPRM since 1994.

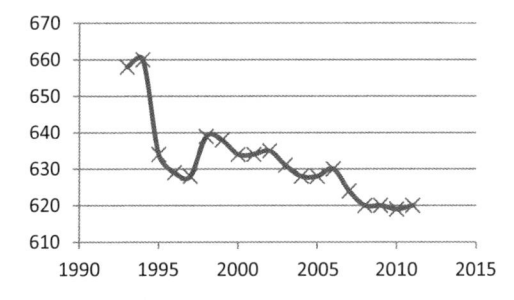

Figure 4. Average math achievement test scores at the UPRM since 1994.

Table 2. Distribution of high school GPAs by math grades received by freshmen at the University of Puerto Rico at Mayagüez from 2008 to 2011.

Grade	Observations	Average High School GPA	Standard Deviation	Minimum	Maximum
A	673	3.87	0.21	2.61	4.00
B	936	3.77	0.29	2.26	4.00
C	1408	3.69	0.33	2.19	4.00
D	388	3.61	0.36	2.43	4.00
F	878	3.54	0.39	2.29	4.00
W	1480	3.62	0.37	2.30	4.00

key to admission to the university and obtaining a better life, secondary school teachers may be giving higher grades for worse performance.

Table 2 shows the average high school grade point average for freshmen who receive an A, B, C, D or F in their freshmen year math courses. The graph indicates that while almost everybody appears to have been an A-grade student in high school, almost half are destined to achieve D's, F's or W's in their freshman year mathematics courses.

High school students tend to focus on getting into the university and to worry about doing well in university courses only later, once they are at university (if at all). The data indicate that this change in admission standards seems to have shifted the incentive for students, teachers and parents away from knowledge and towards higher grades. The data show that high school GPAs have gone up but mathematical knowledge has gone down. It is not the intention of the authors to address the use of standardized tests but rather the consequences that can result when political expediencies rather than long term development motivate mathematics education policies. As a result of its turbulent political background, Puerto Rico can be vulnerable to such phenomena.

Gender Issues

In the United States, the NAEP exam results indicate that males perform better than females in mathematics. This difference, however, is not seen in Puerto Rico where females perform as well as males. In fact, females perform approximately 5 points higher on average in the Geometry and Spatial Sense section. This lack of a gender gap continues to be seen in post-secondary institutions. Based on data from Río Piedras and Mayagüez, approximately 36% of graduates in engineering programs are women and approximately 60% of graduates from other science and mathematics programs are women. While obviously there is still a gap in engineering fields between male and female graduates, it should be noted that this gap is far less than anywhere else in the United States. There appears not to be as significant a gender issue in Puerto Rico with respect to mathematics education as there is in most other countries in the Western Hemisphere.

Mathematical Olympiads

Eugene Francis and the *Asociación Puertorriqueña de Maestros de Matemática* (APMM) introduced Mathematics Olympiads to Puerto Rico through local competitions in the different educational districts of Puerto Rico and a championship in which the finalists from the regional competitions would participate. Francis organized the tests with the Olympiad problems and was in charge of the grading and the results. At the time of his death he was working on a book of problems that were to include all of the championship Olympiad problems with solutions for teachers and students interested in these mathematical activities.

This work has been continued and has, in fact, expanded significantly. In 2001, *Olimpiadas Matemáticas de Puerto Rico* (OMPR) was pioneered by Cáceres and Portnoy as an initiative for identifying and nurturing mathematical talent across the island. The number of participating students has grown steadily since its inception; currently over 6,000 students participate each year.

OMPR has been able to awaken interest in mathematics and STEM disciplines in general. During the last decade, Mayagüez has become a major hub of mathematical activity on the island. Hundreds of students and parents drive for hours across the island to attend the different sessions OMPR holds throughout the year. As a measure of success, besides having earned over 60 international medals and other awards, OMPR prides itself in the fact that many of their former Olympians are accepted by premiere universities in the United States. For example, in 2010, 10 OMPR students were accepted at MIT and in 2013 seven students were granted early admissions. Cherry Gong, who represented both Puerto Rico and the United States at the International Mathematical Olympiad (IMO), graduated from Harvard in 2011 and is pursuing doctoral studies in mathematics at MIT. Apparently, MIT is the preferred institution of the current Puerto Rican generation of Olympians. In 2010, George Arzeno, now also at MIT, earned the first gold medal for Puerto Rico at the IMO. These college students have created a support network to help younger students gain access to some of the best mathematical institutions and minds.

As part of their efforts, OMPR hosted the *Olimpiada Centroamericana de Matemáticas* in 2010. In 2014 they will host the International Kangaroo Mathematics Contest and in 2015 the *XXX Olimpiada Iberoamericana de Matemáticas*. OMPR also has shared its model of competition with other countries in the region. One of the main results in this connection is that Jamaica has begun a program based on OMPR.

The success of OMPR is a harbinger of a new era of "liberating pedagogy" that will increase the number of mathematicians and scientists — future Fields Medalists and Nobel Laureates — from Puerto Rico. This will be the pride of a people.

Math Circles

Although support and admiration for the accomplishments of OMPR are widespread, some educators have expressed philosophical reservations about competitions. Robert Kaplan, for example has written:

> Nothing, you'd think, would be more foreign to a mathematician than transforming his art into a contest. [...] Where has the leisure gone in which young mathematicians used to think through difficult thoughts? [...] Is the goal of understanding to be replaced by merely triumphing over some roomful of other people? [...] Isn't man against the gods a higher drama than man against man?

In this spirit, Héctor Rosario from the University of Puerto Rico at Mayagüez founded the Puerto Rico Math Circle in 2012. This initiative grew out of the need to provide continuous support for students who wish to engage in high-caliber mathematical thinking. In 2013, the University of Puerto Rico at Mayagüez and the Puerto Rico Math Circle hosted *Circle on the Road*, the annual gathering of math circles in the United States. This project is promoted by the National Association of Math Circles and sponsored by the Mathematical Sciences Research Institute.

To the advocates of mathematics competitions, Harvard mathematician Barry Mazur has proposed the following variant, as recorded by Kaplan and Kaplan (2007):

> If there are to be competitions, might they not be in a more fruitful style? Barry Mazur asks why all — or the most telling — questions couldn't be of the form: "What do you think about this?" He calls the present form of competitions "closed": here is the problem we have made for you, solve it. He suggests instead an "open form": tell us something of your own; surprise us.

Rosario has proposed a week-long conjecture-based mathematics competition. According to a former director of the National Association of Math Circles, David Auckley, and the current president of the World Federation of National Mathematics Competitions, Alexander Soifer, no such contest exists. In such a competition — more like a fellowship — students would be introduced to some "open form" topic, and be asked to produce conjectures springing from it. Naturally, they would need to prove or disprove their claims.

Conclusion

In their concluding remarks to *Out of the Labyrinth: Setting Mathematics Free*, Robert and Ellen Kaplan appeal to the human condition: "We have an inborn urge to know, to make, and to enjoy the world around us, that neither fatigue nor fear can long suppress," so let us "rise out of the labyrinth and see the whole world spread below us" (Kaplan and Kaplan, 2007). Puerto Rico has taken some steps to liberate mathematics and its education from the hesitance instilled by centuries of colonialism. Still, much has to be done to improve the curriculum material as well as the teacher preparation programs. In order to be successful, we cannot limit ourselves to replicating the approach of the United States; that attempt at replication has not been effective, as the results of international studies consistently show. Puerto Rico must transcend the effects of imperial hubris if it wishes to advance a culture that values mathematical ideas. That should be the goal of our mathematics education agenda.

Bibliography

Alianza para el Aprendizaje de Ciencias y Matemáticas (AlACiMa), *Sharing Our Journey to Improve Mathematics and Science Education*, San Juan, 2010.

Banuchi, R. (2013). *Gobernador firma la ley que elimina la Junta de Síndicos*, El Nuevo Día, April 30, 2013

Baxter, G. P., Bleeker, M. M., Waits, T. L., & Salvucci, S. (2007). *The Nation's Report Card: Mathematics 2003 and 2005 Performance in Puerto Rico — Highlights* (NCES 2007–459). US Department of Education, National Center for Education Statistics, Washington, DC: US Government Printing Office.

Cordero Avilés, G. (2013). García Padilla Propone Reducir Junta de Síndicos de la UPR. *Elnuevodia.com*. El Nuevo Día, 12 Mar. 2013. Web. 24 Apr. 2013.

Coll y Toste, C. (1910). *Historia de la Instrucción Pública en Puerto Rico hasta el año de 1898*.

Dion, G. S., Kuang, M., & Dresher, A. R. (2008). *The Nation's Report Card: Mathematics 2007 Performance of Public School Students in Puerto Rico — Focus on the Content Areas* (NCES 2009–451). National Center for Education Statistics, Institute of Education Sciences, US Department of Education, Washington, DC.

Educational Survey Commission of the International Institute of Teachers College. (1926). *A Survey of the Public Educational System of Porto Rico*. New York: Columbia University.

García, M. (1964). Two decades of mathematical education in Puerto Rico. *The Mathematics Teacher*, 57(4), 235–239.

Gómez-Diez, F. J. (2000). *La educación jesuita en Puerto Rico (1858–1886): Entre la sustitución del Estado y el Seminario colegio*, Mar Oceana 5, 91–124.

Kaplan, R., & Kaplan, E. (2007). *Out of the Labyrinth: Setting Mathematics Free*, Oxford University Press, Oxford.

López Yustos, A. (1997). *Historia documental de la educación en Puerto Rico*. Hato Rey, Puerto Rico: Publicaciones Puertorriqueñas Editores.

Maryland Council on Economic Education (2006). *Carlisle Indian Industrial School*.

McCoy, A. W., & Scarano, F. A. (Eds.) (2009). *Colonial Crucible*. Madison, WI: The University of Wisconsin Press.

Moscoso, F. (2003). *Sociedad y Economía de los Taínos*, Editorial Edil,

National Education Association (2006). *A Report on the Status of Hispanics in Education: Overcoming a History of Neglect*, Washington DC.

Navarro-Rivera, P. (2009). The imperial enterprise and educational policies in Colonial Puerto Rico, in *Colonial Crucible*.

Osuna, J. J. (1949). *A History of Education in Puerto Rico*, Editorial de la Universidad de Puerto Rico, Río Piedras.

Programa de Matemáticas, Departamento de Educación de Puerto Rico (2007). *Estándares de Contenido y Expectativas de Grado.*

Programa de Matemáticas, Instituto Nacional para el Desarrollo Curricular, Departamento de Educación de Puerto Rico (2003). *Marco Curricular.*

Quintero Alfaro, A. G. (1972). *Educación y Cambio Social en Puerto Rico.* San Juan: Editorial de la Universidad de Puerto Rico.

Via, M., Gignoux, C. R., Roth, L. A., & Fejerman, L., Galanter, J., et al. (2011). History shaped the geographic distribution of genomic admixture on the island of Puerto Rico. *PLoS ONE, 6*(1): e16513. doi:10.1371/journal.pone.0016513.

About the Authors

Héctor Rosario is a professor of mathematics at the University of Puerto Rico at Mayagüez and holds a Ph.D. in mathematics education from Columbia University. He is a member of the advisory board of the National Association of Math Circles, and is the representative for Puerto Rico in the *Red de Educación Matemática de América Central y el Caribe* (REDUMATE) and the Inter-American Committee of Mathematics Education (CIAEM).

Daniel McGee worked 18 years at the University of Puerto Rico at Mayagüez where he directed numerous projects involving developmental mathematics, high school mathematics, pre-service teacher training, and materials for pre-calculus and calculus classes. His materials are used in universities and high schools throughout Puerto Rico. He has recently become director of the Kentucky Center for Mathematics where he continues his tradition of innovation and leadership.

Jorge M. López is a professor of mathematics at the University of Puerto Rico in Río Piedras and has conducted research on commutative harmonic analysis and on the history of mathematics and its relevance to mathematics education. He is the founder of the Regional Training Centers in Mathematics Instruction (CRAIM) and has developed, in collaboration with the Freudenthal Institute and the University of Wisconsin–Madison, a complete Mathematics in Context curriculum for primary and middle schools in Puerto Rico. He has been a visiting scholar at Harvard, UC Berkeley, and Utrecht. Recently, he was the first mathematician nominated to the Academy of Arts and Sciences of Puerto Rico.

Ana Helvia Quintero is professor of mathematics at the University of Puerto Rico in Río Piedras and holds a Ph.D. in the learning of mathematics from MIT. Her main interest is mathematics education. She has been involved in several projects aimed at improving the school system. From January 2001 to June 2002 she was Under Secretary of Education of the Commonwealth of Puerto Rico. Based on that experience, she published a book in 2006, *Muchas reformas, pocos cambios* (Many Reforms, Hardly Any Changes).

Omar A. Hernández has been a professor in the Department of Graduate Studies in the School of Education at the University of Puerto Rico in Río Piedras since 2007. He holds degrees from the *Universidad Pedagógica Nacional* in Colombia, Purdue University, and the University of Puerto Rico, where he earned his doctorate in curriculum and instruction. Originally from Colombia, he has spent over 20 years working at several academic institutions across Puerto Rico.

Chapter 16

TRINIDAD and TOBAGO: Mathematics Education in the Twin Island Republic

Shereen Alima Khan and Vimala Judy Kamalodeen

Abstract: This article reviews the practices and trends in mathematics education in Trinidad and Tobago from pre- to post-colonial periods. It highlights critical factors that shaped these practices and the efforts made at globalizing the curriculum so as to improve student performance.

Keywords: Mathematics education; history of education in Trinidad and Tobago; colonial education; mathematics curriculum; mathematics performance.

Trinidad and Tobago is a twin-island republic with a history of mathematics education couched by a legacy of British colonial education systems. While our development is similar to that of other Caribbean islands, our unique multicultural mix influences educational initiatives and outcomes. In this chapter, we describe the history of mathematics education from colonial times to present day and include a discussion of the impact of post-colonial perspectives on mathematics curriculum development and implementation, assessments and teacher education. We conclude with a discussion of future directions and initiatives to democratize the curriculum, indigenize resources, and develop greater equity in assessment.

Native Peoples, Education Practices, and Colonization

In order to fully appreciate the trends in mathematics and its teaching in Trinidad and Tobago, it is important to understand the history of the education system in this twin-island in the Caribbean. Educational policies have been controlled and shaped by three major forces: the role of religion in delivering education to the masses, the importation of curriculum content, and the social, political and cultural agendas of the providers (MOE, 1998a).

Indigenous People: Pre-Columbus Era

A brief history of Trinidad's early inhabitants is necessary to understand the cultural heritage of the people. This view is supported by London (2002a) who emphasized the need for context and further explained that to reconstruct the past, events must be related to other events.

The first inhabitants of Trinidad were peoples from South America who traveled there about 6,000 years before Christopher Columbus arrived in the West Indies. The two major tribes that settled in Trinidad were the *Caribs* and *Arawaks*. Although there is no documented evidence of an educational system during their reign, the customs and practices of both groups clearly indicate that they were adept in some practices that would have incorporated the use of mathematics. The Caribs were particularly skillful in the crafting of canoes, pirogues and tools. The use of measurement and geometry can be inferred from their construction skills, as evidenced in the structure of their homes and the design of their weapons (bows and arrows). These Amerindians had no knowledge of metals and so their tools were of polished stone, bone, shell, coral, or wood (The Story of the Caribs and Arawaks, 2014). There was no currency but trade was quite common among the tribes. Items of exchange used in bartering comprised pottery, gold and even women. Later on, tobacco cigars were used as a form of currency. These native peoples were unable to withstand the rigor of British rule and soon became a minority in the country.

British Rule and the Christian-dominated Curriculum (1798–1962)

During slavery, there was no formal provision of education for the slaves. Prior to the abolition of slavery in 1823, the Cabildo (the Town Council) opened schools in Trinidad. The curriculum focused mainly on reading, writing, and arithmetic, as well as spelling, but the development of Christian morality and knowledge of the English language were considered most important. The children of the free classes alone received instruction in private schools that were usually headed by clergymen. The more affluent ones were educated privately at home or in the public schools and universities of Great Britain.

The elements of arithmetic were taught but this was done less frequently than writing. In fact, additional fees were levied on children who opted to take subjects other than religion such as arithmetic. The mode of teaching was "enthroned rote learning and, in most cases did not allow for or encourage understanding of what was committed to memory ... the method of *repetez sans cesse* became an established route to what was considered excellence in schooling" (London, 2002b, p. 113). There were periods on the timetable devoted to memorization of arithmetical tables only (Bacchus, 1994).

After the abolition of slavery, religious bodies established private schools. The Roman Catholic Church and the Church of England continued to invest in education. In 1848, missionaries of the Presbyterian faith came from Canada to instruct recent immigrants and were particularly successful in establishing schools among those of East-Indian origin (The Maurice Report, 1959). As such, primary and secondary schools were set up for perpetuating the religions of the religious denomination that administered the school. Curricula were set by Boards of Governors and focused on mimicking the curriculum that dominated British schools at that time and on the furthering of religious knowledge. Thereafter, Muslims and Hindus were allowed to establish schools alongside the existing Catholic, Anglican, and Presbyterian schools.

In 1851, Governor Harris, in contemplating the difficulties arising from differences of race, language and religion, decided that the government should embark on secular education through the establishment of ward schools located in different geographical regions (Williams, 1962). He recognized that education had a crucial role to play in building a society out of the different groups that populated the island. He felt that the descendants of the African slaves and the indentured Indian laborers needed more than the proselytized education that the religious bodies were offering.

In the Harris regime, all children in government institutions received schooling free of expense while the ward met the cost of textbooks and other materials. He also appointed a School Inspector to oversee the schools in the ward, established a training school for teachers of primary schools to be maintained by public funds and a Board of Education to manage the schools in each ward. Selection of textbooks had to be sanctioned by this board.

The pre-occupation with organizational matters in this era left little room to focus on the curriculum. Up to this time, there was no documented evidence of what was taught and the manner in which it was taught. The first report on school practices was prepared in 1869, with the appointment of Sir Patrick Keenan as the secretary of state for the colonies. He was asked to report on the state of education in Trinidad. He visited 76 educational institutions and examined 3,103 pupils and reported on both organizational as well as curricular matters. The Keenan Report was influential in curriculum reform and by 1870 Trinidad adopted a dual system of education. Schools were divided into two classes: (a) Government schools, to be maintained by public funds, and (b) Government-Assisted or denominational schools, partially funded by the government. This dual system still exists today and mathematics teaching and performance varies distinctly across these two systems.

In many ways, the history of education in Trinidad and Tobago is similar to that of other English-speaking countries in the Caribbean, such as Jamaica and Barbados. The development of education was closely tied to British colonial rule with structures of schooling and education following and supporting existing societal structures (Altbach, 1971). Religious groups independently established schools to further perpetuation and dominance of those religious groups and today denominational schools are still actively involved as providers of education to the nation's youth. English is the principal language spoken with bare remnants of heritage languages in icons such as street names and people's names.

Post-Colonial Era

In 1869, Keenan reported on the curriculum by examining every child in every school in the subjects taught in the curriculum. He scored each child using a scale of 0–10 for each subject and explained that the score was a representation of the individual child's answering relative to the proficiency, which might be reasonably expected. He found that the average mark obtained in arithmetic was only 1.6, compared to 3.6 in reading, 3.3 in spelling, 2.5 in dictation, and 3.6 in writing. He further noted that there was no appreciation of the meaning of what was read and little attempt at expression. He was very severe in his remarks on the teaching and learning of arithmetic, perhaps too severe since the questions posed to children and the language of the problems were not always culturally relevant. The example below shows a typical problem given to students in upper primary:

> This is the most extraordinary and disappointing result of all ... My examination was of a most elementary character. It consisted generally of an easy exercise in notation, in each of the simple rules and in proportion. In thirty of the ward schools, I did not find half a dozen children capable of working such a simple sum as the following:
>
> *A quarree contains $3\frac{1}{5}$ English acres; what is the rent of 5 quarrees at £1 6s 6p the English acre?*
>
> Anything so awkward, so destitute of intellectual resource, and so wanting in the power of dealing with numbers, as characterized by the performances of pupils who tried to do this little sum, I never witnessed (MOE, 2001, p. 32).

With respect to the teaching methods, the report continued:

> At first I was perplexed whether to attribute the defects to the instruction of teachers or to the quality of the pupils' intellect, but I soon found that the masters pursued a system of purely mechanical teaching which could not possibly yield satisfactory results, and that pupils were afflicted with a positive inaptitude for arithmetical reasoning. It is not improbable that this inaptitude is only a consequence of rustiness of intellect, produced by neglect and unskillful instruction. Curiously enough, the best answering in arithmetic was made by the pupils in the girls model school ... the mark obtained was 5.4 (MOE, 2001, p. 32).

He commented also on the nature and importance of mathematics in the curriculum when he said:

> From this low proficiency in a subject which is universally regarded as an essential branch of education, and which is the most practical, the most useful, the most disciplinal, and the most popular of all the elementary school lessons, I came to the conclusion that the operation of the ward schools was certainly not promoting the learning or stimulating the intelligence of the people of Trinidad

and Tobago. Geometry, algebra and mensuration were not taught in the ward schools (MOE, 2001, p. 33).

The curriculum of the ward schools was deficient when compared to the denominational schools, in which there was more variety and higher standards, perhaps because these denominational schools were more established and well resourced. For example, at the College of the Immaculate Conception — still known as CIC today — an all-boys Catholic school, mathematics was taught in three courses: a preparatory course, a commercial course, and a classical course. At this school, the boys were sons of proprietors, merchants, professionals, and members of the public service. Those who were destined for mercantile pursuits took the commercial course and the arithmetic was tapered to prepare students to become good clerks or accountants. The classical course extended these areas to the study of great ancient authors and fathers of the Church. Latin, Greek, English, Spanish and German were added and the boys were proficient in translating from one language to another.

Keenan had much to say on the teaching of mathematics at the secondary level, which he felt was a weak point in the colleges. He noticed that the boys were not exposed to sufficient mathematics at the first two levels and felt that there should have been a full course in algebra, plane and spherical geometry, mechanics, and trigonometry at this level. He concluded that although the boys were proficient in calculating rapidly and accurately, their mathematical capacities were kept shallow, or only feebly developed (MOE, 2001). This critique impacted future curricula and the development of primary and secondary curricula followed.

Development of Mathematics Curricula in Primary and Secondary Schools

Primary

The mathematics curriculum grew from the teaching of arithmetic to the teaching of mathematics, thus incorporating newer topics to accommodate changes dictated by the internationally acclaimed mathematics revolution of the seventies. This period, also known as the New Mathematics era, was marked by tremendous advances in research and automation brought about by the introduction of machines and digital computing devices (Nichols, 1968).

The first formal primary school curriculum in Trinidad was created in 1975. The government gained financial and technical assistance from the Inter-American Development Bank to sponsor this project. The philosophy espoused in the introduction was rather brief; it stated:

> Through Mathematics, the child is provided with a fundamental foundation for logical thinking and learns to appreciate science, philosophy and other areas of learning which influences his growth and development in the society in which he lives.

Thus the purpose of mathematics education in schools seemed to be focused on the development of logical thinking for personal development and contribution to society at large. The five strands in this curriculum were number, measurement, money, space and shape, and statistics. In keeping with the thrust towards New Mathematics, topics like sets, transformation geometry, and statistics and probability were introduced for the first time. The treatment of number remained rigorous and topics like fractions were introduced quite early. Extensive work on ratio and proportion, area, decimals and topics in number theory were also features of this curriculum.

The 1975 syllabus can be described as content driven, with no mention of the process goals in mathematics. Emphasis on mathematics as a tool to service the sciences and other occupations was clear. There were no suggestions for teaching and only scanty suggestions for use of resources. The scarcity of instructional materials and the poor quality of textbooks had an impact on the effectiveness of the delivery of the curriculum (Ramsey, 1998).

In 1999, the government once more secured a World Bank loan to review the primary curriculum. The Director of Curriculum Development at the time, Lloyd Pujadas, commented on the major weaknesses of the previous curriculum and an extract of his comments is given below.

> Widespread teaching through the abstract mode with emphasis on drill and rote learning, lack of hands-on activities in the development of mathematical concepts, the inability to deliver the curriculum for each class within the expected time frame and the failure to take into account students' experiences and cultural context in the delivery of the mathematics content (Primary School Mathematics Syllabus, 1999).

The new curriculum emphasized process goals, such as communication, connections, reasoning, and problem solving. The new syllabus blended the psychology of learning Mathematics drawing from the work of Piaget, Dienes, Bruner, Skemp, and Van Hiele to create a constructivist perspective in methodological approach. There were philosophical statements relating to conceptual understanding, developing positive dispositions, and critical thinking. The new curriculum also addressed the content overload. Some of the topics omitted were the area of a circle, ratio, probability, and some transformation geometry (enlargement and shearing). The major change was the slowing down of the abstract topics by adopting a spiral approach. Fractions were spread over the five years and computation of decimals was delayed to the upper levels. Algorithms were introduced in stages using more informal approaches involving the use of concrete materials.

This 1999 syllabus was developed to reflect a more child-centered, constructivist approach, with the teacher providing a learning environment where students search for meaning, appreciate the value of mathematics in all aspects of life and develop a spirit of inquiry and inventiveness (Pariag, 2008). Yet, teachers sampled in the Victoria educational district felt that the mathematics primary syllabus was too wide and that students were not ready to grasp concepts therein (Lalbeharry, 2008). Students' incapability of solving problems and their inability to express themselves clearly and precisely using scientific language were also areas of concern (MOE, 1998b).

Due to ongoing concerns of cultural relevance and the need for indigenous content, a revision of the 1999 Primary Curricula was scheduled to take place in 2007, but due to a number of political hurdles, it actually took place in 2012.

Secondary

The most powerful influence on the secondary mathematics curriculum was the examination system. It is said that assessment drives learning, but this curriculum was derived in response to preparation for external examinations from Britain. The General Certificate in Education (GCE) was originally introduced in England, Wales and Northern Ireland in 1951, replacing the older School Certificate (SC) and Higher School Certificate (HSC). It was intended to cater to the increased range of subjects available to pupils since increasing the school departure age from 14 to 15 in 1947. It is an academic qualification that examination boards in the United Kingdom and a few of the former British colonies or Commonwealth countries like Trinidad and Tobago and the wider Caribbean, conferred to students.

In 1968, the Parliament adopted the first major education plan for the country, which proposed expansion targets for various levels of education (MOE, 1974). The post-independence era saw a shift of focus from primary to secondary education. During 1968–1983, secondary education experienced rapid growth; the secondary enrollment rate increased from 24% in 1970 to 83% by the late 1980s (Tsang, Fryer & Arevalo, 2002). Entry to secondary schools was gained based on the results of the Common Entrance Examination which resembled, in form and effects, the reputed Eleven-Plus Examination in the United Kingdom.

The GCE had two stages, the Ordinary Level, which is taken at the end of the fifth year of secondary school, and the Advanced Level, which is taken two years after the Ordinary Level. Only the highly successful students move on to the Advanced Level. Preparation for the Advanced Level is done in the sixth form, a characteristic feature of the British system of education. This certificate is used for university admission and scholarships were awarded on the results of these examinations.

Around the time of independence in Trinidad and Tobago, the business of all the secondary schools was to prepare candidates for the General Certificate of Education (GCE) exams. Pupils were not allowed to take subjects in which their achievement scores in the earlier years had been low. Hence, those who were not performing well in mathematics at the lower levels of the secondary system were not permitted to take it at the upper levels. Although, most students were allowed to take five or six subjects at one sitting of the examination, mathematics was not a compulsory examinable subject in many schools.

The Thrust of Caribbean Examinations

As more students were gaining access to secondary schools, the number of candidates entering the Cambridge Examinations was growing, not just in Trinidad and Tobago, but in the neighboring Caribbean territories as well. At the Second Conference of Heads of Government of Commonwealth Caribbean Countries held in Jamaica in January 1964, it was agreed that the setting up of a Caribbean Examinations Council (CXC) was a fit and proper subject for regional cooperation. Members of the Council were drawn from the 16 territories and the region's two universities, the University of the West Indies and the University of Guyana. The major role of the Council was to conduct such examinations as it may deem appropriate and award certificates and diplomas on the results of examinations so conducted.

After much deliberation the council was established in 1972, and in 1979 the Caribbean Secondary Education Certificate (CSEC) replaced the existing Ordinary Level GCE Cambridge Examinations. In 1998, the Caribbean Advanced Proficiency Examinations (CAPE) replaced the Cambridge Advanced Levels Examinations. This was a significant milestone in the history of Trinidad and Tobago, especially after becoming a Republic in 1976. The CXC council relied on technical expertise from Cambridge University in the setting up of the examination system. The pioneers of the CXC were trained by Cambridge in the preparation of examination papers and the development of mark schemes. In the early years, a member from Cambridge was present at the CXC marking exercise.

The shift in examination administration resulted in the inclusion of local teachers in the examination process for the first time. Secondary school teachers in the Caribbean participated in the development of their syllabi and in areas such as paper setting and script marking. This process continues to today. CXC now offers examinations in 33 subjects at the CSEC level (formerly O Level) and 25 subjects at the CAPE level (formerly A Level). Subject panels comprising representatives from the various territories oversee all matters relating to syllabus development and examinations. In the 2011 annual report, the chairman made the following comments:

> CXC is one of the shining examples of functional cooperation in the Caribbean. It is a collaborative effort among our Governments, through their Ministers and Permanent Secretaries; our many resource persons — markers, subject panellists; the amazing staff of CXC, at Headquarters in Barbados and at the Western Zone Office in Jamaica; and the people of the Caribbean who have supported our examinations for more than three decades (Harris, 2011).

CXC has opened up opportunities for professional development for teachers of the region who seek to further their studies in areas like Curriculum Development/Design and Measurement and Evaluation in order to take up vacant positions with CXC. Regional integration was also an outcome of the new examination system, as CXC uses the system of "cottage marking" where all scripts are marked by a team assembled in one of the regional venues for a period of at least two weeks. This inclusion allowed teachers to be examiners and to have an important voice in critical decisions affecting syllabus development. It also had a direct impact on their classroom teaching and examination preparation.

The first Caribbean Mathematics syllabus was developed in 1975 and five years later mathematics was examined for the first time as a CXC subject. The Caribbean Mathematics syllabus was wider in scope than the Cambridge syllabus at that time. Topics like reasoning and logic, standard deviation, quadratic inequalities, and application of vectors were present as optional topics. As the years went by, the syllabus was revised and some of these topics were omitted. Further changes saw the removal of earth geometry, plans and elevations, and transformations such as shearing and stretching.

The general trend in revising the curriculum was to reduce the content. The only addition was the inclusion of problem solving by introducing a question on investigations. So far there is no school-based assessment (SBA) in mathematics but all other CXC subjects (except for English language) have an internal assessment. The SBA in mathematics was suspended when the subject panel representative from Trinidad presented a case for its omission signed by 81 teachers from Trinidad (Khan, 1995). The council listened and decided to delay its implementation until the members of the territory were ready. For the first time, Trinidadians understood that they had a voice in curriculum matters, more so, as Trinidadian examination candidates make up 25% of the regional participants.

Development of Textbooks

Up to the 1950s, most of the teaching materials were foreign-based as they were aligned to British curricula. The textbook *A Shilling Arithmetic* (and later, *A New Shilling Arithmetic*), published in 1866, was the first textbook used. The first couple of pages presented a set of tables and it was expected that these had to be committed to memory. There were multiplication tables and tables of conversion of measures for money, weight, length, surface, capacity, and time. Money was measured in pounds, shillings, and pence. This was consistent with the currency used in the colony and all other measures followed the British Imperial System.

The presentation of the material in the text followed a fixed sequence. In each section, there is a brief explanation of the concept/principle to be learnt, followed by one worked example. Exercises consisted of mechanical problems and word problems, many of these requiring higher levels of understanding of the concepts. There was a notable absence of worked examples involving word problems but these were present in the exercises. The language structure in the word problems was quite demanding for pupils who were still learning the English language.

The goals of teaching Arithmetic were expressed in the foreword of the book, in which an extract was quoted from Dr. Whewell's work on *A Liberal Education* (Colenso, 1866) and the focus on boys doing arithmetic is noted.

> As the basis of all real progress in mathematics, boys ought to acquire a good knowledge of arithmetic and a habit of performing the common operations in arithmetic, and of applying rules in a correct and intelligent manner ... arithmetic is a matter of habit, and can be learnt only by long, continued practice.

In later years, adaptations of these textbooks were made to reflect a more Caribbean flavor. A series of mathematics books, written by an Irish immigrant, Cutteridge, was used. *Cutteridge Arithmetic* attempted to localize the content by setting sums involving yams and plantains, cutlasses and hoes, chicken houses and one-room houses (Campbell, 1996). He ruled out sums involving trains, clocks, mixtures, and discount and compound fractions, and replaced them with a wide range of everyday calculations covering local problems. This was an attempt to create textbooks with more indigenous content.

Another approach adopted at the secondary level was to change the name of the text but not the content. *Mathematics for the Caribbean*, a series that was popular for many years was not written with the Caribbean in mind. The examples and situations in the texts were basically the same as the original text written for Irish or English students. Many foreign publishers flooded the market using this approach and so the trend of importing curricula persisted.

Summary — Post Colonial Era

The presence of Britain, during the colonial period and thereafter has had the largest influence on the way mathematics was taught and examined. Curricula were designed to reinforce existing societal structures and were perpetuated even after independence from Britain in 1962 until the late

1970s in terms of textbooks and formal examinations. It can be concluded that during the post-colonial era, mathematics was included in the curriculum so that the mind could be trained, strengthened and improved through exercise. Edward Lee Thorndike is credited with leading this movement. In addition to its contribution to mental discipline, mathematics as a pure and applied science represented a unique accomplishment with which educated people ought to be acquainted (Bishop, *et al.*, 1996). In the next era, there is a greater drive to gain control of curricula and realign goals to that of an emerging country as post-Republic Trinidad and Tobago.

Indigenizing the Curriculum and Restructuring the Examination System

Changes to the curriculum at the secondary level were mainly in response to the need for locally relevant examinations. At the primary level, changes were prompted by concerns for low performance on examinations. Policy makers felt that the only way to change the curriculum was to change the assessment practices. It is therefore fair to say that the curriculum is assessment-driven. An evaluation of the primary school curriculum conducted by Northey, Bennett and Canales (2007) supported this view. These evaluators also noted that teaching methods were still predominantly "chalk and talk" with a focus on recall of facts and application of procedures in routine situations.

From Common Entrance to SEA

The purpose of the Common Entrance Examination (also called the Eleven-Plus Examination) was the selection of students for secondary school. This examination was necessary as the number of places at secondary level was substantially lower than that of primary. Its initial focus was competence in mathematics and language arts, but later on included other subjects. It acted as a "gatekeeper, creating differential test-taker outcomes, and governing access to further educational opportunity" (Sacks, 2001; Zwick, 2002 in DeLisle, 2006, p. 92).

Competition for places at highly ranked or "prestige" schools resulted in intense preparation for the Common Entrance Examination. With thousands of students vying for a few hundred places in the prestige schools, a pass mark of at least 90% was required in mathematics for entry. This competitive system has had a negative impact on the curriculum. More time was spent on testing than on teaching and even those students who performed well, entered secondary school without the basic foundation for secondary school work (MOE, 1998b).

The Common Entrance Examination (CEE) was criticized for using objective testing only and not allowing students to express themselves and explain their thinking. As such, in 2001, the Secondary Assessment Examination (SEA) replaced the CEE utilizing open-ended questions. It was redesigned to better measure students reasoning ability and verbal skills (De Lisle, 2006) in three subject areas: mathematics, language arts, and creative writing. It is still in use today. There is a greater emphasis on problem solving and more coverage of the syllabus content than the previous CEE examination.

The movement towards open-ended items in the examination had a positive impact on performance. The new examination came into effect in 2001, but several interventions had to be made before any changes in results were observed. The teachers had to be re-tooled in writing and assessing open-ended items. In addition, the examination itself was transformed to include more scaffolded-type questions to allow students to have more access to the questions. The scoring system enabled students to gain marks for partial solutions. The impact of this change was seen in the results of the 2004 examination, illustrated in Table 1.

The graph in Figure 1 further explains the improved performance. A significant number of pupils, averaging 20%, were now scoring above 80% on the SEA. It can be seen that while previous performance dipped at the 81–100% range, performance increased considerably in 2004 and this trend was maintained.

However, performance at the lower end of the spectrum remained largely unchanged. The movement to open-ended responses would not have helped those with low reading/comprehension scores and this is one explanation for this trend. This explanation is justified by examining the scores in creative

Table 1. National mean scores in SEA mathematics (DERE, 2013a).

Year	2001	2002	2003	2004	2005	2006
National average in SEA (%)	57.6	52.5	59.4	62.6	63.1	61.6

Percentage of Students — **National Performance in Mathematics at end of primary school**

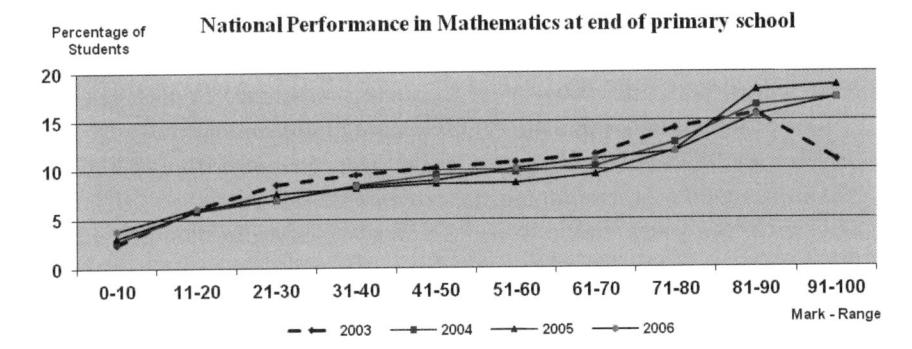

Figure 1. Performance in SEA mathematics by mark ranges, from 2003 to 2006 (DERE, 2013b).

Table 2. Mean marks for mathematics and language arts in SEA (DERE, 2013c).

Year	2007	2008	2009	2010	2011	2012	Avg
SEA Maths %	59.9	60.5	63.1	62.5	67.1	66.1	63.1
SEA LA %	55.5	55.4	57.6	56.8	58.1	55.6	56.5

writing and language arts. The average score over the period 2007–2012 for mathematics is 63.1% compared to 56.5% for language arts. Improvement in literacy is therefore necessary for further improvement in mathematics. Thus examination performance at SEA improved in mathematics, even with an emphasis on problem solving. The data in Table 2 indicates that the trend is maintained.

What is interesting here is that the SEA teachers have been under criticism for teaching to the test, using "chalk and talk" methods and relying on rote memorization. Yet, the performance of students is not unsatisfactory at this level of the system. It would appear that parental involvement and extra lessons are major factors that impact performance at this high stakes examination.

From O and A Levels to CSEC and CAPE

Due to the thrust for Universal Secondary Education, nearly all primary school students move over to the next tier of education, the secondary level. This has been facilitated through a rapid increase in secondary schools throughout Trinidad and Tobago. At the end of five years of secondary education, approximately 17,000–18,000 students currently take the Caribbean Secondary Education Certificate (CSEC) examination in 5–8 subjects and mathematics is compulsory.

The percentage of candidates attaining satisfactory grades of I–III in CSEC examinations ranges from 40% to 47%, according to the most recent statistics (2008–2013). Prior to this period the percentage did not vary considerably. A drop in performance is noted after 2009. At this time, CXC discontinued the Basic Proficiency Examination, which targeted the lower-performing students and all candidates now take the General Proficiency. The results of the entire Caribbean region were affected in a similar manner.

The low performance in the CSEC 2012 Mathematics examination has been cause for alarm in Trinidad and the wider Caribbean region, and resulted in a public call for analysis of reasons for the declining performance. There are approximately 90,000 CSEC Mathematics candidates from 20 Caribbean territories. An editorial in a local newspaper (Chee Hing, 2012) entitled "Math Failings" commented on the results of candidates in the entire Caribbean region:

> ... two-thirds of pupils failed the CSEC Math Exam in June 2012 ... Topics such as the range, perimeter, and profit and loss that should be covered at the lower secondary level were not fully understood.
>
> CXC's Subject Awards Committee (SAC) said it was "deeply concerned about the quality of work produced by candidates at this level." ... and called on the region to address the issue of teaching and performance in mathematics by re-organizing its mathematics

Table 3. Percentage of students obtaining Grades I–III in CSEC Mathematics (DERE, 2013d).

Year	2008	2009	2010	2011	2012	2013	Avg %
Percentage attaining grades I–III at CSEC Mathematics	47	47	45	40	41	42	43.7

Comparison of performance in mathematics CSEC and SEA, 2008-2012

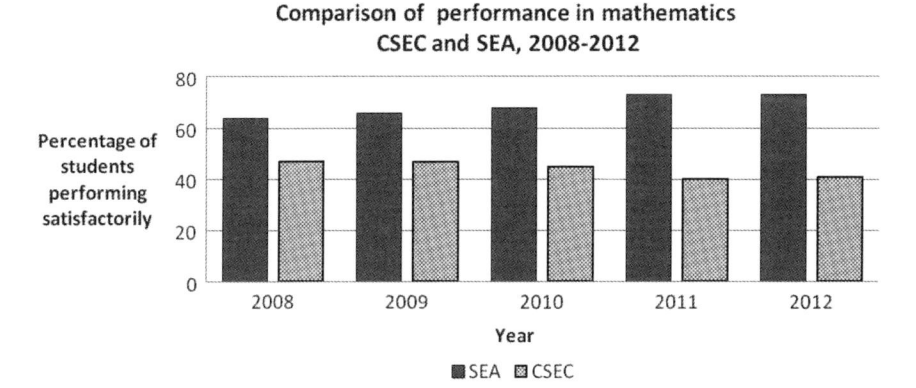

Figure 2. Percentage of students performing satisfactorily in mathematics at the end of primary (SEA) and at the end of secondary (CSEC) school (DERE, 2013a; DERE, 2013d).

programme, supporting teacher training and facilitating access to instructional resources.

Chee Hing (2012) continued, "At a macro-level, it may well be time for a national debate on the ongoing decline in mathematical ability. Do pupils need a reversion to old, traditional ways of learning?"

The above analysis reveals that performance at the upper secondary level in mathematics is much lower than performance at the upper primary level. A comparison of the performance at the end of primary with performance the end of secondary is shown in Figure 2. The data in Figure 2 compares percentage of students scoring above 50% in SEA mathematics with percentage of students obtaining Grades I–III in CSEC mathematics.

This decline in performance as students move from primary to secondary school is disturbing. In analyzing teaching methods at both levels, Lam (2007) noted that teachers at both primary and secondary levels used traditional approaches relying mainly on "chalk and talk". In this study, Lam (2007) also noted that most teachers were unable to see the drawbacks of their methods and claimed that "good explanations" were effective strategies. When quizzed on the non-use of resources, they argued that students were still underperforming in many resource-rich classrooms.

Data from CXC through school reports indicate that as many as 60% of the candidates are struggling with mathematics. Over the years, topics like algebra, measurement, and geometry appear to be quite challenging for students. Performance on higher level thinking skills requiring abstract reasoning and problem-solving also pose problems for students.

A system of ongoing National Tests for diagnosis

The concern for low performance in mathematics measured by summative tests at the end of both levels of the system prompted the Ministry of Education to introduce a series of National Tests to monitor student performance before exiting the system. According to De Lisle (2008), national assessments in the Anglophone Caribbean are administered to serve different purposes, including measuring student achievement, for selection, and to hold schools accountable for the implementation of the national curriculum by year level, where the main priority is on closing the achievement gap of underperforming schools. One of the commendable objectives of the National Test is to use the data to make informed decisions about interventions to improve the educational quality and help policy makers monitor trends in the nature and quality of student learning over time.

Approximately 17,000 to 20,000 students take the National Test in five core areas of the primary school curriculum every year. The results in mathematics indicate that levels of numeracy are still below the expected standards. Over the period 2008–2012, an average of 47% of students in Standard 1 and 48% of students in Standard 3 are not meeting the required standard in mathematics. Tests are designed along three cognitive levels: knowledge/recall, algorithmic thinking, and problem-solving. Performance is lowest on items requiring higher levels of thinking.

At secondary level, a National Certificate in Secondary Education (NCSE) was launched in 2006 at the third form level for 14–15 year-old students. The assessment is made up of an internal component of 60% and a standardized written test worth 40%. Analysis of data over the period 2008–2012 indicates that at the end of Form 3, an average of 53% of students are performing below satisfactory levels in mathematics.

The above comparison of performance in these National Tests continue to reveal that approximately half of the student population is not performing at the expected levels in mathematics at both primary and lower secondary levels (see Figure 3). The data also reveals that students at government secondary schools consistently perform lower than students at government-assisted (denominational) secondary schools.

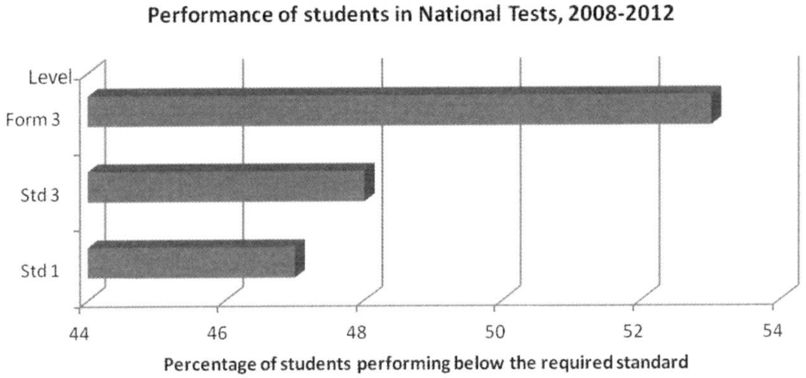

Figure 3. Percentage of students performing below the required standard in National Tests at primary (Standard 1 and 3) and secondary levels (Form 3) (DERE, 2013e; DERE, 2013f).

This trend continues at the upper secondary or Form 5 level in the CSEC examinations. The spike in student performance in mathematics at SEA and at CAPE (17+) is not comforting enough for what is now considered a critical issue in education and has future implications for National Development. Critics of the assessment mechanisms in Trinidad and Tobago recognize that primary gatekeeping mechanisms in many international school systems occur at 16+ to 18+, while in the Caribbean, the role of the Eleven-Plus Examination as an early gatekeeper has been inherited from the colonial era (Jules, 1994; Payne & Barker, 1986 in De Lisle, 2006, p. 92). It can be concluded that its continued retention continues to perpetuate an elitist education system that was introduced and enforced by British colonials.

Teacher Education in Mathematics

Teacher preparation, education, and in-service training inform the way a curriculum is enacted and the following discussion develops a chronology of teacher education for mathematics teachers at both primary and secondary levels. The next sections detail the primary experience and secondary experiences in teacher education.

The primary experience

Primary school teachers were drawn from the system itself and under a pupil-teacher mentoring arrangement, new teachers gained entry into the primary school system. Later on, teachers were required to pass a qualifying examination — the Teachers' Certificate Examination — before they can practice. Failure to pass the examination after two attempts resulted in dismissal. There were no examinations in the practice of teaching in the pupil-teacher system. Training colleges were set up at different campus locations to formally train teachers and teachers exited with a two-year Certificate in Teacher Education. In 2001, the Bachelor of Education was formally introduced as a four-year degree program for in-service teachers. Entry requirements were passes in five subjects at CSEC including mathematics and English language. Teachers exit as primary school specialists.

The curriculum of teacher training institutions has always been one of concern, and a major issue is the balance between pedagogy and content in initial teacher preparation. Since recruits were originally taken from the primary schools, it was necessary to have a strong content base in the program to build on their primary base. At CSEC level, a grade III is accepted, but it suggests that teachers with the weakest mathematical ability can enter the teaching service with minimum content proficiency. This problem is compounded when prospective entrants repeat the course just to make the minimum grade level for entry. As such, the content deficiency still exists and colleges have been mandated to strengthen the academic background of teachers (MOE, 1993).

The issue of provision of specialist teachers in mathematics to provide instructional support for those teachers who lack confidence in the teaching of mathematics has been raised. A need for specialist teachers in mathematics for diagnosis and remediation was suggested (Pujadas, 1984), although data from a 1968 study on primary school organization did not indicate significant differences with the use of a semi-specialist approach (Gerberich, 1974). However, increased camaraderie among staff was noted as an unexpected outcome of the study. In spite of these data, the Ministry of Education initiated a one-year teacher education course in both mathematics and reading at the University of the West Indies. The purpose was to train teachers as facilitators to plan interventions for primary school teachers so as to improve levels of literacy and numeracy. A total of 106 trained primary school teachers were deployed and given specialized training in the teaching of mathematics (58) and reading (48). They were assigned to districts and worked under school supervisors (in district teams) as well as the curriculum officer (mathematics) at the head office of the Ministry of Education.

They were given a range of responsibilities, including implementation of the new syllabus, diagnostic testing/remediation, conducting professional development workshops, assisting supervisors and the curriculum officer in the monitoring of teachers, preparation of programs of work for teachers, and designing of instructional materials.

This system of primary school facilitators was initially planned for three years but continued with periods of pauses followed by renewals for just over two decades. A committee was appointed to conduct a study on the impact of the facilitators on the system. Areas such as the management of the program, the extent to which teachers were using strategies acquired in their training sessions, and the levels of competence of students in mathematics were measured. Approximately 2,000 pupils at three levels (Standard 1, 3 and 5) were tested from a sample of 60 schools. Observations and interviews were carried out among 176 teachers and 43 principals. Documents such as Facilitators and Principals' reports were analyzed as well.

The findings of this study, extracted from the report prepared by QuanSoon (1994), suggested that group-work was not emphasized and that teachers were not implementing teaching strategies learnt in the training workshops. Data from teacher interviews indicated that teachers improved in some areas, mainly test construction, planning for instruction, and the use of "hands-on approaches". Other areas of implementation were difficult because they did not always understand what was presented in the workshops and much of what was learned was forgotten. The absence of resources prevented them from using active learning. In addition, when they returned to their classrooms, principals made no provisions for the transmission of workshop knowledge and skills to the rest of the staff. They also indicated that facilitators gave helpful advice, but did not visit often enough. With respect to attending to students with special needs, teachers felt they were insufficiently trained and were incapable of devising strategies to effectively address these needs.

The study also showed that mathematics performance was lower than expected. Performance on the content strands was also poor with only Standard 5 students performing at a barely satisfactory level in number and measurement. The most significant finding in the study was the weakness in management and structures at curriculum and supervisory levels and the lack of effective feedback mechanisms to the central office of the Ministry of Education.

The major recommendations of the evaluation committee were:

- The setting up of a formal management structure to achieve self-sustainability and accountability.
- The introduction of semi-specialist teachers in the schools to provide linkages to the facilitators.
- The provision of opportunities for ongoing professional development for the facilitators.

The Ministry of Education responded to the above concerns by employing new officers to the Curriculum and Supervision Divisions. However, a number of facilitators left the program because of lack of incentives. They were not replaced and in 2001 there were only ten facilitators to service some 500 primary schools.

The Curriculum Division, as part of its monitoring process, collected data on the implementation of the 1999 primary curriculum. Although a move towards curricula that employed constructivist approaches was being made, observation of teachers and detailed analysis of facilitators' school visit reports revealed the following about primary teaching:

> Constructivist methods were not fully understood and teachers were still unfamiliar with the new approaches in the revised curriculum. Another deterrent to successful implementation of the revised curriculum was the content background of the teachers at the primary level. A large number of teachers had surface knowledge of the mathematics and were unable to make it meaningful and relevant to their pupils (Khan, 2001).

The Curriculum Division organized large-scale training of 8,000 teachers in more than 500 primary schools over the period 2002–2006. Training was conducted each year for five years until every teacher in all levels were trained. This was a new model of training, as previous "train the trainer" models were deemed ineffective. This particular model of training had the advantage of ensuring that each teacher received the new strategies first hand. In addition, facilitators followed up the training by holding site-base workshops in which principals and the entire staff were updated with the new methods. Materials used in the workshops were also provided to each school so that teachers could implement the strategies when they returned to their classrooms. The content of the workshops were based on needs assessment using data from National Tests, observation of teachers and syllabus demands.

Teachers were introduced to a range of instructional strategies that were part of the current curriculum. They returned to their classrooms and tried out the strategies. The facilitators were able to monitor their classroom practices and provide the required support. The workshops were deemed successful after analysis of teacher evaluations. Teachers felt that they gained insight into new methods related to teaching of algorithms, use of manipulatives, understanding of slow learners, alternative assessments, curriculum integration, and the teaching of reasoning (MOE, 2006).

Workshops came to an end in 2007 when the facilitators' program was discontinued. Although the training program was successful, it took a long time to reach all the teachers. For example, a five-year period will elapse before a teacher is called for re-training. In addition, the absence of facilitators rendered it impossible to monitor the teachers to determine if the new strategies were being implemented.

The secondary experience

Secondary teacher education has not been as formalized as that of primary education. Entry to secondary mathematics teaching has been a tertiary level degree in mathematics from an established university. The establishment of government schools in the 1970s propelled teachers with a Teacher's Certificate in education to move from primary to secondary to fill a large number of vacancies. The more established denominational schools continued its practice of hiring non-graduate teachers selected from among their sixth form students.

In 1962, of the 709 teachers in the government and assisted schools, only 85 (about 11%) were graduates of universities with professional training. Campbell (1997, p. 102) suggests "... traditionally it was felt that a university graduate was already equipped to teach secondary school without any professional training as a teacher." The shortage of secondary teachers ruled out initial teacher training and Campbell (1997, p. 102) suggests that:

> ... the expansion of secondary education in the late 1950s and early 1960s was characterized by a low ratio of university graduate to non-graduate teachers, especially in the new government schools which sought to follow the older, better established

denominational secondary schools into the path of academic subjects and Cambridge external examinations.

In the early seventies, the Ministry of Education (MOE) commissioned the School of Education (SOE) of the University of the West Indies (UWI) to provide and deliver a one-year Diploma in Education (Dip.Ed.) program on its behalf to practicing teachers. Teachers entering the program have a content first degree in mathematics. A 2012 study of teacher experiences on the Dip.Ed. program (James, et al., 2013, p. 87) reveals generally positive views:

- Intensive, enlightening, inspiring. I learned new methods/philosophies about teaching/learning. A great opportunity to reflect on my practice as well as get feedback ...
- My perspectives on teaching, learning student ability [sic] and my relationships with my students have been widened and challenged for the better.
- Another benefit is appreciating the business of education as evolving — being open to various changing methods of teaching to reach students. Exposure to the vast amount of research being done in the field of education was another definite benefit. I am now encouraged to research and look at new ways of developing my work and myself.

While these views are not specific to mathematics teachers, they are typical of sampled groups. There are deficiencies in the training program, however, and these weaknesses continue to impact classroom teaching competence. Furthermore, some teachers indicated that there was an inadequate focus on innovative methods of teaching that would cater to students with different learning abilities (James, et al., 2013, p. 91):

- The teaching methods that I learnt were for above average students and I have below average and learning-disabled students so I was only able to use some aspects of the methods.
- The implementation of the strategies learnt — they are not designed for large classrooms; students at various intellectual levels in one class ...
- There was not enough focus on catering for different learning abilities/styles.

To date, more than 50% of teachers have been trained through the Dip.Ed. program but as this program is voluntary, there are still large numbers of untrained teachers in school. Continued professional development opportunities continue through a number of tertiary level providers in the form of M.Ed., M.Phil., and Ph.D. However, for those not pursuing those programs, there are limited opportunities for growing professionally. James, et al. (2012) note that while some measure of collaboration between MOE and SOE may be incorporated into the delivery of the Dip.Ed. program, there are no formalized networks for facilitating future professional development.

In very recent years, the newly-formed University of Trinidad and Tobago (UTT) began to offer pre-service B.Ed. teacher preparation programs for secondary mathematics, among other subjects. Although the program is in its infancy, much progress is being made to overcome the hurdles that are present.

The Curriculum Development Division (CDD) of the MOE has offered occasional training for teachers when a new syllabus is being introduced (national or CXC) or when students are performing poorly on a particular topic. Data from the CXC reports or from classroom tests are used to identify training needs. Secondary school teachers also participate in regional workshops sponsored jointly between the ministry and the examining body, CXC, but these are ad hoc at best.

In 2004, the CDD initiated training of heads of departments, but this was short-lived due to changing policy initiatives. As such, professional development for secondary mathematics teachers remains highly unstructured and non-mandatory. This remains one of the challenges facing secondary mathematics teachers.

Summary

Trinidad and Tobago has made attempts to formalize assessment processes, streamline syllabi and localize curricula for cultural relevance. Strides in the last fifty years have seen the expansion of access to mathematics education in schools with a significant amount of teacher preparation and continuing professional development. Decline in performance at the national level is a cause for concern and it is receiving attention from education stakeholders.

Future Trends and Implications

In an era of global competition, efforts were made by the Ministry of Education to develop curricula that were aligned to both national and global directives. In particular, the adoption of Essential Learning Outcomes such as problem-solving and technological competence became more important to curriculum developers. Current initiatives in mathematics education include the promotion of students' thinking skills in culturally-relevant curricula, promotion of gender equity in assessments, enhancement of numeracy and literacy across all education levels and continued emphasis on competitions for promoting academic excellence. These areas are explored in some detail in the next sections.

Globalizing the curriculum

In an attempt to democratize the education process in Trinidad and Tobago, new curriculum documents were created in 2002 in eight subject areas at the secondary level, including mathematics. This movement was evident throughout the Caribbean and, indeed, throughout the world (Papadopoulos, 1998 in Worrell, 2004, p. 1). The mathematics curricula were designed to promote students' thinking skills in a culturally relevant context and aimed to promote multicultural understandings. Yet, the curriculum needed to respond to global trends and directions in mathematics as well. As such, the curriculum was located in what Robertson (1995, p. 477) describes as the "local in the global". Challenges to the design of these curricula include:

> ... a relationship between the canonical knowledge of Europe and North America, which has traditionally constituted mainstream academic knowledge in once-colonized Caribbean countries (London, 2003), and other forms of knowledge that may be of equal importance to students, and to the society (Worrell, 2004, p. 1).

These curricula are being currently used in schools in the first three years of secondary schooling. The advent of new syllabi and curricula allowed for rapid growth in the local textbook industry. Caribbean publishers emerged and enticed writers to produce local textbooks tapered to the syllabus. This has resulted in a surge of these textbooks on the local market as subject-area experts responded to this initiative. CXC also held regional workshops in areas that were new to the teachers and participants produced resource modules for distribution to the territories. Research is needed on how these curricula have affected the teaching-learning environment at secondary level and their role in improving problem-solving skills among secondary students.

Achieving excellence through mathematics competitions

Trinidad and Tobago is the only English-speaking Caribbean territory to participate in the International Mathematical Olympiad, and it has been doing so every year for over 20 years. Although the country is yet to secure a gold medal, it has won one silver and four bronze medals. Participants have been largely male students coming from denominational schools.

The Curriculum Division has planned and implemented two tiers of competition to promote interest and performance among mathematics students. The Mental Mathematics Marathon has been successfully run for the past twenty years at the primary level, while the Junior Mathematics Triathlon has been run for students at lower secondary level. The focus is on problem-solving involving the use of efficient strategies in quantitative, spatial, and logical reasoning.

At the primary level, there is keen competition among districts to win the top awards. The Mental Mathematics Marathon is a resounding success bringing together key stakeholders in the system engaged in promoting and rewarding excellence in mathematics. The Junior Mathematics Triathlon focuses on performance tasks on a variety of content areas and relies on oral and written communication of responses. These competitions also provide a springboard for students to enter other competitions at the secondary level and create opportunities for them to recognize their talents in mathematics. Winners of this competition go on to win national scholarships and other prestigious awards. In fact, many Trinidad and Tobago students gain entry into very prestigious international universities such as MIT, the University of Waterloo, and Oxford University.

A thrust for gender equity

There is a lament that women do not access high stakes positions in science and mathematics in spite of increasingly comparable performances in mathematics and science between boys and girls in North American studies. In Trinidad and Tobago, there are interesting trends in performance between boys and girls. Generally, girls outperform boys at several levels including

SEA and CSEC, but not at more competitive levels such as the CAPE or in mathematics competitions described in the earlier section.

Studies on the impact of gender at the SEA level show negligible differences between boys and girls at higher levels of performance but increased differentials at the lower levels (De Lisle and Smith, 2004, p. 41). These authors argue that the focus on verbal-linguistic skills seems to favor girls at the SEA level and further marginalize boys with poor reading skills. Because SEA performance leads to secondary school placement, the placement of low-performing boys at certain types of schools can lead to increased disadvantages at the CSEC level. Performance in the Programme for International Student Assessment (PISA) (OECD, 2010) reveals that Trinidad and Tobago ranks at the 52nd position in mathematics out of 67 countries studied and shows poor performance in both reading and mathematics among 15–16 year olds. A possible explanation for this is that PISA emphasizes problem-solving, a consistent area of weakness among students. Relationships between reading and mathematics need further exploration, and consideration of assessments that promote gender equity made. A strategy that is currently being explored nationally is the numeracy intervention at primary level. The latter is discussed below.

A need for improved levels of numeracy and literacy

The country is now introducing a Numeracy Strategy alongside the new curriculum. The Numeracy Strategy is a work in progress, and takes into consideration all past issues. Five components have been identified as areas for strengthening as shown in Figure 4 (Khan, 2012).

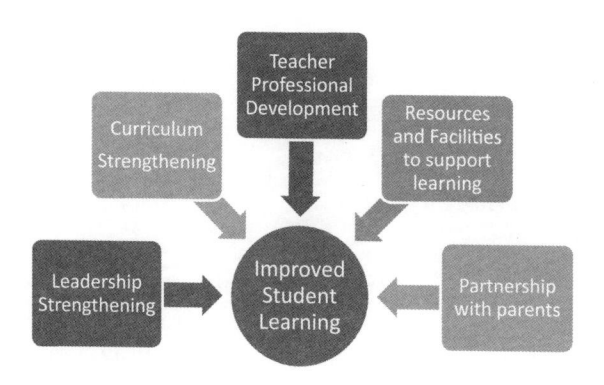

Figure 4. Areas to strengthen in the Numeracy Strategy.

Embedded in the numeracy strategy is the notion of district teams who will assume leadership of their districts by determining specific needs and planning interventions to address them. Mathematics coaches who will be trained in pedagogy and leadership skills to implement district initiatives will provide instructional strengthening. The inclusion of parents in providing support for initiatives is also critical to the success of the strategy. The curriculum review is already in progress and the shift to more depth and less breadth is a critical issue to be addressed. The strategy initially targets the primary level of the system but will eventually reach the secondary as more resources become available.

Final Thoughts

A historical perspective has allowed this paper to highlight significant events and milestones in mathematics education in Trinidad and Tobago. Mathematics is valued at all levels of the education system, and exceptional performances by students are seen in mathematics competitions, upper end of SEA, and at CAPE levels. The continuation of the dual system of schooling, inherited from colonial times, continues to dictate student success and polarize performance. Poor performance in mathematics examinations elicits public outcry, yet examination of critical issues in mathematics remains pedantic. The gains made by localizing the curriculum remain minute as teachers continue to be voiceless in a traditional top-down approach to administration of education policy. National teacher preparation and education programs remain inadequate. Mechanisms to support mathematics teachers in the classroom need to be promoted through systematic and sustained research-driven approaches. Potential useful initiatives — such as inclusion of numeracy and democratization and indigenization of curricula — can benefit from analysis through socio-political lenses (Gutiérrez, 2013), as the influence of political directorates in mathematics education remains prime.

Bibliography

Altbach, P. (1971). Education and neocolonialism. *The Teachers College Record*, 72(4), 543–558.
Bishop, A. J., Clements, M. A., Keitel, C., Kilpatrick, J., & Laborde, C. (Eds.) (1996). *International Handbook of Mathematics Education, Part 2*. Netherlands: Kluwer Academic Publishers.

Bacchus, M. K. (1994). *Education as and for Legitimacy: Development in West Indian Education between 1846 and 1895*. Waterloo: Wilfred Laurier University Press.

Campbell, C. C. (1996). *The Young Colonials: A Social History of Education in Trinidad and Tobago. 1834–1939*. Barbados: University of the West Indies.

Campbell, C. C. (1997). *Endless Education. Main Currents in the Education System of Modern Trinidad and Tobago. 1939–1986*. Barbados: The Press University of the West Indies.

Chee Hing, K. (2012). Math Failings. *Trinidad Newsday*. Editorial August 20, 2012. Retrieved from http://www.newsday.co.tt/editorial/0,165056.html

Colenso, J. W. (1866). *A Shilling Arithmetic*. London: Longmans, Green & Co.

De Lisle, J. (2006). Dragging eleven-plus measurement practice into the fourth quadrant: the Trinidad and Tobago SEA as a gendered sieve. *Caribbean Curriculum, 13*, 91–129.

De Lisle, J. (2008). Can standards-referenced, large-scale assessment data lead to improvement in the education system? Judging the utility of student performance standards in the primary school national assessments of educational achievement. *Caribbean Curriculum, 15*, 71–114.

De Lisle, J., & Smith, P. (2004). Reconsidering the consequences: gender differentials in performance and placement in the 2001 SEA. *Caribbean Curriculum, 11*, 23–56.

Division of Educational Research & Evaluation (DERE) (2013a). *Retrieved from database — Mean score of students on SEA 2001–2012*. Port of Spain, Trinidad and Tobago: Author (DERE).

Division of Educational Research & Evaluation (DERE) (2013b). *Retrieved from database — National performance of students by mark-range, SEA 2003–2006*. Port of Spain, Trinidad and Tobago: Author (DERE).

Division of Educational Research & Evaluation (DERE) (2013c). *Retrieved from database — National performance of students in mathematics and language arts, SEA 2007–2012*. Port of Spain, Trinidad and Tobago: Author (DERE).

Division of Educational Research & Evaluation (DERE) (2013d). *Retrieved from database — Performance of students on CSEC 2008–2013*. Port of Spain, Trinidad and Tobago: Author (DERE).

Division of Educational Research & Evaluation (DERE) (2013e). *Retrieved from database — Performance of students on National Tests (primary) 2007–2012*. Port of Spain, Trinidad and Tobago: Author (DERE).

Division of Educational Research & Evaluation (DERE) (2013f). *Retrieved from database — Performance of students on NCSE (secondary) 2007–2012*. Port of Spain, Trinidad and Tobago: Author (DERE).

Education System of Trinidad and Tobago (2013). Retrieved July 10, 2013 from http://photos.state.gov/libraries/port-of-spain/223843/PDFs/EDUCATIONAL%20SYSTEM%20T_amp_T%202011.pdf

Gerberich, J. R. (1974). *The Carnegie project in primary school organisation, Trinidad and Grenada — 1968–1970. An Evaluation*. Jamaica: UWI.

Gutiérrez, R. (2013). The sociopolitical turn in mathematics education. *Journal for Research in Mathematics Education, 44*(1), 37–68.

Harris, N. (2011). *Chairman's statement. Annual report 2011*. Barbados: Caribbean Examinations Council. Retrieved November 26, 2001 from http://www.cxc.org/SiteAssets/AnnualReports/AnnualReport2011Final.pdf

James, F., Phillip, S., Herbert, S., Augustin, D. S., Yamin-Ali, J., Ali, S., & Rampersad, J. (2013). Is anybody listening? Teachers' views of their in-service teacher professional development programme. *Caribbean Curriculum, 20*, 77–100.

Jules, V. (1994). A study of the secondary school population in Trinidad and Tobago: Placement patterns and practices — A research report. St. Augustine, Trinidad: Centre for Ethnic Studies, UWI, cited in J De Lisle, Dragging eleven-plus measurement practice into the fourth quadrant: The Trinidad and Tobago SEA as a gendered sieve. *Caribbean Curriculum, 13*, 91–129.

Khan, S. A. (1995). *Views of secondary school teachers in Trinidad and Tobago on the introduction of SBA in mathematics*. Unpublished Memo.

Khan, S. A. (2001). *Report on the status of primary mathematics*. Trinidad and Tobago: Ministry of Education, Division of Curriculum Development.

Khan, S. A. (2012). *A numeracy strategy. A model for improving numeracy in Trinidad and Tobago*. Unpublished, August 22, 2012.

Lam, E. (2007). Mathematics education in Barbados and Trinidad: challenges and progress, in D. Kucherman (Ed.), *Proceedings of the British Society for Research into Learning Mathematics, 27*(3), November 2007.

Lalbeharry, D. (2008). *Investigation of the problems primary teachers face in the teaching of Mathematics in the county of Victoria*. Retrieved November 8, 2013 from http://uwispace.sta.uwi.edu/dspace/handle/2139/1909

London, N. (2002a). Curriculum convergence: an ethno-historical investigation into schooling in Trinidad and Tobago. *Comparative Education, 38*(1), 53–72.

London, N. (2002b). Curriculum and pedagogy in the development of colonial imagination: a case study. *Pedagogy, Culture & Society, 10*(1), 95–121.

London, N. A. (2003). Ideology and politics in English-language education in Trinidad and Tobago: The colonial experience and a postcolonial critique. *Comparative Education Review, 47*(3), 287–320.

Ministry of Education (1974). *Draft Plan for Educational Development in Trinidad and Tobago, 1968–1983*. Trinidad and Tobago: MOE.

Ministry of Education (1993). *Report on the National Task Force on Education (1993) Green Paper*. Trinidad and Tobago: MOE.

Ministry of Education (1998a). *Report of the Curriculum Task Force, Part I — The Proposed Primary Education Curriculum*. Trinidad and Tobago: MOE.

Ministry of Education (1998b). *Review of the Primary School Curriculum*. Trinidad and Tobago: Curriculum Division.

Ministry of Education (2001). *The Keenan Report of 1969 — Report upon the State of Education in the island of Trinidad*. Trinidad and Tobago: School Publication Unit.

Ministry of Education (2006). *Primary Level Mathematics*, pp. 37–40. Report of the Curriculum Development Division 2003–2005. Trinidad and Tobago: Curriculum Development Division.

Nichols, D. E. (1968). In W. C. Seyfert (Ed.) *The Continuing Revolution in Mathematics*. Washington, DC: NCTM.

Northey, D., Bennett, L., & Canales, J. (2007). *Support for a seamless education system: curriculum and instruction, testing and evaluation and Spanish as the first foreign language: final report*. USA: Simon Fraser University.

OECD (2010). *PISA 2009 Results: Executive Summary*. Retrieved July 8, 2013 from www.pisa.oecd.org

Pariag, J. (2008). *An evaluation of the implementation of the primary mathematics curriculum (extended literature review)*. Trinidad: UWI. Unpublished paper.

Papadopoulos, G. (1998). Learning for the twenty first century: Issues, in *Education for the twenty first century: Issues and prospects*, pp. 25–46. Paris: UNESCO Publishing, cited in Worrell, P. (2004). Addressing cultural diversity in a creole space: the SEMP language arts curriculum. *Caribbean Curriculum, 11*, 1–22.

Payne, M. A., & Barker, D. (1986). Still preparing children for the 11+: Perceptions of parental behaviour in Barbados. *Educational Studies, 12*(3), 313–325, cited in J. De Lisle, Dragging eleven-plus measurement practice into the fourth quadrant: the Trinidad and Tobago SEA as a gendered sieve. *Caribbean Curriculum, 13*, 91–129.

Primary School Mathematics Syllabus (1975). Trinidad and Tobago: An IDB-Ministry of Education Publication.

Primary School Mathematics Syllabus (1999). Trinidad and Tobago: An IDB-Ministry of Education Publication.

Pujadas, L. (1984). *Towards the promotion of a curriculum for schools in Trinidad and Tobago for the late 80's and beyond*. Report of the Curriculum Unit, Ministry of Education. Trinidad and Tobago: Curriculum Development Division.

QuanSoon, S. (1994). *Report of Committee Re: Impact of remedial reading and mathematics programmes* on *the development of reading and mathematics among primary school students* for submission to Minister of Education. Trinidad and Tobago: UWI.

Ramsey, Y. (1998). *Report on the status of mathematics education in Trinidad and Tobago*. Paper presented at the CASTME Regional Conference on Educational Technology for Science and Mathematics Education in the Caribbean: Social and Cultural relevance for the 21st Century, 12–15 July, 1998, St. Vincent and the Grenadines, West Indies.

Robertson, R. (1995). Glocalization, in B. Ashcroft, G. Griffiths and H. Toffin (Eds.), *The Post-Colonial Studies Reader*, 72nd Edition. NY: Routledge.

Sacks, P. (2001). Pseudo-meritocracy: A response to "The future of affirmative action" by Susan Sturm & Lani Guinier. *Boston Review*, cited in J. De Lisle, Dragging eleven-plus measurement practice into the fourth quadrant: the Trinidad and Tobago SEA as a gendered sieve. *Caribbean Curriculum, 13*, 91–129.

The Maurice Report (1959). *Education Report of the Committee on General Education.* Trinidad and Tobago: Ministry of Education.
The Story of the Caribs and Arawaks (2014). Retrieved July 1, 2014 from http://www.raceandhistory.com/Taino/Caribs.htm
Tsang, M. C., Fryer, M. & Arevalo, G. (2002). *Access, Equity and Performance. Education in Barbados, Guyana, Jamaica and Trinidad and Tobago.* Washington: IADB.
Williams, E. (1962). *History of the People of Trinidad and Tobago.* New York: Frederick A. Praeger Inc.
Worrell, P. (2004). Addressing cultural diversity in a creole space: the SEMP language arts curriculum. *Caribbean Curriculum, 11,* 1–22.
Zwick, R. (2002). *Fair Game?: The Use of Standardized Admissions Tests in Higher Education.* New York: Routledge Falmer.
Zwick, R., & Schlemer, L. (2004). SAT validity for linguistic minorities at the University of California, Santa Barbara. *Educational Measurement: Issues and Practice, 23*(1), 6–16, cited in J. De Lisle, Dragging eleven-plus measurement practice into the fourth quadrant: the Trinidad and Tobago SEA as a gendered sieve. *Caribbean Curriculum, 13,* 91–129.

About the Authors

Shereen Alima Khan has 12 years of experience as a Curriculum Coordinator at the Ministry of Education, where she was involved in developing and assessing mathematics curriculum and conducting in-service training for teachers at primary and secondary levels. She served as a mathematics teacher educator for 13 years and currently works at the University of Trinidad and Tobago. Her service to the Caribbean Examinations Council in developing regional examinations and curricula spans 30 years. She is now a doctoral student in mathematics education at Columbia University.

Vimala Judy Kamalodeen is a lecturer of mathematics and information technology education at the University of the West Indies, St. Augustine, Trinidad and holds an Ed.D. from the University of Sheffield, UK. She has worked as a mathematics educator for many years and then as a Curriculum Supervisor at the Ministry of Education, where she was involved in a number of projects related to secondary and primary mathematics as well as teacher training. Her focus is on improving the school system by enhancing the skills and attitudes of teachers.

Chapter 17

VENEZUELA: Signs for the Historical Reconstruction of Its Mathematics Education

Fredy Enrique González
(Translated by Nathan C. Ryan)

Abstract: This chapter will serve as the foundation upon which a panoramic view of the historical evolution of Mathematics Education in Venezuela will be constructed. It reviews the history — through different periods — and current status of mathematics education in Venezuela. Moreover, a periodization of the trajectory taken by Mathematics Education in Venezuela will be given.

Keywords: History of mathematics education; mathematics in Venezuela; history of mathematics; mathematics education; national identity.

Introduction

This chapter will serve as the foundation upon which a panoramic view of the historical evolution of Mathematics Education in Venezuela will be constructed. To fully understand the topic at hand requires an elucidation of the various meanings, different but complementary, attached to the phrase "mathematics education". The intended meaning when one writes the phrase "mathematics education" (in lowercase) is that of the mathematical instruction that every citizen of a country should receive, be it through formal means (classroom education) or informal ones (instructional settings other than the classroom). On the other hand, Mathematics Education (with the first letters capitalized) refers to a discipline that creates knowledge and understanding related to the methods of teaching and learning mathematics. A more general treatment of the relationships and links between these two phrases can be found in Belisario and González (2012).

The Mathematical Practices of the Aboriginal People of Venezuela

As a result of the research in ethnomathematics carried out in the country, a great deal is known about the mathematics of the aboriginal people of Venezuela, as one can see in the contributions made by the following authors:

(1) D. Sánchez (2009) studied the number systems used by the aboriginal people of Venezuela; this "research concerns the number system and counting methods of the aboriginal peoples of Venezuela; a summary of their mathematical knowledge is given. Specifically, the names for numbers in their respective languages are provided; references are made to tools, still used to this day, with which one can count, for example, the number of days one has been traveling and has yet to travel; an account is given of how they used their mathematical and geometrical understanding in engineering and construction — for example, how the Ye'kuana people built their churuata or "atta" is described — these are large circular buildings with conical roofs designed to house several extended families."

(2) C. Longart and C. Tirapegui (2008). "A field study is described in which interaction between students was used to promote geometric reasoning, a particular component of mathematical reasoning. The research began because of observations in a classroom of students of the second level of basic education, students who generally do not have exposure to the basic objects of study in geometry and the relationships between them. The evolution of their geometric reasoning was studied according to the theory of Van Hiele: when they were in contact (socializing) with other children who, despite not being enrolled in formal schooling, were able to design geometrically intricate, beautiful and useful baskets made of palm (moriche) fiber. In the crafts made by Warao boys, girls and mothers, the presence of the basic elements of geometric reasoning are immediately recognized; the moriche basket is a spiral of palm fiber that starts off as a perfectly flat circle and then starts to curve upward, depending on the particular piece; the fiber is tied with fine threads that head out radially from the center and open equidistantly like a ray of radii, periodically there are sections of different colors inserted into the fiber spiral, in order to create decorations consisting of figures of people or of elements inspired by their surroundings. Observing

elements that reveal the geometric reasoning of these indigenous people (who, as mentioned above, have never gone to school), the experience of weaving the baskets occurs in and fosters a 'communicative environment' in which the messages flow constantly and in multiple directions, framed in a social and cultural context that gives them meaning and value."

(3) O. Villalobos and D. Ruiz (2009). "This research has as its goal the reporting of the collection of geometric ideas and reasonings that arise in three autochthonous activities among the Pemón people in the Wonken region. These activities are: the construction of living spaces, the practice of basketry and the cultivation of farms known as *conucos*. This study is supported by the theory of 'universal' activities due to Bishop (1999). This theory states that in every ethnic or cultural group in which its members do at least one of the following: locate, design, count, measure or explain, evidence of the existence of mathematics can be found. All these activities are related to their need to manage their environment and make up the foundation for the development of mathematics throughout a culture. To carry out this study, a qualitative ethnographic methodology was used, a methodology characterized by its treatment of problems related to 'population-individual' and 'cultural interaction-language' dynamics. Villavicencio (2001) indicates that this methodology allows for the recognition of phenomena in the various cultures being considered, as well as how these phenomena are manifest and also the various representations the individuals form of these phenomena. The units of analysis were the geometric ideas present in the activities carried out by members of the Pemón community in Wonken: the construction of living spaces, the practice of basketry and the cultivation of farms known as *conucos*. The general techniques used in carrying out this research were (a) interviews, (b) participatory observation, (c) discussion groups. The tools used to collect data were (a) research journals and (b) audio recordings. The results obtained from the aforementioned interviews, discussions, telling of stories, research journals, and sketches by the researchers allow for the presentation of a description of the geometry of the Pemón people found in the Wonken region. This geometry includes basic ideas such as roundness, needles, space, shapes and territorial vision."

(4) O. González (2002). Ethnomathematics among the indigenous Kurripako (Arawak) people and programs of intercultural bilingual education. "We are aware that the inquiry we have been carrying for years

among the indigenous ethnicities in Venezuela has already been discussed in classical works of linguistics such as those of Berlin and Kay (1969) but for other groups such as the Inuit in Alaska and the Tzeltal in Central America. The current situation though, suggests that what is essential to carry out these important reflections is more than just what is being done academically; the new legislative climate offers the indigenous people of Latin America the opportunity to educate ethnic groups previously excluded from the educational system, in a truly self-determined way, in coordination with the non-indigenous or *criollo* educational systems, and with a truly intercultural profile."

(5) R. Luque (2009). The presence of mathematical elements in the Wayúu people. "This research had as its general goal bringing to light the mathematical elements in the culture of the Wayúu people. The research was done with the goal of identifying and interpreting the collective meaning of the mathematics currently used by the Wayúu people. To this end, the mental representations of the Wayúu were studied, in particular the external representations were examined from two perspectives: the anthropology of Chevallard (1992) which asserts that mathematical objects are cultural inventions and the semiotics of Pierce which is concerned with the meaning of the symbols present in the culture. Moreover, the theory of mental representations of Johnson-Laird (1993) was applied to obtain theoretical constructs regarding collective meanings in the culture of the Wayúu people. After finding the units of meaning, the mathematical categories present in the Wayúu people were discovered. These theoretical approximations will contribute to the teaching-learning process of institutional mathematics. Additionally, it will aid in the creation of written material in the autochthonous language. The methodology fits in an ethnographic framework, was complemented by mathematical anthropology, and as such, went to the heart of the Wayúu culture and extracted their worldview from it by means of the characteristics of the collective thinking and metaphorical language. Unstructured interviews with key knowledgeable subjects were carried out. Later, these were submitted to a semiotic study of their mathematical elements. As a result of these studies, after the application of the phenomenological method, mathematical structures such as the use of pattern, counting and time were brought to light.

(6) A. Abreu and R. Luque (2007). In this study the use of area measurements of certain materials used in the creation of various kinds of crafts by the members of the Wayúu people located in the State of Zulia, in the western part of Venezuela, was examined.

The Teaching of Mathematics in Venezuela During the Colonial Period

The colonial period in Venezuela begins with the arrival of Christopher Columbus to the Venezuelan coasts (1498); the works of Yajaira Freites (2000) and Walter Beyer (2010) present interesting information regarding mathematics and its teaching in Venezuela during the colonial period. For example, Freites (2000, p. 9) indicates that:

> Venezuelans came late to acquiring mathematical knowledge ... the first faculty of the Royal and Pontifical University of Caracas had little interest in mathematics; this was not a part of the institution.

Consistent with Freites (*op. cit.*), the first department of mathematics to be created in Venezuela was at the Royal and Pontifical University of Caracas and was created by royal decree of Carlos II on October 21, 1765; the chair of this department, however, remained vacant because the university did not have the financial resources to maintain it.

Beginning in 1788, Baltazar de los Reyes, regent of the department of philosophy at the Royal and Pontifical University, introduced the study of the basic notions of arithmetic, algebra and geometry because he considered them to be indispensable not only in the study of physics, but also in the study of theology. The boldness of Reyes Marrero caused him so many problems that he was brought before the king, who ultimately decided "that the ideas of algebra, geometry, and arithmetic were to be taught to students who wanted them to be taught to them."[1] As these mathematical ideas were covered in courses on philosophy, there was no mathematics chair *per se*; the idea to re-establish the chair proposed by Reyes Marrero was trumpeted once again in 1790 by Juan Agustín de la Torre, who, at the time, was the rector of the University of Caracas. Nevertheless, his proposal

---
[1] Unless otherwise noted, these quotes come from Freites (2000).

(which considered mathematical understanding limited to only that which was required for surveying) was poorly received.

The two main problems that prevented the creation of positions in mathematics and various disciplines other than theology, in addition to a lack of understanding by the people who were making such decisions, were the lack of people qualified to fill them and the lack of financial resources to guarantee their sustainability.

Not surprisingly, the building of roads, piers and harbors, the immediate need to take advantage of the natural resources provided by the land and, additionally, the construction of forts and other notable public works, made the teaching of mathematics indispensable. This teaching, as indicated above, was done in an "extramural way". This was how the royal engineers and officials received permission to teach mathematics: first to the military itself and gradually, to some civilians.

This is the context in which the Caracas Academy of Geometry and Fortification was established (1760–1768) — "the first Academy about which anything is known" (Freites, 2000, p. 15) — under the command of Lieutenant Colonel Nicolás de Castro. Castro stayed in this position until he was transferred to Panamá. At the same time as the founding of Castro's academy, in 1761, the Military Academy of Mathematics in the province of La Guayra (this place nowadays is known as La Guaira and is the capital of the State of Vargas, next to the city of Caracas) was founded by Artillery Captain Manuel Centurión. Another department of mathematics, whose existence was brief, was located in Caracas and had Simón Bolívar as an attendee; it was created by Capuchin Monk Francisco Andújar. The eighteenth century then came to an end.

In 1808 there were two active mathematical academies, one in Caracas and the other in Cumaná. The Cumaná Academy was directed by Royal Engineer Juan Pires and was attended by Antonio José de Sucre. De Sucre later came to be known as the Grand Marshall of Ayacucho, and christened as the Abel of America by Simón Bolívar upon hearing of his assassination in the Berruecos Mountains (Colombia) in 1830. The Caracas Academy was overseen by Colonel José Tomás Mires and in it "were taught arithmetic, algebra, geometry and trigonometry; civil engineering as well as line and topographic drawing"; from this academy the first Venezuelan engineers graduated, according to Arcila Farías (1961, p. 253 cited in Freites, 2000, p. 16).

April 19, 1810 marks the founding of Venezuela as an independent republic; the country then begins a new phase (that concludes in 1821, with victory at the Battle of Carabobo, a battle which sealed the permanent independence of Venezuela from the Spanish empire) as a result of which all social structures would be affected. Certainly, this applies to education in general and mathematics in particular. Walter Beyer (2009, 2010) has carefully examined the state of mathematical education during this period.

The Teaching of Mathematics in Venezuela during the Republican Period

In 1827, the Liberator Simón Bolívar enacts a decree that frees the University of Caracas from the control of the Church and awards it "economic and political autonomy"; additionally, and importantly, the election of the administration was to be done by the faculty.

With this reform, impelled by the Liberator himself by means of the new statute that he himself enacted, the chairs of Mathematics and Physics were created, along with those of Chemistry and Botany. The problem was finding suitable people to fill these positions and their corresponding departments.

The mathematical studies of students were completed before enrolling in a university's degree programs. They lasted three years and were very elementary (1st year: mathematical readings; 2nd year: geography and chronology; 3rd year: readings on arithmetic, algebra, topography and applied geometry).

In order to learn mathematics during the early years of the independent republic, Venezuelans relied on the Universities of Caracas and Mérida, the national schools and the Academy of Mathematics. The national schools, established at the behest of the central government, created a system that covered much of the country and were intermediate institutions that awarded the title of Bachelor (bachiller). The mathematics taught in these schools was the same as was taught at the universities before enrollment in their degree programs.

With regard to the Academy of Mathematics, it was originally conceived as "a school of military mathematics"; nevertheless, in 1831 it was founded not as a school but, rather, as an academy, an "institution aimed at teaching the civic and military applications of mathematics" (Freites, 2000, p. 20). Its founding regent was Juan Manuel Cajigal and it began offering courses in 1834; the operations of the academy were full of vicissitudes, financial limitations, a lack of understanding by the rulers of the country, and racial prejudices.

After the death of Bolívar, there was a period of readjustment in Venezuela during which groups of people became united by common political affiliations or by common social and cultural affiliations. Among the latter were the scientific societies, one of which was the Economic Society of the Friends of the Country which, given its interest in raising the knowledge base of the citizenry, created a chair in Arithmetic and Elementary Geometry, in addition to some in other areas such as Music and Grammar.

Another important scientific society was the College of Engineers of Venezuela, founded in 1860, whose mission included "the fostering of the natural and exact sciences in the country" (Freites, 2000, p. 24).

With the end of the so-called Federal War (1868) and the rise to power of General Antonio Guzmán Blanco — who effectively was president (either directly or through surrogates) from 1870 to 1888 and developed a plan to modernize Venezuelan society by implementing a series of economic, political, social, cultural and educational initiatives — the educational initiatives included "demanding that public education affirm in 1870 that all elementary education will be free and compulsory".

The enactment of the *Decree of Public, Compulsory and Free Education* (http://www.analitica.com/bitblio/aguzman/publica.asp) by General Antonio Guzmán Blanco, on June 27, 1870, marks the beginning of a new phase in all things related to Venezuelan education in general and mathematical education in particular. The Guzmán Blanco decree was so significant that the history of Venezuelan education can be divided into two parts: one before the decree and one after the decree. Among the many consequences of this law was the need to build and endow more schools, thus making it possible that "more young women and men, in addition to learning to read and write, would learn their numbers and would begin to learn about the four basic operations of arithmetic" (Freites, 2000, p. 29).

The presidency of Guzmán Blanco paid attention to the national schools, which became federal schools. Two of these (the one in Maracaibo and the one in Valencia) reached such a high academic level that they became, respectively, the University of Zulia (1891) and the University of Carabobo (1892).

In 1878, Guzmán Blanco decided that the chair of mathematics that had previously been in the Academy of Mathematics was to be transferred to the University of Caracas. This transfer led to the founding of the College of Exact Sciences at the University of Caracas. As a result of this, Engineering was now deemed to be a discipline worthy of university study and the

Academy's work in the military sphere came to an end. The engineers, organized in the College of Engineers of Venezuela founded in 1861, made a significant contribution to the evolution of science and mathematics in Venezuela and, in fact, played an important role in the adoption of the metric system by the country.

In the latter parts of the nineteenth century, the basic ideas of mathematics continued to be taught at an introductory level before enrollment into a university degree program while the so-called advanced ideas of mathematics were being taught at the College of Engineering.

The Teaching of Mathematics in Venezuela in the Twentieth Century

There are those who assert that, in terms of education, Venezuela did not reach the twentieth century until the 1940s since the intervening period "between 1899 and 1936 were years of barbarity, darkness, denial of freedom and of almost complete ignorance" (Fermón, 1989 cited in Beyer, 2010). Thus the 20th century of Venezuelan education began in 1936 with the founding of the National Pedagogical Institute (IPN) whose graduates in the area of mathematics, along with the engineers graduated from the universities, took control of the teaching of mathematics. The period starting in 1936 with the founding of IPN would culminate with the arrival of New Math in 1969. This new period would last about ten years, until 1979, right up to the enactment of the Education Act.

The broad impact of the IPN is reflected in the following statement:

> The IPN and its graduates played a supremely important role in curricular changes, teacher training, and the creation of educational policies. Additionally, they carried out various studies and diagnostic tests related to the situation in the elementary and secondary schools. In terms of the development of mathematical education, their intervention was key, particularly in terms of its relation to the CIAEM (Inter-American Committee of Mathematics Education) and the establishment of New Math in Venezuela (Beyer, 2010).

Other examples include two IPN graduates who specialize in mathematics: Boris L. Bossio Vivas (Bolívar, 2005) who was a prolific writer of text books, and Reimundo Chela (Planchart, 2000), who was the first Venezuelan to earn a doctorate in mathematics (he received this academic degree from the University of London in 1961, under the supervision of Albrecht Frölich) and was the driving force behind the creation of the College of Sciences in the Central University of Venezuela. The IPN also is credited with having the first graduate program in Mathematics Education in Latin America — it was at the level of a master's degree and was founded in 1974 (Orellana Chacín, 1980, p. 128).

In 1958 the College of Sciences was established at the Central University of Venezuela, with a department of Physics and Mathematics, which awarded the first bachelor's degrees in mathematics in 1962 to Mauricio Orellana Chacín and Jesús Salvador González (Beyer, Orellana Chacín, & Rivas, 2009). From this moment on, other colleges of science at other universities were created. In 1970, the first graduate program (at the master's level) in mathematics in Venezuela was founded at the University of Carabobo (Orellana Chacín, 1980, p. 131).

Another important event was the signing at the Central University of Venezuela, of an agreement of cooperation between the College of Science and the College of Humanities and Education to train secondary school teachers.

The events describe the development of the teaching of mathematics in Venezuela from its pre-Columbian past until 1992 when the Venezuelan Association for Mathematics Education (ASOVEMAT) was founded.

The Teaching of Mathematics in Venezuela Following the Founding of ASOVEMAT

The founding of the Venezuelan Association for Mathematics Education (ASOVEMAT) was a watershed moment in the history of mathematics education in Venezuela. At this point in time mathematics education is considered both an area of practice and a disciplinary field since the actions of this association had a significant impact on those who taught mathematics professionally. It allowed them to develop a sense of personal identity as well as a sense of belonging to a community that distanced itself from other groups working in mathematics; namely, other groups that focused on either creating or developing mathematics (such as pure mathematicians) or applying it (as is the case for engineers, chemists, biologists, statisticians and professionals in other fields who use mathematics in their professional responsibilities).

From December 4 to 9, 1961 the Inter-American Conference on Mathematical Education (I CIAEM)[2] took place in Bogotá, Colombia (Barrantes & Ruiz, 1998). This meeting was attended by a Venezuelan delegation consisting of 12 faculty members. These scholars were dedicated to the promotion of the New Math. This style of mathematics was taught only at the university level and had yet to reach elementary and secondary education.

Orellana Chacín gives an account of the Venezuelan participation in the I CIAEM:

> In the I CIAEM, Professor J. A. Rodríguez presented the placement test given in Venezuela and Manuel Balanzat made another presentation about the "Teaching of Mathematics in Venezuela" which was mostly concerned with teaching after secondary school. It was at this conference that the need to change the nature of mathematical teaching in Venezuela, as much in terms of content as in methodology, was formulated. The faculty that attended the I CIAEM, especially Jesús González, José Giménez Romero, Evelia de Anzola, Luis J. Marcano and José Alejandro Rodríguez, upon returning from the conference, diffused, via professional presentations or their teaching activities, several of the ideas they considered necessary in order to change the nature of mathematics teaching at the secondary school level as well as at the post-secondary school level. These faculty members remained engaged for many years in activities related to the reform of teaching in secondary school.

Venezuela has continued its involvement with subsequent Inter-American Conferences on Mathematics Education, including the most recent meeting, the thirteenth — commemorating the fiftieth anniversary of the 1961 meeting — that took place in June of 2011, in Recife, Brazil.

The work by the Venezuelans who attended the I CIAEM bore fruit and in 1969 the so-called "Education Reform" was created (Rojas de Chirinos,

---

[2] CIAEM also stands for Comité Interamericano de Educación Matemática (Inter-American Committee of Mathematics Education), the sponsoring organization of this conference.

2006). Its adoption was driven by the government, at the time controlled by the Christian Social Party, and, consequently, the teaching of the New Math was extended to the elementary and secondary schools of Venezuela. This created the need to prepare the teachers[3] who would be entrusted with teaching this "new" mathematics.

Concern about the particular details of the teaching of mathematics in our country was already present at the beginning of the 1960s. In 1961, in the eastern part of Venezuela, the Seminar in the Teaching of Physics and Mathematics was held; this is considered to be "the first activity of its type to be developed in Venezuela" (Beyer, 2001).

The 1970s, preceded by the educational reform of 1969 — the one that extended New Math to elementary and secondary schools — contained important events that would have an impact on mathematics education (as a field of application) and Mathematics Education (as a disciplinary field) until this day.

Indeed, in 1970, a graduate program in mathematics was started at the University of Carabobo by means of partnerships between the University of Madrid and the University of Oklahoma. Before this time,

> ... in some of the country's institutions such as the Central University of Venezuela, the University of the Andes, the University of Zulia, and the University of Carabobo itself, graduate courses had been offered, as well as some seminars, but all of it occurred in fits and starts; it did not adhere to any existing structure. These activities occurred at the whim of the faculty working at these institutions, or were due to a visit from or contact with a professor who offered to teach a course. These activities gave the faculty awareness of certain mathematical topics but did not accomplish any kind of systematic training of students or scientific production (Orellana Chacín, 1980, p. 131).

The graduate program at the University of Carabobo was established by a resolution of the University Council in 1970, under the initiative of the Secretary of the University of Carabobo, Freddy Mulino Betancourt. Graduate study formally began with a graduate program in mathematics at the College of Engineering (Sáenz Palencia, 2011, p. 17).

---

[3] A thorough history of the initial and continuing training of mathematics teachers in Venezuela can be found in León, Beyer, Serres & Iglesias (2012).

With the passing of time, this graduate program, after first leaning towards becoming a master's in Mathematics with a certification in teaching, became the current master's in Mathematics Education.

Freddy Mulino Betancourt and Emilio Medina were instrumental in developing the graduate programs related to mathematics. Betancourt was the first Venezuelan to receive a doctorate in mathematics with a dissertation in the field of Mathematics Education. Emilio Medina's dissertation which he defended in 1975, examined the program of study for future teachers, as implemented in the pedagogical institutes in Venezuela. This study led him to the conclusion that it was absolutely necessary to reform these institutes because they failed to address the reasons for which they were created (the training of teachers of mathematics for secondary education).

Emilio Medina put several of his ideas into practice at the University of Carabobo in its master's of science degree with a certification in mathematics, whose popularity was dwindling. This program of study was modified under his guidance — his idea was to refocus it on the teaching of mathematics and it was because of this decision that the advanced degree program continued and became a master's in mathematics with certification in teaching.

The educational reform of the early 1970s was accompanied by curricular changes and by the opening of new post-secondary educational institutes. These, in turn, brought with them the need to train the academic personnel to handle both the demands generated by the changes in subject content — brought on by the curricular changes — as well as the student demands for enrolling in these recently-created institutes.

To meet these demands, two strategies were put in effect: one of them was the opening of graduate programs and the other was the founding of the National Center for the Improvement of the Teaching of Science (CENAMEC) in 1973. CENAMEC played an important role in the evolution of Mathematics Education in our country. This institution became a center for those interested in putting into effect educational innovations whose implementation would contribute to improving general scientific training for Venezuelan citizens, mathematics training, in particular.

The Mathematical Coordinating Committee of this organization was made up of important teachers in the discipline, some of whom were closely tied to the reform movement. Not surprisingly, this was a time of reflection, search and agreement regarding the model for teaching mathematics in Venezuela. It's basis was the work by CENAMEC starting in 1975.

There were many problems that needed solutions. These included: the content and methodology for the teaching of mathematics in secondary schools; the quality of the initial training of teachers; bringing current teachers up to date, whether they had graduated or not; and resources for the teaching of mathematics. With a view towards discussing these issues in an appropriate setting, it was decided that an event should be designed whose explicit purpose was to consider the aforementioned issues. Thus, the First Meeting of Teachers of Mathematical Teaching from the Educational Institutes was held in 1982. The idea of this meeting was to bring together those who were involved in the teaching of future secondary school teachers, regardless of whether the meeting's attendees were from the institutes, from schools affiliated with universities, or from the universities themselves. Five meetings addressing the same issues were held (from 1982 to 1986, one per year). The meetings were not restricted to faculty involved in post-secondary teaching, but were open to all teachers of mathematics interested in improving their skills. To accomplish this, mini-courses were taught and workshops were held at these meetings. It was in this way that the number of mathematics teachers at elementary schools and secondary schools who attended these meetings increased.

Due to this gradual increase, beginning in 1987, the name of the meeting changed and it became known as the Meeting about the Teaching of Mathematics. This change was not merely a semantic one, but it also signaled a qualitative change in the make-up of the community of Venezuelan mathematics educators. Every year, around two hundred teachers from a variety of places in the country headed towards Caracas to discuss, study and consider various topics related to the processes of mathematical teaching and learning.

Nevertheless, teachers from the interior of the country began to express the need to have these meetings in other regions of the Republic. Hence, in 1991, the First Midwestern Meeting of Mathematics Education was held, promoted by a group of teachers associated to the Mathematics Department at the Barquisimeto Pedagogical Institute. This meeting was one of the first large meetings about mathematical teaching held outside of Caracas. At these centers teachers of mathematics felt empowered to express themselves, to give an indication of their successes in the classroom, and to inform the younger participants of professional development opportunities.

After the meeting in Barquisimeto there was a flurry of activity in Venezuela. This activity led to the promotion and organization of similar events in different parts of the country. Of particular consequence were those actions carried out by teachers in the Northwestern region of the country, in the areas known as Insular and Guayana. There the work emanated from the Department of Mathematics at the Maturín Pedagogical Institute. This work included the organization of the Mathematics Education Meetings for that region. During one of these meetings, the Venezuelan Association for Mathematics Education (ASOVEMAT) was formed. It currently has a national presence and chapters in more than half the states of the Venezuelan nation.

The founding of ASOVEMAT in Maturín in 1992 marks the period of the greatest development of Mathematics Education in Venezuela. Since then, mathematics educators in Venezuela have been part of a formal collective. This association is the driving force behind the Venezuelan Conference on Mathematics Education (COVEM), the flagship national conference, of which eight meetings have been held (Maturín, 1994; Valencia, 1997; Maracaibo, 2000; Trujillo, 2002; Barquisimeto, 2004; Maracay, 2007; Caracas, 2010; Coro, 2013).

An Attempt at the Periodization of the History of Mathematics Education in Venezuela

Establishing historical periods for any social process is problematic (Semo, 1977). It is rare that the proposed periods will be universally accepted due to the challenges in establishing the events that frame the periods. Running this risk, a periodization of the trajectory taken by Mathematics Education in Venezuela will be given.

The *Pre-Colombian Period* spans from the oldest history of our original inhabitants to August 2, 1498: the date of arrival at the mouth of the Orinoco River of the expedition commandeered by Christopher Columbus. The *Colonial Period* begins with the arrival of the Spanish and comes to a head with the Declaration of Independence on July 5, 1811. The *Republican Period* goes from 1821 (with the signing, in Carabobo, of the permanent Independence of Venezuela from the Spanish empire) until present day. Three stages can be identified in this period: III-a, III-b, III-c corresponding, respectively to the nineteenth, twentieth, and the part of the twenty-first century that has transpired so far.

Six separate segments can be identified in the twentieth century (Stage III-b of the Republican period). The first of those (1900–1958) is highlighted by the passing of a series of codes for public education on top of the creation of the Ministry of Public Education in 1880, which provided schools with some regulatory guidance. This stage ended in 1958 with the end of the dictatorship of Marcos Pérez Jiménez. This attracted many immigrants who came to our country to take advantage of the many opportunities created by the arrival of democracy. The second segment began with the I CIAEM which, as described above, provided an opportunity for a group of Venezuelan teachers of mathematics. It ended with the Educational Reform of 1969, which brought the New Math to our elementary and secondary schools.

The third segment (1970–1980) has its beginning with the establishment of graduate studies in mathematics at the University of Carabobo and ended with the meeting organized in Barquisimeto (State of Lara), in which questions were asked about the way in which the teaching of mathematics was being done — this questioning was related to the failure of the New Math in other countries.

The fourth segment of Stage III-b began in 1980 with the 1st Meeting of Instructors of Mathematics Teaching from the Institutes of Post-secondary Education. This meeting began a series of regular gatherings, in which teachers of mathematics met not only to reflect upon their experiences but also to learn about novel approaches to teaching. Occasionally, they would also attend visits from foreigners or fellow Venezuelans in notable positions related to mathematics in the Ministry of Education, in CENAMEC or in a university or pedagogical institute. The end of this segment is situated in 1990 with the 1st Meeting of Coordinators of Graduate and Research Programs which brought forward issues related to the need for links, nonexistent at the time, between research in Mathematics Education and the programs providing advanced degrees (master's). Gradually, the idea that the research being carried out in the pursuit of advanced degrees should be based in formal theories and approaches became better received.

The start of the fifth segment of Stage III-b is identified with the founding of ASOVEMAT, the natural by-product of the ongoing process of community development begun at the 1st Meeting of Instructors of Mathematical Teaching from the Institutes of Post-secondary Education and a series of other institutional, local, regional and national events. Some of these events had been attended by internationally renowned mathematics educators who would often point to the need to form an organization that would give all of these activities an intentional direction. Among these mathematical educators were Ubiratan D'Ambrosio and Claude Gaulin.

The end of this segment occurred with the 3rd Latin American Congress on Mathematical Education (III CIBEM), held in July of 1998, on the grounds of the Central University of Venezuela. This was the second (the IV CIAEM being the first, in 1975) large international event arranged by the community of Venezuelan mathematics educators.

With the III CIBEM the twentieth century of Mathematics Education in Venezuela came to an end. To start the next millennium, the ALIEM XXI (Latin American 21st Century Program of Research in Mathematics Education) project was developed. This project was conceived as

> a conceptual tool presented to people, institutions, and organizations interested in improving the quality and level of mathematical competency of the citizens of Latin America (González, 2000).

ALIEM XXI provided paths for research in Mathematics Education in Latin American countries and, in this way, also promoted interaction and collaboration between communities of mathematics educators in different countries. Since it was likely that these communities would confront common problems, it seemed best to try to resolve these common problems collaboratively.

The start of Stage III-c of the Republican Period is at the beginning of the first decade of the twenty-first century. The event that marked its beginning is the ratification of the Constitution of the Bolivarian Republic, approved by referendum in 1999. This new agreement on social coexistence along with other aspects of the political character of the day would have — and continue to have — a profound impact on all the economic, social, political, cultural, and educational processes that would come to pass in the country. In educational matters, there were far reaching changes such as the installation of the Bolivarian Educational System, which would be reflected in the activities related to school mathematics in primary and secondary education, as well as at the university level carried out by the instructors of mathematics, especially at those institutions of post-secondary education with a so-called "Bolivarian focus" (Ministerio del Poder Popular para la Educación, 2007).

The end of Stage III-c coincided with the approval by the National Council of Universities of the Doctorate in Mathematics Education at the Libertador Experimental and Pedagogical University (Pedagogical Institute of Maracay, UPEL), the first such degree program in Venezuela.

The efforts to create a Doctorate in Mathematics Education in Venezuela can be traced back to 1997 and to the 1st Venezuelan Symposium in Research in Mathematical Education held in Valencia (State of Carabobo). The project presented there by Fredy González was known as the Venezuelan Program for the Doctorate in Mathematics Education (PROVEDEM). A second version of this project was announced and considered at the 3rd Latin American Congress on Mathematics Education (III CIBEM; Caracas, Central University of Venezuela) (González, 1998). Later, lines of research were created in general doctoral programs in Education. The first of these was proposed by Blanca Quevedo in the doctoral program in the College of Social Sciences and Education offered at the University of Zulia (LUZ) and the second was proposed by David Mora in the doctoral program in Education at the Central University of Venezuela (UCV). In 2003, Fredy González and Mario Arrieche presented a proposal to create the Doctoral Research Area in Mathematics Education to the administration of the doctoral program in Education at the UPEL Maracay. This last effort was the most immediate precedent for the Doctoral Project in Mathematics Education at the UPEL (DEM-UPEL).

So, after a long process which started in 1997 and which was related to the so-called PROVEDEM Project (Venezuelan Program for the Doctorate in Mathematical Education), a doctoral program in Mathematics Education came to fruition.

Indeed, the National Council of Universities (CNU) in 2012 approved the proposal for the creation of this doctoral program. This doctoral program at UPEL (DEM-UPEL) is an option for graduate study that will train a professional researcher in the field of Mathematics Education. Its goals are (1) to train researchers at the highest academic level who will be able to form research groups committed to the creation and application of theories, methodologies and techniques to the field of Mathematics Education; (2) to significantly increase the number of Venezuelan researchers who have attained a doctorate; and (3) to positively impact the quality of Mathematics Education in the Venezuelan educational system.

The DEM-UPEL is run by professors and other professionals who teach mathematics, with considerable experience teaching in the discipline at various levels. Some faculty members are connected to research projects in the field of Mathematics Education. Students can take courses either full time or part time.

It is expected that those who complete the program will be well-suited to:

(a) Evaluate and investigate problems linked to the processes of teaching and learning mathematics, contributing results that will enable the evolution of Mathematics Education as a scientific discipline.
(b) Propose and validate innovations in teaching with the use of technology. These innovations will make it viable to overcome the problems faced in mathematics classrooms at the various levels of Venezuelan education, taking into account the economic, social, historic, cultural and political contexts of the country.
(c) Work independently and collaboratively in order to create, as a result of research activities, new understanding in the field of Mathematics Education.
(d) Propose theories and models that describe, explain and improve the current organization and functioning of Mathematics Education, as a scientific endeavor, not only in Venezuela but also in other countries in Latin America.
(e) Develop innovations in the processes of teaching and learning mathematics with an eye towards application, with educational goals in mind and with an emphasis on the possibilities opened up by electronic tools and technology — not only those tools that can contribute to the understanding of actual mathematics (e.g., software) but also those related to the creation of new means of interaction between participants in educational situations linked to mathematics.

All this makes us think that, with the opening of the DEM-UPEL, Mathematics Education in Venezuela is taking a qualitative leap forward, whose consequences will be felt in the near future.

Bibliography

Abreu, A., & Luque, R. (2007). *Concepciones de la medida de área en los alumnos Wayúu en la Tercera Etapa de Educación Básica*. La Universidad del Zulia, Facultad de Humanidades y Educación. Maestría en Enseñanza de la Matemática. Master's Thesis, unpublished.

Arcila Farías, E. (1961). *Historia de la Ingeniería en Venezuela*, 2 Vols. Colegio de Ingenieros de Venezuela, Año Centenario 1861–1961. Caracas: Editorial Arte.

Barrantes, H., & Ruiz, A. (1998). *Historia del Comité Interamericano de Educación Matemática*. Bogotá, Colombia: Academia Colombiana de Ciencias Exactas, Físicas y Naturales, Barry University, International Commission on Mathematical Instruction. Retrieved from http://www.cimm.ucr.ac.cr/ciaem/?q=es/node/37

Belisario, A., & González, F. (2012). Historia de la Matemática, Educación Matemática e Investigación en Educación Matemática. *UNION, Revista Iberoamericana de Educación Matemática, 31*, 161–182. Retrieved from http://www.fisem.org/www/union/revistas/2012/31/archivo_16_de_volumen_31.pdf

Berlin, B., & Kay, P. (1969). *Basic Color Terms: Their Universality and Evolution*. Berkeley: University of California Press.

Beyer, W. (2001). Pasado, presente y futuro de la Educación Matemática venezolana. Parte II. *Enseñanza de la Matemática 10*(2), 3–20.

Beyer, W. (2009). *Estudio evolutivo de la enseñanza de las matemáticas elementales en Venezuela a través de los textos escolares: 1826–1969*. (Tesis Doctoral). La Paz, Bolivia: Edición del Instituto Internacional de Integración Convenio Andrés Bello; Grupo de Investigación y Difusión en Educación Matemática (GIDEM). ISBN: 979-99954-817-6-6

Beyer, W., Orellana Chacín, M., & Rivas, S. (2009). Esbozo biográfico de un insigne matemático venezolano: Jesús Salvador González (1930–2008). *Boletín de la Asociación Matemática Venezolana, XVI*(1), 39–50. Retrieved October 31, 2013 from http://www.emis.de/journals/BAMV/conten/vol16/Biografia.pdf

Beyer K., W. O. (2010). Senderos, caminos y encrucijadas de las matemáticas y la educación matemática en Venezuela. *UNION, Revista Iberoamericana de Educación Matemática, 23*, 15–44. Retrieved October 30, 2013 from http://www.fisem.org/www/union/revistas/2010/23/Union_023_008.pdf

Bishop, A. (1999). *Enculturación matemática: La educación matemática desde una perspectiva cultural*. Barcelona, España: Paidós.

Bolívar, W. (2005). *Boris Bossio Vivas: su obra, aportes e impacto*. Trabajo Especial de Grado (Tutor: Walter Beyer). Unpublished. Caracas: Universidad Central de Venezuela.

Chevallard, Y. (1992). Concepts fondamentaux de la didactique: Perspectives apportées pau une approche anthropologique. *Recherches en Didactique des Mathematiques, 12*(1), 73–112.

Fermín, M. (1989). *Momentos Históricos de la Educación Venezolana*. Caracas: Editorial Romor.

Freites, Y. (2000). Un esbozo histórico de las matemáticas en Venezuela. Parte I: Desde la Colonia hasta finales del siglo XIX. *Boletín de la Asociación Matemática Venezolana* (Caracas), *VII*(1y2), 9–43. Retrieved from http://www.emis.de/journals/BAMV/conten/vol7/yfreites.pdf

González, F. (1998). *Proyecto PROVEDEM (Programa Venezolano de Doctorado en Educación Matemática)*. Presentation made at III Congreso Iberoamericano de Educación Matemática (III CIBEM). Caracas: Universidad Central de Venezuela.

González, F. (2000). Agenda latinoamericana de investigación en educación matemática para el siglo XXI. *Educación Matemática, 12*(1), 107–128.

González, O. (2002). Etnomatemáticas entre los Indígenas Kurripako (Arawak) y los Programas de Educación Intercultural Bilingüe. *Equisángulo*. Retrieved from http://www.saber.ula.ve/bitstream/123456789/20292/1/articulo5.htm

Johnson-Laird, P. N. (1993). La théorie des modèles mentaux, en M. F. Ehrlich, H. Tardieu, & M. Cavazza (Eds), *Les modèles mentaux. Approche cognitive des representations*. Masson, pp. 1–22.

León, N., Beyer, W., Serres, Y., & Iglesias, M. (2012). *Informe sobre la Formación Inicial y continua del docente de Matemática: Venezuela*. Presented at the Escuela-Seminario Internacional Construcción de Capacidades en matemáticas y Educación Matemática.

Longart, C., & Tirapegui, C. (2008). *Aprendiendo Geometría en Ambientes Interculturales: El caso de escolares criollos y tejedores warao (Venezuela)*. Universidad Nacional Experimental de Guayana (UNEG), Master's thesis. Unpublished. Retrieved from http://www.bibliodar.mppeu.gob.ve/?q= node/143624&backtocolection=repo/Universidad%20Nacional%20Experi mental%20de%20Guayana%20(UNEG)&page=0%2C0%2C0%2C0%2C0% 2C0%2C0%2C0%2C0%2C0%2C0%2C0%2C0%2C0%2C0%2C0%2C0%2C0 %2C0%2C0%2C5

Luque, R. (2009). *La presencia de los Elementos Matemáticos en el Pueblo Wayúu*. La Universidad del Zulia (LUZ). College of Humanities and Education. Doctorate in Social Sciences. Doctoral Thesis, unpublished. Retrieved from http://servidor-opsu.tach.ula.ve/7jornadas_i_h/paginas/doc/JIHE-2011-PN53.pdf

Ministerio del Poder Popular para la Educación (2007). *Currículo nacional bolivariano: Diseño curricular del sistema educativo bolivariano*. Caracas: Autor (Educere, Mérida, v. 11, n. 39, dic. 2007. Retrieved from http://www.scielo.org.ve/scielo.php?script=sci_arttext&pid=S1316-49102007000400020 &lng=es&nrm=iso

Orellana Chacín, M. (1980). *Dos décadas de Matemática en Venezuela*. Caracas: Universidad Nacional Abierta.

Planchart, E. (2000). Raimundo Chela. *Boletín de la Asociación Matemática Venezolana, VII*(1y2), 53–57. Retrieved from http://www.emis.de/journals/BAMV/conten/vol7/planchart.pdf

Rojas de Chirinos, B. (2006). Modernización y regionalización: Un estudio de la reforma educativa en la escuela primaria venezolana 1969–1979. *12*(Extraordinario), 127–147. *Journal of the Libertador Pedagogical and Experimental University* (Venezuela). Retrieved from http://www.redalyc.org/pdf/761/76109908.pdf

Sáenz Palencia, L. (Ed.) (2011). *Prospecto de los Programas de Postgrado UC 2012*. Valencia, Venezuela: Ediciones de la Universidad de Carabobo, Vicerrectorado Académico, Dirección General de Postgrado. Depósito legal: 1f04120123784168. ISBN: 978-980-12-6206-0.

Sánchez, D. (2009). El Sistema de Numeración y algunas de sus aplicaciones entre los aborígenes de Venezuela. *Revista Latinoamericana de Etnomatemática*,

2(1), 43–68. Retrieved from http://www.revista.etnomatematica.org/index.php/RLE/article/view/15

Semo, E. (1977). Problemas Teóricos de la Periodización Histórica. *Dialéctica*, *II*(2), 11–21. Retrieved from http://es.scribd.com/doc/81665353/Dialectica-n%C2%BA-02-enero-1977

Villalobos, O., & Ruíz, D. (2009). *Aproximación a la geometría Pemón en el sector Wonken*. Universidad Nacional Experimental de Guayana (UNEG); Master's thesis in Educational Sciences.

Villavicencio, M. (2001). El aprendizaje de las matemáticas en el Proyecto Experimental de Educación Bilingüe de Puno y en el Proyecto de Educación Bilingüe Intercultural del Ecuador: Reflexiones sobre la práctica y experiencias relacionadas (Martha Villavicencio), en A. E. Lizarzaburu & G. Zapata Soto (Comps.) *Pluriculturalidad y aprendizaje de la matemática en América latina. Experiencias y desafíos* Madrid, España: Ediciones Morata, S. L. Conjuntamente con PROEIB-Andes (Cochabamba, Bolivia) y DSE (Bonn, Alemania), pp. 167–190.

About the Author

Fredy Enrique González is a professor of mathematics at the Universidad Pedagógica Experimental Libertador (UPEL Maracay) and holds a Ph.D. in education from Universidad de Carabobo (Venezuela). He has recently (2013) become Coordinator of UPEL's doctoral program: *Doctorado en Educación Matemática*. He is the founder of the *Núcleo de Investigación en Educación Matemática "Dr. Emilio Medina"* (NIEM). He has been involved in several projects aimed at improving the education of pre-service mathematics teachers in the Venezuelan school system.

EPILOGUE

A Half Century of Progress

Patrick Scott

As indicated in the Preface, this volume is dedicated to the memory of Howard Franklin Fehr, a founding trustee of the Inter-American Committee on Mathematics Education (IACME)[1] at the first Inter-American Conference on Mathematics Education in 1961. He edited the proceedings of the first two Inter-American Conferences and served formally as the Vice President of IACME from 1972 to 1975. IACME celebrated its fiftieth anniversary during its 13th Conference in Recife, Brazil, in June 2011. It serves as a useful framework for discussing the "Half Century of Progress" in mathematics education in the Southern Americas.

There have been three stages in the history of the Inter-American Committee on Mathematics Education. The first conference in 1961 elected Marshall Stone as president and he continued in that position until 1972, and as honorary president until 1987. The focus of the first stage was essentially to implement curricular reforms related to the New Math. Luis Santaló (a Spaniard established in Argentina) continued the work of Stone. In the second stage, under the leadership of Ubiratan D'Ambrosio, the focus shifted somewhat from *what* should be taught to *how* it should be taught. This has reflected what has happened in the international mathematics education community. Among the topics that have received special attention are teacher preparation, socio-cultural influences, and the use of technology.

---

[1]For a detailed account of this organization, consult *History of the Inter-American Committee on Mathematics Education*, by Barrantes and Ruiz (1998). It is available in both Spanish and English (http://www.centroedumatematica.com/aruiz/libros/La%20Historia%20del%20Comite%20Interamericano%20de%20Educacion%20Matematica.pdf).

In the first decade of the twenty-first century, IACME moved into a third stage under the leadership of Ángel Ruiz, and continues to be the leading mathematics education organization in the region. Its conferences are still held every four years and bring together mathematics educators from across the Americas with renowned international guests. Plenary and parallel presentations present cutting edge topics in the teaching and learning of mathematics; oral communications and poster sessions give mathematics education researchers from throughout the hemisphere an opportunity to present and discuss their results; and mini-courses and workshops present practical applications particularly relevant to local teachers. In this stage, the scientific rigor in the selection of conference presentations has increased and the use of technology for communication concerning the organization of the conference has reached a high level. The Luis Santaló Medal has been created "to honor individuals who have contributed significantly to the development of IACME across their lifetimes" and the relationship with the International Commission on Mathematical Instruction (ICMI)[2] has intensified. Moreover, IACME was the first regional organization to be recognized by ICMI as a "Multinational Mathematical Education Society Affiliated to ICMI" in 2009.

As research in mathematics education has increased in the region, the IACMEs of every four years have not been sufficient to meet the needs of researchers to present their results. The interest of Spanish and Portuguese mathematics educators attending IACME V (1985) and IACME VI (1987) led to the first Iberoamerican Congress on Mathematics Education (CIBEM)[3] in Seville in 1989. Since then, there has been a CIBEM every four years with a format very similar to that of IACME's, giving regional researchers and local teachers an opportunity for relevant professional development. The CIBEMs are held two years after the IACMEs in the year after the International Congress on Mathematical Education (ICME). Essentially, a CIBEM in Spain or Portugal is followed by two in Latin America.

A development related to both CIBEM and progress in mathematics education in Iberoamerican countries has been the creation of the Iberoamerican Federation of Mathematics Education Societies (FISEM).[4] FISEM's stated goal is to "promote the interchange of experiences and

---

[2] http://www.mathunion.org/ICMI
[3] http://www.cibem.org
[4] http://www.fisem.org

information that will permit the improvement of the teaching and learning of mathematics at all levels." Apparently, 50 years ago none of the 14 Latin American countries that now have mathematics education associations federated with FISEM had such organizations.

Another regional organization that partly began to give researchers in mathematics education an opportunity to present their results at conferences is now called the Latin American Committee on Educational Mathematics (CLAME).[5] It began annual conferences in 1987 with the first "Central American and Caribbean Meeting on Teacher Preparation and Research in Educational Mathematics". In 1997, with the formal creation of CLAME, the meeting name was changed to the Latin American Meeting on Educational Mathematics (RELME). It continues to meet every year in a Latin American country.

The most recently established regional mathematics education group is the Central American and Caribbean Mathematics Education Network (REDUMATE).[6] REDUMATE was created by the participants of the ICMI-sponsored Capacity and Networking Project (CANP) that was held in Costa Rica in August of 2012. REDUMATE sponsored its first Central American and Caribbean Congress on Mathematics Education (CEMACYC)[7] in the Dominican Republic in November of 2013. It followed the same format as the IACMEs, thus providing plenary presentations, as well as opportunities for regional researchers and local teachers. There is a close relationship between REDUMATE and IACME.

The chapters in this volume clearly show the progress in the preparation of teachers of mathematics. Fifty years ago almost all elementary teachers were trained in high school level normal schools. Over the years, most of those normal schools have been converted into institutions of higher education.[8] At first, the tendency was to offer special three-year teaching degrees and then to transition to regular four or five-year university bachelor's degree programs. A similar process has usually taken place for the more specialized preparation of secondary mathematics teachers. Also, most

---

[5] http://www.clame.org.mx
[6] http://www.centroedumatematica.com/redregional
[7] http://i.cemacyc.org
[8] A similar process happened in Europe and the United States. As late as the 1970s, there were still some elementary teachers in some states of the United States who had graduated from high school level normal schools and did not hold university degrees.

countries reported the influence of the New Math movement. It usually led to many specific professional development efforts, mostly for secondary mathematics teachers, as attempts were made to implement the new content and methods.

Fifty years ago, although in the Southern Americas there were some master's degree and doctoral programs in pure mathematics — and a few professors had earned doctoral degrees in mathematics education in the United States or Europe — it was apparently not until 1974 that the first graduate program in mathematics education was established (in Venezuela). As of 2013, most countries now have one or more master's degree programs in mathematics education. In fact, at least eight countries now offer doctoral programs in mathematics education: Argentina, Brazil, Chile, Colombia, Cuba, Jamaica, México, and Venezuela. Other countries have doctorates in education in which students might emphasize mathematics education.

It is perhaps important to note that most countries report with pride their involvement in the International Mathematical Olympiad (IMO), as well as in local and regional mathematics competitions at various levels. This may be the one activity that has been most successful in bringing mathematicians, mathematics educators, and classroom mathematics teachers together in an effort to support and improve the mathematical abilities of students.

Some of the regional olympiads may have also served to foster camaraderie among the Spanish and English-speaking Caribbean nations, integrating the latter into the regional discourse. Since its inception, IACME has had Spanish, Portuguese and English as official languages for its conferences, but the reality has been that Spanish — and Portuguese to some extent — has become the dominant language. Thus, integration and support of mathematics education in the English-speaking Caribbean, to say nothing of French and Dutch speaking areas, remains a challenge.

There has been significant progress in mathematics and its teaching in the Southern Americas during the past 50 years, but the promise of a mathematics education that facilitates the life choices and possibilities of every citizen is still an unrealized dream. The challenge for mathematicians to apply advances in their discipline to help solve many of the problems faced by all sectors in their countries also has not been met as successfully as it might. We must all build upon the positive accomplishments of the past 50 years so that every year the joys of solving problems will be shared by more and more of the citizens in the Southern Americas.

Index

1991 National Constitution, 133

abacus, 93
abolition of slave trade, 119
abstract algebra, 69
abstract courses, 121
Academia Colombiana de Ciencias Exactas, Físicas y Naturales, 130
Academia Militar y Escuela Politécnica, 120
Academia Superior de Ciencias Pedagógicas, 96
academic program, 326, 335
Academy of Medical, Physical and Natural Sciences, 212
Academy of Military Engineers, 117
Academy of Sciences of Cuba, 211–213
access to educational system, 73
accumulation of capital, 241
achievement maps, 106
active learning, 129
adult education, 145
Advanced Level, 413
Advancement of the Teaching of Mathematics (IPEM), 373–376
African, 407
Agrarian Reform, 45
algebra, 62, 63, 410
algorithmic thinking, 422
Alianza para el Aprendizaje de Ciencias y Matemáticas (AlACiMa), 394

Alianza para el Fortalecimiento en el Aprendizaje de las Matemáticas (AFAMaC), 393
Alliance for Progress Project, 133
Amautas, 33
American occupation of Haiti, 247
Amerindians, 406
Anales de ingeniería, 121
analytical reading, 135
Andrés Bello Inter-American Prize for Education, 50
anthropology, 129
Apologetic Summary History of the People of These Indies, xii
application problems, 14
Applied Mathematics, xiv, 142
Arango, R., 329
Arawaks, 406, 441
Argentina, xxi, 1–29, 48, 120, 122, 128, 135, 160, 351, 358, 375, 461, 464
Argentina Scientific Society, 3
argumentation strategies, 16
arithmetic, 62, 407
arithmetical tables, 407
Aritmética Social, 388
Arosemena, J., 328
Arosemena, M., 327
Arosemena de Tejeira, O., 333
Astronomical and Meteorological Observatory, 122
Atcon Report on Latin American Universities, 141
Augustinians, 326

Autonomous University of Chiriquí, 335
Avelino Siñani–Elizardo Pérez Law, 45
Aymara, 32, 33
Aztecs, 296, 297

Bachelor's in Computer Science, 209
Bachelor's in Education, 207
Bachelor's in Mathematics, 207–210
Baldor College, 201
Barbados, 408
Barranquilla, 120
Barrios, J. R., 272
bartering, 406
basic numeracy, 127
Bazelais, B., 244
Begle, E. G., 389
Belgrano, M., 2
Belize, 266
Bellegarde, D., 248
Bentham, J., 119
Bergeron, A., 121
Bernard Reform, 240, 254
bilingual schools, 127
blended courses, 185
Bogotazo, 131
Bolívar, S., 117
Bolivia, 31–56, 118
Bolivian Education Code, 40, 41
Bolivian Society of Mathematics, 47, 48
Bolivian Society of Mathematics Education, 49
Bonaparte, J., 117
Bouchereau, C., 246–249
Bourbaki, 372
Bourbaki group, 140
Boussingault, J.-B., 118
Boyer, J.-P., 243
Bravo, M. V., 329
Brazil, xi, xii, 48, 57–87, 135, 143, 287, 358, 359, 370, 371, 375 464
Brígida Álvarez, 390
British, 406
Bruner, 412

budget for education, 130
Buenavista College, 196
Bulletin of the Cuban Society of Mathematics and Computer Science, 219
bullying, 144

Caballero, A. N., 127, 128
cacique, 382
Calcul différentiel, 121
Calcul infinitésimal, 121
calculating skills, 8
Calculus of Quaternions, 121
Camacho, C., 368, 370
Cambridge Examinations, 413
Capacity and Networking Project (CANP), 463
Caribbean, 382, 406
Caribbean Advanced Proficiency Examinations (CAPE), 414
Caribbean Basin Initiative, 253
Caribbean Examinations Council (CXC), 413
Caribbean Institute of Mathematical Sciences (CIMS), 259
Caribbean Secondary Education Certificate (CSEC), 414, 420
Caribs, 406
Carlisle Indian Industrial School, 384
Carney, M., 251
Cartagena, 116, 120
cartography, 118
Casa K'inich, 269, 289
catedrático, 383
Catholic Church, 116, 382
Catholic values, 119
Center for Mathematical Modeling, 109
Center for Research and Development in Education (CIDE), 101
Central America and the Caribbean (OMCC), 285
Central American and Caribbean Congress on Mathematics Education (CEMACYC), 463

Index

Central American and Caribbean Mathematics Education Network (REDUMATE), 463
Central Bank or *Banco de la República*, 124
Central University of Las Villas, 202
centralist government, 122
centralized educational system, 91
centrist government, 119
Centro Residencial de Oportunidades Educativas de Mayagüez (CROEM), 390
Centros Regionales de Adiestramiento en Instrucción Matemática (CRAIM), 391
"chalk and talk", 417
Charcas, 34, 35
Charles III, 116
child-centered, 412
Chile, 48, 89–114, 135, 143, 160, 279, 375, 464
Chorographic Commission, 120
Christian Brothers, 123
Christian indoctrination, 116
Christophe, H., 242, 261
chroniclers, xvii
Chuquisaca, 36
Church hierarchy, 119
Church of England, 407
Church's domination, 119
CINVESTAV, 308, 311, 312
Ciudad Universitaria, 129
civil engineering, 61
classic high schools or *bachilleratos*, 127
classroom guides, 133
Codazzi, A., 120
coffee, 119, 124
Colegio Balboa, 322
Colegio de Panamá, 327
Colegio de San Bartolomé, 328
Colegio Mayor de San Bartolomé, 116
Colegio Mayor del Rosario, 116
Colegio Militar de Ingeniería, 120
Colegio Propaganda Fidae, 327
Colegio Seminario, 327

Colegio Tridentino de San Agustín, 269, 272, 273
Colegio de Salvador, 196
colegios menores, 116
College of Agriculture and Mechanic Arts, 384
College of Physical Sciences and Mathematics, 347
College of the Immaculate Conception (CIC), 410
Colombia, ix, 115–149, 230, 287, 323, 327, 330, 375, 464
Colombian Institute for the Promotion of Higher Education (ICFES), 134
Colombian Mathematical Society, 141
Colombian Science Academy, 120
Colombian Society of Engineers, 121
colonial Colombia, 116
colonial period, 58, 115, 194, 195
colonial science, xx
colonialism, 58, 241
colonialist powers, 241
colony, 381
Columbia University, 386
Columbus, C., 406
Comenius Research and Development Center, 107
Comentarios a las Ordenanzas de Minas, xv
Commercial School of Languages, 323
Commission of Study of National Education, 333
Committee of Mathematical Education of Paraguay (CEMPA), 354
Common Core standards, 106
Common Core State Standards, 391
Common Entrance Examination, 413, 417
Commonwealth, 413
communication, 412
communism, 128
community leaders, 145
concept of country, xi
concordat with the Vatican, 122

468 Index

confederate government, 122
Conferencias Iberoamericanas de Educación Matemática, xxiii
connections, 412
conquest, xi
Consejo Nacional de Ciencia y Tecnología, CONACYT, 354
Conservatives, 119
constitution of 1801, 241
Constitution of the Republic of Colombia in 1886, 122
Constitution of the United States of Colombia in 1863, 120
Constitutional Reform of 1936, 128
constructivist, 312, 315, 412
contextualization of the mathematics, 211
continuing education, 75
Coordinating Commission of National Education, 324
Copán, 266–269, 289
Copernican system, 116
corrupt political practices, 144
Costa Rica, 48, 151–192, 271
cottage marking, 414
Cours d'analyse, 138
criollos, 298, 442
Cuba, 143, 193–222, 226, 277, 287, 383, 464
Cuban native cultures, 194
Cuban Revolution, 203
Cuban Society of Mathematics, 216, 217
Cuban Society of Mathematics and Computer Science, 216, 218
Cuban Society of Physical and Mathematical Sciences, 202, 216
cultural contexts, xxii
cultural events, 145
cultural practices, 58
cultural relevance, 412
culturally relevant curriculum, 80
currency, 406
curricular revision, 133
curriculum, 225, 226, 231, 304, 306, 310, 311, 313, 316, 406

curriculum development, 414
curriculum officer, 424
Curriculum Plan C, 207
curriculum reform, 57
Curso de cálculo diferencial, 121

D'Ambrosio, U., 69, 461
Dartigue, M., 249, 250
Dartigue Reform, 240
de Caldas, F. J., 117
de Gamboa, F. X., xv
de Guadalupe, F. A. L., 269
de Las Casas, B., xii
de Luna Victoria y Castro, F., 322
de Mosquera, T. C., 120
de Paula Santander, F., 118
decimals, 127
Decroly, O., 128
democratic education, 70
denominational schools, 408
Dewey, J., 63
diagnosis, 424
Dienes, Z. P., 310, 412
Dieudonné, J., 140
dioceses, 116
Diploma in Education, 428
disabilities, 73
displaced population, 144
distance education, 79
Dominican Republic, 463
Dominicans, 116, 326
drill, 411
Dubois, E., 244
Dumouchel, 245
Dupuy, A., 240
Durán, F. M., 329

e-learning, 90
EAFIT (*Escuela de Administración, Finanzas y Tecnología*), 142
Earthquake of January 12, 2010, 259
East-Indian, 407
Ecole Normale Superieure, 252, 256
economic protectionism, 128

Economic Society of Friends of the Country, 195, 196, 198
educational deprivation, 71
educational planning, 131
educational reform, 63, 333
educational standards, 134
El Salvador, 266, 271
elementary education, 61, 197, 199
elementary school, 203
Eleven-Plus Examination, 413, 417
Encomienda, 194
end-of-college examination, 135
engineering, 63
engineering schools, 61
Enlaces, 90
England, xii, 225, 226, 242, 412
entrepreneurial sector, 144, 146
Epsilon, 219
equity, 71
Escalante Gutiérrez, J. A., 33, 49
Escuela Nacional Unificada, 99
Estándares Básicos en Competencias en Matemáticas, 134
Estimé, D., 252
ethnomathematics, xvii, 441
Euclidean geometry, 93
Europe, 464
European ideals, 119
external examinations, 412
extra-mathematical problems, 16
extracurricular activities, 145

facilitators, 425
Faculté des Sciences (FDS), 251
Faculty of Mathematics and Computer Science, 215
Faculty of Science, 197
fascism, 128
Father of Public Education, 329
Federación Iberoamericana de Sociedades de Educación Matemática, xxiii
Federal Education Law, 21
Federal War, 37
Federici, C., 141
Fehr, H. F., ix, x, 461

Fejér, L., 140
Fery, H., 243
Fields Medal, 140, 143
Filloy, E., 311, 312
First Military Occupation, 199
Fondo Universitario, 141
Foraker Act, 384
Foreign Scholarships Program, 90
foreign teachers in science and mathematics, 158
Fourier Series, 141
fractions, 127
France, 277, 314
Francis, E., 390
Franciscans, 326
Francisco Morazán National Pedagogical University (UPNFM), 279, 280, 282, 283
free trade, 119
Freire, P., 68
French Canal construction, 323
French Guiana, 224
French *Lycée*, 127
Freyle, J. D., xiv
Friedman, M., 92

garage universities, 131
Garavito, J., 121
García, G., 365, 366
García, M., 388
Garfield High School, 50
Géffrard, F.-N., 245
gender equity, 430, 432
gender issues, 398
General Certificate in Education (GCE), 412
General Inspection of Elementary School, 332
General Plan of Studies for the Island of Cuba, 197
generic competencies, 135
geometry, 62, 63, 69, 410
German Academic Exchange Service (DAAD), 141
German Democratic Republic, 226
German Pedagogical Mission, 133

German teachers, 331
German texts, 204
Gimnasio Moderno, 127
goals of education, xiii
gold, 119
Government schools, 408
Government-Assisted Schools, 408
graduate programs, 70
Gran Colombia, 322
Great Britain, 128, 314, 407
Great Depression, 66, 128
gross enrollment ratios, 131
Guatemala, 266, 270–273
Guerrier, P., 243
guerrilla movements, xxiv
guerrilla warfare, 144
Guilbaud, T., 246
Guyana, 223–238
Guzmán Blanco, A., 446

Haiti, xi, 239–264
Haitian Institute of Advanced
  Mathematics, 259
Haitian-American Sugar Company
  (HASCO), 247
hands-on activities, 411
Hay–Herrán Treaty, 123
Héraux, H., 246–249
Herbart teaching methodology, 332
Hibbert, L., 251
high school, 200, 201
high-end research, 145
higher cognitive abilities, 184
Higher Education Reform in Cuba,
  207
Higher Normal School (ESP), 275,
  279
Higher School for Boys, 323
Hilbert Spaces, 141
Hilbert, D., 143
Hindus, 407
Hispanic America, 383
Home School, 201
Honduran Committee on
  Mathematics Teaching (CHEM),
  283

Honduran Mathematical Olympiad,
  286
Honduras, 265–293
Horváth, J., 140
Hurtado, M. J., 328
hydraulics, xiv

Iberoamerica (OIM), 285
Iberoamerican Congress on
  Mathematics Education (CIBEM),
  462
Iberoamerican Federation of
  Mathematics Education Societies
  (FISEM), 462
Iberoamerican Mathematical
  Olympiad, 287
ICETEX (*Instituto Colombiano de
  Crédito Educativo y Estudios en el
  Exterior*), 142
illicit crops, 144
illiteracy, 129
in-service, 423
inalienable rights, 119
Inca, 33
Inca rule, 115
incentives, 426
independence, 413
India, 224
Indian Citizenship Act of 1924, 384
indigenous, 416
indigenous groups, 80
indigenous population, 73
Industrial Revolution, 116
inequities, 73
INICIA Test, 108
initial teacher training, 172
initiatives, 74
innovation and creativity, 144
innovations, 74
Institute of Basic Sciences (ICB), 352
Institute of Cybernetics, Mathematics
  and Physics (ICIMAF), 213
Institute of Mathematics and
  Computer Cybernetics, 212
Institute of Mathematics and Related
  Sciences (IMCA), 369–371

Institute of Sciences, 349
Institution of Saint Louis of Gonzague, 245
Instituto Técnico Central, 123
Institutional Educational Project (PEI), 134
institutions of higher education, 58, 60
instructional technology, 133
integrated curriculum, 66
intellectual instruments, xiv, xv
Inter-American Committee on Mathematics Education, xxii, 308, 461
Inter-American Conference on Mathematics Education, 373, 461
Interdisciplinary Program for Research in Education (PIIE), 101
International Commission on Mathematical Instruction (ICMI), xxi, 64, 304, 462
International Congress of Mathematicians, 142
International Mathematical Olympiad (IMO), 285, 431, 464
International Mathematical Union (IMU), xxi, 142, 306
international online discussions, 145
interventions, 424
intra-mathematical problems, 9
Introduction to Modern Mathematics, 138
INURED, 259

Jamaica, 408, 464
Japan, 283, 289
Jesuits, xx, 116, 299, 326, 383
Jones Act of 1917, 384
journal, 66
Journal of Mathematical Sciences, 219
Journal of the Cuban Society of Physical and Mathematical Sciences, 202
journal *Operations Research*, 219

Kandel, L., 251
Keenan Report, 408
Kemmerer Mission, 124
Kindergarten Normal School, 201
Klein, F., 64, 303, 304
Kline, M., 311
knowledge/recall, 422
Konder, P., 141
KOWIZUKO, 208
Krasner, M., 140

La Escuela Normal, 329
La Gran Colombia, 117
laboratory schools, 68
Láinez Law, 3
Lancaster, J., 271, 279
Lancastrian method, 327
Lancastrian system of education, 242
land surveying, 118
Laserna, M., 138
Latin, 118
Latin American Committee on Educational Mathematics (CLAME), 463
Latin American Meeting on Educational Mathematics (RELME), 463
law, 63
Law of Public Instruction, 196, 198
law schools, 61
l'Ecole des Sciences Appliquées (ESA), 246
Le Petit Séminaire Collège Saint Martial, 245
learned societies, 130
learning of mathematics, 317
Lefschetz, S., 140
Les Filles de la Sagesse, 245, 261
Les Frères de l'Instruction Chrétienne, 245, 261
Lescot, E., 252
Lherisson, L. C., 239, 244, 245
Liasson, 208
Libby, F. E., 332
Liberal hegemony, 120
Liberal ideas, 118

Liberal party, 122
Liberal principles, 119
Liberal Reform, 265, 272–274, 278
Liberal regime, 122
Liberal Republic, 128
Liberals, 119, 240, 244
Licenciatura Académica en Matemática, 97
licensure, 74
liceus, 62
Lindo, J., 273
Linnaeus, C., 116
literacy, 127, 424
Llopis, S., 277
Logan, R. W., 239, 241, 243
logic, 69
Lutz, E., 332
Lycée Pétion, 243

Madiou, T., 241
Madrid, J. R., 287
Magdalena River, 120
mandatory education, 15
Mandelbrojt, S., 140
manipulatives, 427
manumission at birth, 119
Marco Curricular, 385
Marist College, 201
Martínez, C. J., 332
Massive Open Online Courses, 186
Master of Mathematics, 215
master teachers, 211
Masters in Mathematics, 142
Math Circles, 400
mathematical achievement, 73
mathematical aptitudes, 125
mathematical concepts, 411
mathematical content knowledge, 74
mathematical curriculum, 90
mathematical modeling, 90
Mathematical Olympiads, 136, 371
Mathematics Camp, 336
mathematics competitions, 204
mathematics curriculum, 152
mathematics education as a discipline, xxi

mathematics education in engineering programs, 210
Mathematics Education in Venezuela, 439, 453, 455, 457
mathematics for social justice, 80
mathematics instruction, 156
mathematics instructors, 210
mathematics learning, 226
mathematics methodology, 207
Mathematics Olympiads, 399
Mathematics of Planet Earth, xxvi
mathematics teacher training, 205
mathematics teaching, 226, 312
mathematics textbooks, 60, 62, 200
Mathématiques, 121
May 1968, xxiv
Maya, 265–269, 289, 296, 297
Mayagüez, 386
Mazur, B., 401
Measure Theory, 141
measurement, 411
Medellín, 120
medical schools, 61
medicine, 63
Medina, E., 451
MEMI, 52
Memoirs of 1924 and 1926, 333
memorization, 127, 407
Méndez Pereira, G., 332
Méndez Pereira, O., 333
mensuration, 410
Mental Mathematics Marathon, 431
mentoring, 423
mercantile system, 240, 241
mestizos, 298
Mexican Society of Mathematics (SMM), 307
México, xiv, 48, 120, 128, 135, 143, 266, 270, 277, 295–320, 375, 464
Michelow, J., 97
military academies, 60
military dictatorships, xxiv
military engineers, 116
MINEDUC, 91
mining, xiv

mining engineers, 118
mining school, 118
Ministry of Education, 206
Ministry of Higher Education, 206
Ministry of Public Works, 124
modern communications, 145
Modern Mathematics, 67, 280
Modern Mathematics Movement, 69
modern pedagogy, 129
money, 411
Monroe, P., 386
Montessori, M., 128
Morazán, F., 271, 272
Moreau de Saint-Méry, M. L., 241
Morgan, H., 327
moriche basket, 440
multicultural, 430
Muñoz Marín, L., 385
Music and Reciting School, 323
Muslims, 407
Mutis, J. C., 116

Napoleon's invasion of Spain, 117
National Assessment of Educational Progress (NAEP), 394
National Association of Math Circles, 309, 401
National Association of Teachers of Mathematics (ANPM), 307, 309
National Autonomous University of Honduras (UNAH), 265, 273–275, 277, 278, 282, 283
National Autonomous University of México (UNAM), 306
National Basic Curriculum (CNB), 281, 284
National Computing Center (CNC), 352
National Congresses on Mathematics Education (CONEM), 377
National Council of Teachers of Mathematics (NCTM), 317
national curriculum, 66
national education, 151
National Education Law, 21
national identity, 381

national initiatives, 74
National Institute, 93, 323
National Institute of Physics and Chemistry, 348
National Institute of Pure and Applied Mathematics (IMPA), 287
National Network of Mathematics, 336
National Panel for Scientific Degrees, 215
National Pedagogical University (UPN), 311
National Polytechnic Institute (IPN), 308
national ranking system (*escalafón*), 129
national reality, 129
National School Mathematical Olympiads (ONEM), 371
National Science Foundation (NSF), 372, 388
National Secretariat of Science, Innovation and Technology (SENACYT), 336
national team, 205
National Tests, 422
National University of Asunción (UNA), 347
nationalist movements, 128
Nationals, 240, 244
Native American, 382
naval and military academies, 61
Nazis, 365
NCTM Math Standards and Expectations, 134
Neogranadine, 322
Neumann, R., 332
new forms of knowledge and behavior, xvii
New Granada, 116
New Math, 133, 308–311, 372, 389, 410, 447, 464
Nicaragua, 270, 273
non-denominational public education, 118
non-denominational schools, 127

non-Euclidean geometry, 138
Normal Home Schools, 201
Normal School for Boys, 323
Normal School for Girls, 329
Normal School for Teachers, 201, 302
Normal School for Teachers and Private Tutors, 44
Normal School of Juan Demóstenes Arosemena, 324
Normal School of La Paz, 44
Normal School of Santiago de Veraguas, 324
Normal School of Warisata, 44, 45
normal schools, 122, 127, 384
North Korea, 226
number, 411
numeracy, 424, 432
Nutrition in Schools, 324
Núñez, R., 122

obstacles, 73
OECD, 135
Olimpiadas Matemáticas de Puerto Rico (OMPR), 399
OMAPA, 353
online, 75
open-ended items, 418
Operation Bootstraps, 396
Order Manuel José Hurtado, 329
Ordinary Level, 413
Organic Law for Public Education N°11 of March 23, 1904, 330
Organic Law N°1 of 1877, 328
Organic Law of Elementary School Education, 331
Organization of American States, xxii
organized crime, 144
Ospina, P. N., 124
Osuna, J. J., 381

Padilla, R. C., 275
Panamá, 116, 123, 124, 135, 321–345, 444
Panamá Canal, 121, 342
Panamá Canal indemnification, 128
Panamá Isthmus railroad, 120

Panamá Normal School, 322
Panamanian Mathematical Olympiad, 335
Papy, G., 69
Paradigma, 288
Paraguay, 2, 347–361
Paraguayan Mathematical Society (SMP), 350
parish priest, 126
pasantías, 90
Paul, E., 261
pedagogical institutes, 205
pedagogical instruction, 126
pedagogical theories, 129
pedagogy, 424
pediatrics, 129
Pemón, 441
peripheral science, xx
permanence in educational system, 73
Perry, G., 144
Perú, ix, xviii, 32, 48, 118, 128, 135, 143, 358, 363–380
Peruvian Mathematical Society (SMP), 369–371, 377
Peruvian Mathematics Education Society (SOPEMAT), 377
Pestalozzi, J., 122, 305
Pestalozzi system, 329
Pétion, A., 242, 243, 261
Ph.D. in research mathematics, 215
physical punishment, 126
Piaget, 412
Plan C, 207
Plan D, 207, 209
Plan Remos, 200
Plan Varona, 199
Political Constitution of the Republic of Panamá, 323
political nature, xxiv
political strife, 143
Polytechnic School of Haiti and the College of Sciences, 252
Poma de Ayala, xix
Pontifical Xavierian University, 116
Popayán, 120
Porto Rico, 387

practical classes, 210, 211
pre-Columbian educational model, xviii
pre-Columbian times, 115
pre-service training for teaching, 173
pre-university, 203
pre-university education, 200, 201
preparation of teachers, 77
Preparatory School for Teachers, 198
Presbyterian, 407
Presidential Medal for Excellence in Education, 50
primary education, 122
primary school attendance, 119
primary school teachers, 126
primary schools, 407
private education, 197
private schools, 91, 407
private secondary schools, 127
pro-US schools, 127
probability, 411
Problem Solving with an Emphasis in Real Contexts, 184
problem-solving, 4, 176, 305, 313, 314, 412, 422
problem-solving skills, 125
profesorado, 275, 278, 280–282
professional, 72
professional development, 414
professional programs, 73
professional schools, 63
professorial career, 130
program accreditation, 142
Programme for International Student Assessment (PISA), 92, 135, 316, 432
progress maps, 89
Progressive Education, 63
Project for National Development, 325
Project for the Popularization of Science, 336
pron, 93
psychology, 129
public education, 197
public educational system, 128

public schools, 91
public secondary schools, 127
Puerto Rico, 196, 381–404
Puerto Rico Statewide Systemic Initiative (PRSSI), 391
Pumarejo, A. L., 128

QS ranking of Latin American Universities, 142
quadrivium, 93
qualifying examination, 129
quantitative reasoning, 135
quaternary sector, 146
quinine, 119
Quintero Alfaro, A. G., 389
quipu, xix, 3, 93
Quipucamayos, 33

railroad, 121
railroad lines, 120
railways, 124
Raleigh, W., 223
rates of retention, 73
reasoning, 412
Reclus–Wyse–Sosa route, 121
Reconquista, xx
reform, 62, 64
Reform Bellegarde, 246
Reform of Modern Mathematics, 175
religion, 406
religious neutrality, 122
religious orders, xvii, 123
remediation, 424
Republic of Colombia, 117
República de la Nueva Granada, 119
Republican Period, 200
research, 226, 234
research in mathematics education, 70, 181
Research Institutes on Mathematical Education (IREM), 377, 378
Research Triangle Institute, 254
resources, 411
Résumé des leçons de calcul infinitésimal, 121

Revista Colombiana de Matemáticas, 141
Revista de Instrucción Pública, 329
Reyes, R., 124
rigorous mathematics courses, 123
Roman Catholic Church, 407
Romberg, T., 391
Rosa, R., 272
rote learning, 407
Rouma, G., 39
Royal Academy of Medical, Physical and Natural Sciences of Havana, 211
Royal Academy of San Luis, 93
Royal and Literary University of Havana, 196
Royal and Pontifical University of San Javier, 326
Royal Botanical Expedition (*La Expedición Botánica*), 116
Royal College, 242
Royal Conciliar San Carlos and San Ambrosio Seminary College, 195
Royal Court of Charcas, 34
Royal Pontifical University of Saint Jerome of Havana, 195
Rudin, W., 256
Ruiz, Á., 462
rural areas, 126

Saber 3, 135
Saber 5, 135
Saber 9, 135
Saber 11, 134
Saber Pro, 135
Salgar, E., 122, 328
Salomon, P., 249
San Carlos University, 270
San Cristóbal de la Habana College, 196
San Juan, 386
Santa Fe de Bogotá, 116
Santaló, L., 20, 461, 462
Santiago College, 196
Sarmiento, D., 122
Saunders, P., 242

scholarship, xvii,
scholarships, 127, 413
school administrators and staff, 145
School for the Indigenous People, 323
school inspections, 129
school inspectors, 126
school materials, 126
school mathematics curriculum, 175
School Mathematics Study Group, 69
School of Applied Sciences, 248, 252
School of Arts and Trades, 323
School of Education, 428
School of Education (*Escuela Normal Superior*), 130
School of Engineering, 120
School of Mathematics, 169
School of Mathematics and Engineering, 121
School of Mines, 121
School of the Isthmus (*Colegio del Istmo*), 327
school-based assessment (SBA), 415
schooling, xvii
Schwartz, L., 140
Schwartz, M. H., 140
science and technology parks, 145
Scientific and Applied Computing Laboratory, 359
Second Latin American Mathematics Symposium in Argentina, 141
Secondary Assessment Examination (SEA), 418
secondary education, 60, 61, 197, 199, 200
secondary private schools, 196
secondary schools, 203, 407
Secretariat of Public Education (SEP), 306
secular education, 407
semi-specialist, 424
separated by genders, 126
separation of church and state, 119, 128
separation of Panamá, 123
Service of Education, 324
Serviez, C. M., 117

set theory, 69, 137
shape, 411
Simón Patiño Pedagogical and Cultural Center, 41
Sisters of St Joseph of Cluny, 245
situations-problems, 18
Skemp, 412
slavery, xvi, 406
Smith, D. E., 386
SOCHIEM, 101
social inequality, 144
social interventionism, 128
social justice, xxiii
social mobility, 144
Sociedad Colombiana de Ingeniería, 138
Sociedad Colombiana de Matemáticas, 142
Societé Haitiano-Americaine de Développement Agricole (SHADA), 248
socio-cultural perspective, 69
Sosa, P. J., 121
Soto, M. A., 272–274, 278
South America, 406
space, 411
Spain, xi, 47, 48, 59, 60, 93, 115, 117, 194, 198, 241, 270–272, 322, 381–383
Spanish colonial culture, 119
Spanish colonies' independence, 117
Spanish Inquisition, 115
Spanish invasion, 240
special needs, 425
specialists, 424
Specialization in Higher Mathematics, 140
Springer, G., 69
Sputnik 1, 388
Staden, H., xii
Standard Fruit Company, 247
standardized college entrance examination, 134
standardized tests, xxiii
statistics, 411
steam navigation, 120

STEM, 395
Stone, M., 69, 461
student movement, 128
Study Plan A, 206
Study Plan B, 206
subsidized schools, 91
Sumario compendioso de las quentas, xiv
summer courses, 208
Suriname, 224
sustainability, 321
Switzerland, 314
syllabus, 411

Taíno Genome Project, xii, 382
Taínos, 382
talented students, 205
Tapia, L., 332
Tardieu-Dehoux, C., 241
teacher conferences, 126
teacher education, 57, 128, 131, 405
teacher harassment, 144
teacher preparation, 71, 73, 226
teacher standards, 129
teacher training, 227, 235, 301, 305, 307, 316
Teacher's Assistant, 332
Teachers College Commission, 386
teaching mathematics, 315
teaching of mathematics, 210, 322, 337
Technical-Professional Schools, 100
Technological Center for Education, 324
Technological University of Panamá (UTP), 335
technologically globalized world, 146
technology, 173, 318
Tegucigalpa, 270–273, 287, 288
Teresian College, 201
Territorial Clause, 381
textbooks, 133, 407
Thorndike, E. L., 417
Thousand Days War, 123, 323

threats to the survival of civilization, xxv
Tiwanakota culture, 32
tobacco monopoly, 119
Tola, J., 366, 367, 369, 372–376
Trade Professional Schools, 202
transformation geometry, 412
Treaty of Ryswick, 241
Trends in International Mathematics and Science Study (TIMSS), 92, 135, 316
tribal cultural practices, 80
tributary system, 382
trigonometry, 63
Trinidad and Tobago, 183, 405–438
tuition, 127
two-party system, 119
types of problems, 14

UNESCO, xxii
United Provinces of New Granada, 117
United States, 48, 50, 63, 66, 69, 93, 104, 123, 124, 128, 225, 226, 242, 301, 311, 314, 317, 381, 464
United States Constitution, 381
Universidad Central, 118
Universidad Antonio Nariño, 143
Universidad de Antioquia, 121, 142
Universidad de Cartagena, 118
Universidad de Córdoba, 143
Universidad de los Andes, 139, 141
Universidad de Popayán, 118
Universidad del Norte, 143
Universidad del Valle, 142
Universidad Metropolitana de las Ciencias de la Educación, 96
Universidad Nacional, 121
Universidad Nacional de Colombia, 120
Universidad Pedagógica de Colombia, 130
Universidad Tecnológica de Pereira, 143
university accreditation, 142

University des Antilles et de la Guyane (UAG), 259
university education, 198
University Gardens, 390
University of Buenos Aires, 3
University of Chile, 96
University of Guyana, 225, 226, 229, 234
University of Havana, 199, 200
University of Leon, 270
University of Panamá, 333
University of Puerto Rico at Mayagüez (UPRM), 390, 397
University of Saint Thomas Aquinas, 116
University of Salamanca, 36, 116
University of San Javier, 322
University of the West Indies (UWI), 428
University of Trinidad and Tobago, 429
upper secondary education, 203
upward mobility, 131
urbanization, 120
USSR, 226

Van Hiele, 412, 440
Vasco, C., 133
vector geometry, 90
vector spaces, 69
Venezuela, 116, 119, 143, 279, 439–460, 464
Venezuelan Association for Mathematics Education (ASOVEMAT), 448, 453
Venezuelan Conference on Mathematics Education (COVEM), 453
Venezuelan Program for the Doctorate in Mathematics Education (PROVEDEM), 456
Vera, F., 137
Vespucci, A., 223
viceroyalties, xi
Viceroyalty of New Granada, 116
Villarreal, F., 364, 365

Vincent, S., 239, 244, 245, 252
Vocational Pre-University Institutes of Exact Sciences, 205
Vogeli, B., 97
von Neumann Algebras, 141
von Neumann, J., 140
voucher system, 91

Warao, 440
ward schools, 407, 409
Wars of Independence, 198
Washington, B. T., 249
Water War, 38
Wayúu, 442
whole numbers, 127

Wilberforce, W., 242
Wirsing, O., 329
workshops, 425
world scenario, xxv
written composition, 135

XIX Congreso Colombiano de Matemáticas, 143

yucayeque, 382
yupanas, xix, 3

Zapata, F., 122
Zúñiga, A., 274